EVOLUTION OF GENES AND PROTEINS

EVOLUTION OF GENES AND PROTEINS

Edited by Masatoshi Nei

CENTER FOR DEMOGRAPHIC
AND POPULATION GENETICS
UNIVERSITY OF TEXAS, HOUSTON

and Richard K. Koehn

DEPARTMENT OF ECOLOGY
AND EVOLUTION
STATE UNIVERSITY OF NEW YORK,
STONY BROOK

SINAUER ASSOCIATES INC. • PUBLISHERS
Sunderland, Massachusetts   01375

EVOLUTION OF GENES AND PROTEINS

Library of Congress Cataloging in Publication Data

Main entry under title:
Evolution of genes and proteins.

 Bibliography: p.
 Includes index.
 1. Chemical evolution.   2. Deoxyribonucleic acid.
3. Proteins.   4. Genetic polymorphisms.
I. Nei, Masatoshi.   II. Koehn, Richard K.
QH371.E925   1983   574.87'328   83-477
ISBN 0-87893-603-3
ISBN 0-87893-604-1 (pbk.)

Printed in U.S.A.

6 5 4 3 2

CONTENTS

PREFACE

The study of evolution at the molecular level has experienced two periods of exciting development during the past two decades. The first period started when the techniques of amino acid sequencing and protein electrophoresis were introduced in evolutionary studies in the early 1960s and lasted for about 10 years. During this period the approximate constancy of the rate of amino acid substitution in evolution and the extensive protein polymorphism in natural populations were discovered. These new findings led to the proposal of various new evolutionary theories, some of which were quite controversial. Controversy was particularly heated over Kimura's neutral mutation theory, which proclaimed that most amino acid substitutions and protein polymorphisms are caused not by Darwinian selection but by neutral mutation and random genetic drift. At the same time, the discovery of approximate constancy of amino acid substitution provided new methods for dating the evolutionary history of organisms as well as methods for constructing phylogenetic trees from molecular data. In the 1970s evolutionary geneticists were extremely busy examining the validity of the new evolutionary theories and applying the new methods of constructing phylogenetic trees.

The second period started only a few years ago and is not yet over. This period was initiated by introduction of a new set of biochemical techniques: DNA sequencing, recombinant DNA, and restriction enzyme methods. These new techniques have already uncovered many unexpected properties of the structure and organization of genes (e.g., exons, introns, flanking regions, repetitive DNA, pseudogenes, gene families, and transposons) and of their evolution. It is now clear that most genes do not exist as a single copy in the genome but, rather, in clusters, and that the number of genes in a cluster varies extensively from cluster to cluster. Comparison of nucleotide sequences from diverse organisms indicates that the rate of sequence change in evolution varies considerably with the DNA region examined and that the more important the function of the DNA region the lower the rate of sequence change. Furthermore, the extent of genetic variation undetectable by protein electrophoresis is enormous. These discoveries again have led a number of evolutionists to propose new evolutionary theories such as concerted evolution and horizontal gene transfer.

Although many other exciting discoveries will undoubtedly be made in the near future, it now seems appropriate to examine and summarize in book form the general implications of these new findings, together with those in the first period of development. Such a book is needed not only for students and scholars of evolution but also for biologists in general. However, it is not an easy job for one or two persons to write such a book, because the subject is so diversified and the progress in each area is so rapid. We therefore decided to produce a book on molecular evolution with the help of experts from various specialized areas of the subject. This was facilitated by a symposium on "Evolution of Genes and Proteins" which we organized for the joint meeting of the Society for the Study of Evolution and the American Society of Naturalists held at the State University of New York at Stony Brook, June 23–24, 1982. This symposium was attended by more than 500 scientists and students from the United States and abroad. In this symposium we attempted to cover all important areas of study in molecular evolution, inviting both molecular biologists and population biologists. Each symposium speaker was then asked to write a chapter of textbook, rather than a usual symposium paper, in order to make the book understandable to a wide range of readers, from students to active researchers.

While a book written by multiple authors has a definite advantage in covering the latest developments in diverse areas of a subject, it suffers from such deficiencies as heterogeneity of writing style, repetitions, and inconsistencies. In the present book we have worked to minimize these deficiencies by making extensive editorial changes of original manuscripts; in some cases we even rewrote parts of the original text. However, we have tried to avoid any change of an author's opinions even if they contradicted another author's views. In the forefront of research, scientists do not always agree with each other, and this disagreement often becomes an impetus for further progress. The reader will notice that there are some inconsistencies in the use of scientific terminology. For example, a noncoding DNA sequence between two coding regions of a gene is called an intervening sequence (IVS) in Chapter 1 but an intron in the other chapters. We have left such inconsistencies, because no consensus has yet been achieved in the scientific community for these terms.

The chapters of this book can be divided loosely into three groups. The first group includes the first four chapters, all of which are concerned with the long-term evolution of DNA. In Chapter 1 Edgell and his associates discuss the evolution of globin gene clusters as a model case of gene evolution. Globin genes have been very important in elucidating the evolutionary change of gene structure in the last few years. In Chapters 2 and 3 the evolutionary significance of gene duplicaton and the mechanism of concerted evolution are discussed in the light of new findings at the DNA level. Chapter 4 deals with the evolutionary change of mitochondrial DNA. Mitochondrial DNA evolves much faster than nuclear DNA and thus is very useful for studying the phylogenetic relationships of closely related species.

The second group consists of Chapters 5 to 9, which are mainly concerned with the genetic variation within species. The major issue in these chapters is the maintenance of genetic polymorphism in natural populations, and data on both protein and DNA polymorphisms are examined. The controversy over the neutral mutation theory is still alive. However, unlike a decade ago, neutralists and selectionists are no longer hostile to each other, and the gap between the views of the two groups of scientists has narrowed substantially. Data on DNA polymorphism are still scanty compared with those on protein polymorphism but clearly show that the genetic variability at the DNA level is enormous. In Chapter 8 Avise and Lansman show that mitochondrial DNA is a useful genetic material for tracing back the evolutionary history of populations.

The last four chapters, which make up the third group, deal with several current evolutionary theories. (Chapter 9 can be included in this group as well as in the second group.) In Chapter 10 Jukes presents an interesting theory on the evolution of the amino acid (genetic) code, taking advantage of recent discoveries of non-universal amino acid codes in mitrochondrial genes. In Chapter 11 Kimura discusses recent developments in the neutral theory of molecular evolution. In Chapter 12 Hall describes experimental observations on the evolution of new metabolic functions in microorganisms. The last chapter is concerned with transposons, i.e., genetic elements which move within and between chromosomes. The evolutionary significance of transposons is still largely speculative, but they are potentially important in explaining the existence of repetitive DNA in higher organisms.

We would like to express our hearty thanks to the contributors of this book for writing excellent chapters and for being tolerant of our editorial suggestions and changes. We hope our joint enterprise will be successful in bringing the latest knowledge of molecular evolution to both students and scientists who are interested in the diversity and evolution of organisms.

MASATOSHI NEI
RICHARD K. KOEHN
November 17, 1982

CONTRIBUTORS

Stylianos E. Antonarakis, Department of Pediatrics, Genetics Unit, Johns Hopkins University School of Medicine, Baltimore

Norman Arnheim, Department of Biochemistry, State University of New York, Stony Brook

John C. Avise, Department of Molecular and Population Genetics, University of Georgia, Athens

Betty Brown, Department of Bacteriology and Immunology, University of North Carolina, Chapel Hill

Wesley M. Brown, Division of Biological Sciences, University of Michigan, Ann Arbor

Frank Burton, Department of Bacteriology and Immunology, University of North Carolina, Chapel Hill

Allan Campbell, Department of Biological Sciences, Stanford University, Stanford

Aravinda Chakravarti, Department of Biostatistics, University of Pittsburgh, Pittsburgh

Mary Comer, Department of Bacteriology and Immunology, University of North Carolina, Chapel Hill

Marshall H. Edgell, Department of Bacteriology and Immunology, University of North Carolina, Chapel Hill

Barry G. Hall, Department of Microbiology, University of Connecticut, Storrs

John G. Hall, Department of Ecology and Evolution, State University of New York, Stony Brook

Stephen C. Hardies, Department of Bacteriology and Immunology, University of North Carolina, Chapel Hill

Alison Hill, Department of Bacteriology and Immunology, University of North Carolina, Chapel Hill

Clyde A. Hutchison, III, Department of Bacteriology and Immunology, University of North Carolina, Chapel Hill

Thomas H. Jukes, Space Sciences Laboratory, University of California, Berkeley

Haig H. Kazazian, Jr., Department of Pediatrics, Genetics Unit, Johns Hopkins University School of Medicine, Baltimore

Motoo Kimura, National Institute of Genetics, Mishima, Japan

Richard K. Koehn, Department of Ecology and Evolution, State University of New York, Stony Brook

Robert A. Lansman, Department of Molecular and Population Genetics, University of Georgia, Athens

Wen-Hsiung Li, Center for Demographic and Population Genetics, University of Texas, Houston

Masatoshi Nei, Center for Demographic and Population Genetics, University of Texas, Houston

Stuart H. Orkin, Department of Pediatrics, Boston Children's Hospital, Harvard Medical School, Boston

Sandra Phillips, Jackson Laboratory, Bar Harbor, Maine

Robert K. Selander, Department of Biology, University of Rochester, Rochester

Charlie Voliva, Department of Bacteriology and Immunology, University of North Carolina, Chapel Hill

Steven Weaver, Department of Biological Sciences, University of Illinois, Chicago

Thomas S. Whittam, Department of Biology, University of Rochester, Rochester

Anthony J. Zera, Department of Ecology and Evolution, State University of New York, Stony Brook

ACKNOWLEDGMENTS

This volume is based on a symposium, "Evolution of Genes and Proteins," organized by R. K. Koehn and M. Nei, and sponsored by The Society for the Study of Evolution at Stony Brook, New York, June 23–24, 1982. We gratefully acknowledge financial sponsorship of the symposium and this volume by the Offices of the Vice Provost for Research and Graduate Studies and the Dean of Biological Sciences of the State University of New York, Stony Brook, the Stony Brook Foundation, and Grant DEB 8118404 from the National Science Foundation. Kathleen Ward, Center for Demographic and Population Genetics, University of Texas at Houston, compiled and checked the Bibliography. Her careful and diligent work is greatly appreciated.

Chapter 2/Wen-Hsiung Li
The author thanks Takashi Gojobori for his help in the preparation of the manuscript and Gregory Whitt for discussions. This study was supported by grants from the National Science Foundation and the National Institutes of Health.

Chapter 6/Richard Koehn, Anthony Zera, and John Hall
Preparation of the manuscript was supported by USPHS Grant GM 21131 and NSF Grant DEB 7908802. This is contribution 431 from Ecology and Evolution, State University of New York, Stony Brook.

Chapter 8/John C. Avise and Robert A. Lansman
This work has been supported by NSF Grants DEB 7814195 and DEB 8022135. The authors thank Charles Aquadro and Berry Greenberg for supplying unpublished data. Charles Aquadro also critically reviewed the manuscript.

Chapter 9/Masatoshi Nei
The assistance from Takashi Gojobori and Dan Graur in the preparation of the manuscript is acknowledged. The author's research was supported by grants from the National Institutes of Health and the National Science Foundation.

Chapter 10/Thomas Jukes
Support from NASA Grant NGR 05-003-460 and the assistance of Carol Fegté are acknowledged.

EVOLUTION OF GENES AND PROTEINS

CHAPTER 1

EVOLUTION OF THE MOUSE
β GLOBIN COMPLEX LOCUS

M. H. Edgell, Stephen C. Hardies, Betty Brown, Charlie Voliva, Alison Hill, Sandra Phillips, Mary Comer, Frank Burton, Steven Weaver, and Clyde A. Hutchison III

The globins are an exceptionally well studied gene system and hence represent an excellent molecular data base with which we can articulate and challenge assumptions concerning the evolution of DNA sequences. Our interests, as molecular geneticists, have been primarily to recognize the regulatory elements within the gene system and secondarily to understand the biological mechanisms that control genome organization. As such, we are newcomers to evolutionary analysis, and hence what we would like to do in this opening chapter is to use the mouse β globin genes as an opportunity to raise issues. Those issues will be addressed more fully either elsewhere in this volume or in the technical literature.

One can imagine using comparative biochemical genetics as a method of recognizing important components of a complex gene system like the β globins. Utilizing such a strategy, one might compare the β globin gene clusters of various species such as the mouse, rabbit, goat, and human in order to identify conserved features in the loci. In such a manner, assuming that the sequence features important to globin metabolism will have changed less than other features, one would expect to find the important regulatory elements relatively conserved. This approach seems threatened by the considerable divergence in gene organization actually found when one compares the

1

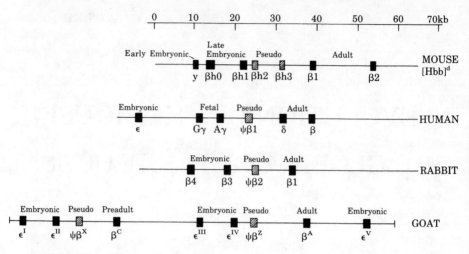

FIGURE 1. Mammalian β globin complex loci.

characterized complex globin gene loci (Figure 1). The gene clusters do not in fact share the same number of gene-like structures, and therefore it is not a trivial matter to decide which genes are to be compared to which (i.e., which are the evolutionarily homologous or "orthologous" sequences in the various species). That there has been considerable divergence in the organizational features of the gene family when one compares different species seems to be a general property of the genome and is not just a special property of the globins. Hence, the identification across species of the gene pairs that are truly orthologous becomes a serious issue. Generally, we feel that the regulation of a gene system is intimately tied into the structural features of the gene cluster. This unexpected degree of structural divergence must, therefore, cause us to at least consider the possibility that the regulatory features of the various globin clusters may not, in fact, be identical.

MOUSE β GLOBINS

The mouse β globin haplotype, $[Hbb]^d$, specifies four β-like proteins: two adult β globins and two nonadult β globins (Russell and Mc-Farland, 1974). However, the 65 kilobases (kb) of DNA cloned from this locus contains seven gene-like structures (Tilghman et al., 1977; Jahn et al., 1980; Edgell et al., 1981) with sequence homology to the adult β globin genes (Figure 1). Three of these were shown by sequence analysis (Jahn et al., 1980; Konkel et al., 1978, 1979; Hansen et al., 1982) to correspond to known β globin proteins: the adult proteins d-major (dmaj) and d-minor (dmin) and the nonadult protein y. In order

2

to have unique gene names and to retain the traditional gene/allele nomenclature, we have renamed the adult genes β1 and β2 to replace the β previously used for both adult genes (β^{maj} and β^{min}, respectively). The four additional genes were given the designation βh, which refers to "beta homologous."

We have done extensive cross-hybridization analyses to define the locations of repetitive sequences within the complex locus (Figure 2). At least six repetitive DNA families have been defined in the locus. The fraction of sequence that is repetitive is not uniform within the β globin locus. At the resolution of *Hae*III fragments probed with labeled genomic DNA, the embryonic region contains only 15% repetitive DNA as compared to 50% in the adult region (F. Burton, pers. comm.).

We have examined the other β globin haplotype prevalent in *Mus musculus*, [Hbb]^s, by library construction, cloning and characterization (Weaver et al., 1981). Given the considerable degree of organizational divergence between species, we were interested in determining the degree of homology existing between two haplotypes. At the level of electron-micrographic heteroduplex analysis we have found very few differences between the haplotypes (Figure 3). There is an interesting pair of insertions in [Hbb]^d near the β2 gene, but in general the haplotypes are quite homologous. However, not all of the structures within the locus are evolving at the same rate. For example, as we will demonstrate later, βh2 and βh3 are evolving much more rapidly than the functional genes. Therefore, the observed extent of homology implies either that the two haplotypes have only recently become distinct or that there is nonreciprocal sequence exchange between the two haplotypes on a quite large scale.

The [Hbb]^s haplotype gives rise to only a single adult β globin. However, both β1 and β2 are transcribed in this haplotype (S. Weaver

FIGURE 2. The β globin complex locus in *Mus musculus* haplotype [Hbb]^d. Functional genes are indicated by the large filled-in blocks and pseudogenes by the hatched blocks. The smaller filled-in blocks of various shapes mark the location and size of the repetitive sequence family which we have most extensively characterized. The stick/flag symbols tag the locations of at least five other repetitive sequences. These latter sequences are probably quite short.

3

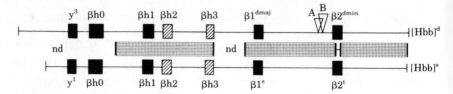

FIGURE 3. Homology relations between two globin haplotypes of the mouse (*M. musculus*). The stippled regions between the two maps indicate where "perfect" heteroduplexes form when examined by electron microscopy. The A and B inserts are each approximately 1.5 kilobases in length.

and B. Brown, pers. comm.). Sequence data from the first coding block of the two genes (S. Weaver, pers. comm.) suggest that they have identical coding sequences. Presumably, these adult genes have been subjected to a recent gene conversion event. The *s* allele of β1 is very homologous to the dmaj sequence, but the $\beta 2^{dmin}$ gene is quite different from the other three adult genes.

HOMOLOGY WITHIN THE COMPLEX β GLOBIN LOCUS

Globin genes consist of three coding blocks (exons) and two intervening sequences (introns). A comparison of the large intervening sequences (IVS2s) of $\beta 1^{dmaj}$ and $\beta 2^{dmin}$ indicates considerable divergence (Figure 4). It is usually concluded that because the intervening sequences (IVSs) are more divergent than their associated coding sequences, they must be under less selective constraint than the coding blocks (Chapter 11 by Kimura). This interpretation is quite consistent with our observation that the nucleotide sequences of the IVSs are more divergent than those found in coding blocks. It has been known for a long time that mutations do not distribute themselves uniformly within a sequence (Benzer and Champe, 1961; Drake, 1970). Mutational hotspots and differences in rates for transversions and transitions have been identified in many different systems. Although some of this may be due to selection, these effects are usually attributed to the nature of the nucleotide sequences and mutational processes. Clearly the number of observed differences we see between two sequences is a complex function of nucleotide sequence and depends on both mutational susceptibility and selective constraints.

Generally, the number of observed mutations is considered to be equal to the intrinsic mutation rate times fixation processes. In the absence of selection, the observed mutation rate within a sequence is dependent on both sequence susceptibility and repair. For example, a poly(T) sequence will accumulate more changes due to ultraviolet

4

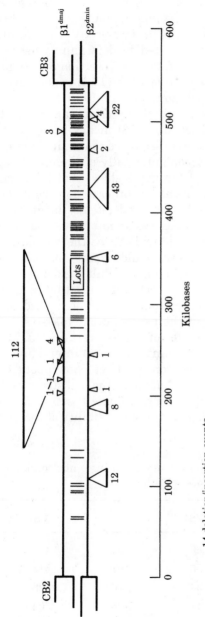

FIGURE 4. A comparison of the large intervening sequences (IVS2s) from two adult β globin genes. CB2 and CB3 refer to the second and third coding blocks, respectively. The vertical lines indicate point mutations and the triangles insertions of the indicated length in nucleotides.

irradiation than will poly(G) sequences. The number of changes that survive will then be determined by the repair mechanisms present in the species. If appropriate, these changes will finally be acted on by the selective constraints peculiar to the species. Therefore, observed differences in rates of divergence can be due to differences in the underlying sequence, which determine the mutation rate, differences in repair processes, and differences in the selective constraints acting on that sequence. Hence, we wanted to determine whether it was possible that the large number of deletion/insertion events in β1 and β2 might be due to the susceptibility of the IVSs to such events.

A large fraction of the deletions found within the *lac* operon in *Escherichia coli* occur between short direct repeats present in the parent sequence (Farabaugh et al., 1978). It has been proposed (Efstratiadis et al., 1980) that similar sequence constraints apply to deletion events in eukaryotes. A computer tally of the direct repeats in the various regions of the β globin genes shows that IVS2 has in fact five times as many direct repeats as any other region (Table 1). Hence, if direct repeat pairs are the substrate for deletions, the many deletions in the large IVS are really not in excess. In fact, if we take the number of deletions per direct repeat pair in IVS2 as the standard deletion frequency for this gene, then we would expect approximately one deletion per coding block (Table 2) for the adult β globins on the basis of repeat density. If one takes into account the actual distribution of sequence, the distribution of deletions can be accounted for on the basis of that sequence without the need to evoke massive differences in selective constraints. We are, of course, certain that there are differences in the selective constraints acting on intervening and coding sequences, but that is based on what we know about biological mechanisms. Without knowing what the actual substrate really is for deletion/insertion events, one cannot evaluate the size of the contribution made by either selection or the susceptibility of the underlying sequence.

TABLE 1. Numbers of perfect direct repeats of three nucleotides or longer that fall within 40 nucleotides of each other.

Gene	CB1	IVS1	CB2	IVS2	CB3
β1^{dmaj}	48	60	99	486	55
β2^{dmin}	54	56	101	516	63
βh2	42	34	110	528	66
βh3	39	41	36*	604	54

* This figure is from the truncated coding block of the βh3 pseudogene.

6

TABLE 2. Predicted deletion/insertion
events for the mouse adult β globin gene.

Gene region	Number of events*
CB1	0.69
IVS1	0.86
CB2	1.43
IVS2	7†
CB3	0.79

* These values were generated by attributing one-
half of the 14 events in IVS2 to each gene and
taking the number of events per direct repeat pair
as the "standard" event frequency.
† Normalizing value.

There is a family of repetitive sequences that occurs at nine locations within the β globin locus (Figure 2). These repetitive sequences
have an interesting organization in that they are 90% homologous to
each other but appear to terminate at random points, relative to each
other. What is the source of this extensive homology, and should that
homology be attributed to selection? These sequences occur 20,000–
50,000 times in the mouse genome. We find it hard to imagine that a
mouse is put at a serious disadvantage if one of these repetitive sequences diverges by more than a few percent. We know from other
investigations that there are processes in the cell that allow two
sequences to become more homologous (see Chapter 3 by Arnheim).
Unexpected homologies between different genes are often attributed
to gene conversion as in the case of the two adult α globin genes in
mammals or the fetal γ globin genes in humans. The molecular mechanism of gene conversion is unknown, but we can be certain that the
process is an enzymatically mediated event; and thus, there must be
sequence constraints on the process. The constraints might be quite
broad (such as simply requiring homology) or quite narrow (requiring,
for example, a precise recognition sequence). Whatever the constraints
are, it would seem quite possible that a set of sequences might come
to satisfy those constraints and therefore be maintained by gene conversion. This would result in a high degree of homology whether or
not it conferred an advantage to the organism. Therefore, in the case
of these repetitive elements, we cannot yet determine whether the

7

high level of homology is due to the properties of the sequence itself or to consequences of the sequence, which are acted on by selection. However, we are inclined to the former.

In the molecular literature, homology is often attributed to selection. It is, however, clear that nucleotide sequence can itself significantly influence the rate of sequence divergence and must, therefore, be examined as an additional factor. We argue that it would be quite useful to try to extract from the sequence data the rules that govern intrinsic divergence rates in the absence of selection in order to assess the contributions of sequence per se versus selection in any given case.

THE β HOMOLOGOUS STRUCTURES

The βh3 gene is an illegitimate recombination product between two genes of differing degrees of homology to the current adult genes (Figure 5). The complete nucleotide sequence of βh3 is known (C. A. Hutchison, pers. comm.) and is quite divergent from the other β globin genes. Besides the large number of point mutations, there is a 160-base pair deletion (presumably due to unequal crossing-over) and an insertion of one base in codon 90, causing a frameshift. There are terminators in all three reading frames, and the normal initiation and termination codons are absent. βh3 is therefore not capable of producing a β globinlike protein and has the common features of a pseudogene. The gene is present as a recombinant in both the [Hbb]^d and [Hbb]^s haplotypes of *Mus musculus*. There are also sequences homologous to βh3 in all of the other species of *Mus* that we have tested and in the deer mouse, *Peromyscus maniculatus*. However, we do not know as yet whether the βh3 sequences present in the other species are in a recombinant or parental gene.

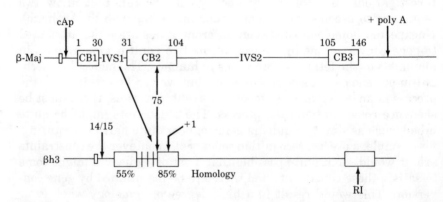

FIGURE 5. The structures of β-maj (β1^{dmaj}) (functional gene) and βh3 (pseudogene). CB and IVS refer to the coding block and intervening sequences, respectively.

8

The βh0 and βh1 structures appear to be normal embryonic genes. The complete sequence of βh0 and a partial sequence of βh1 is known (A. Hill, pers. comm.). These genes do not appear to have any defects that would preclude normal gene function, and both give rise *in vitro* to transcripts using the Manley extract system. Large amounts of transcript also occur in the GM979 erythroleukemic cell line in which the fetal y gene is expressed (Brown et al., 1982). Analysis of poly(A) RNA indicates that at least βh1 is expressed in 13- to 14-day-old embryos.

The βh2 gene is the most divergent of the mouse β globin genes with only 70–74% homology to the adult $\beta 2^{dmin}$ coding blocks. Although the complete sequence of βh2 is known (S. Phillips, pers. comm.), its analysis does not allow us to conclude whether or not βh2 can be transcribed or translated. We can identify problems with the promoter and with the normal IVS1 donor splice site, but it is not clear whether the problems are severe enough to inactivate the gene.

We have examined a variety of other species for the presence of sequences homologous to βh2. We can detect specific hybridization to *Eco*RI fragments in *M. musculus, M. castaneous, M. caroli, M. cervicolor,* and *M. pahari.* However, we see no hybridization to βh2 sequences in *Peromyscus bairdii,* a rodent species that has been estimated to have diverged from *Mus* approximately 65 million years ago. Hence, βh2 must have been present in *Mus* prior to subspeciation of the genus. A similar structure may, of course, exist in *Peromyscus* but be undetectable with our probes due to rapid divergence.

βh2 has been accumulating nucleotide changes at a rate comparable to that of pseudogenes. Despite this rapid rate of divergence, most of the structural features for producing a protein are still present. We have asked whether the nucleotide changes have been accumulating randomly and whether the features that confer globin function have changed as rapidly as the remainder of the sequence. To answer this question we compared βh2 to a "prototype" β globin that we constructed from the most frequent amino acids found at each position in a set of nine functional β globins (S. Phillips and S. Hardies, in preparation). Differences from the prototype sequence were scored (Table 3) for heme interacting residues, cooperative residues, and "other" residues (Eaton, 1980). As would be expected, the functional genes (e.g., $\beta 2^{dmin}$), analyzed in this fashion, have been accumulating more changes in their less important "other" residues than in the two functional residue classes. However, all three residue classes have changed equally in βh2. Apparently βh2 evolves rapidly without evidence of the same sort of constraints imposed on the functional β globin genes.

9

TABLE 3. The distribution of mutations within the βh2 and β2^{dmin} genes.

Residue class*	βh2		β2^{dmin}	
	Conserved	Nonconserved	Conserved	Nonconserved
Heme interacting	13	7 (35%)	19	1 (5%)
Cooperative	19	14 (42%)	31	2 (6%)
Other	61	32 (34%)	82	11 (12%)

* The functioning classes of residues are defined in Eaton (1980). Other residues represent "nonfunctioning" residues in that they are not involved in heme binding or have contact with the α chains or the other β chains within the tetramer.

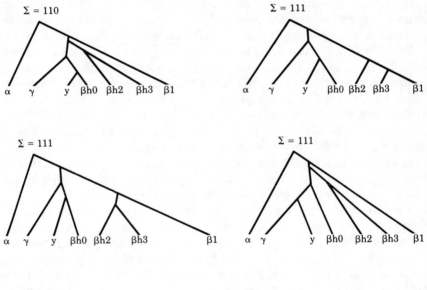

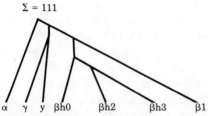

FIGURE 6. Phylogenies derived by parsimony for various β globin genes. All of the globin genes are from mouse except for the human γ globin gene. Coding blocks 2 and 3 are used. The total number of mutations within each tree is indicated by Σ. Branch length is not relevant in these diagrams.

10

The general organization of βh2 is that of a canonical β globin gene. It has retained the recognized structural features that are required to produce a protein but does not appear to be evolving as a functional gene. We have found no transcript either *in vivo* (Brown et al., 1982) or in the relaxed Manley *in vitro* transcription system (B. Brown, pers. comm.). All of these features suggest that βh2 does not code for a functional β globin. It does not, on the other hand, satisfy most of the conditions usually used to define a pseudogene.

THE MOUSE β GLOBIN GENES AS A FAMILY

We have used a minimum distance tree algorithm (W. Fitch, pers. comm.) to examine the phylogeny of the mouse β globin genes and the human γ globin genes. This procedure, which is an attempt to unravel the history of the β globin genes, will of course be imprecise due to the large degree of divergence within the family giving rise to many obscuring parallel and backward mutations. Several trees are generated by this program that are within one base change of each other (Figure 6). The βh2 and βh3 structures are particularly difficult to localize within the phylogeny, and their variability accounts for most of the differences between the various trees. Presumably this is due to the higher rate of divergence experienced by these genes and hence the larger number of obscuring backward and parallel mutations. An ancient origin for these genes would also contribute to the difficulty in constructing a reliable tree. All of the various topologies can be summarized in a single tree with ambiguities (Figure 7).

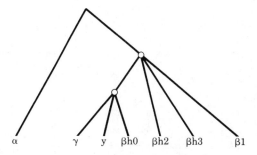

FIGURE 7. A mouse β globin phylogeny. All of the globin genes are from the mouse except for the human γ globin gene. Coding blocks 2 and 3 are used. The two circles represent nodes with ambiguity.

11

Our analysis of the sequences 3′ to (downstream from) the coding blocks indicates that the mouse βh3 3′ sequences are quite homologous to the analogous sequence associated with the human δ gene (Figure 8). We conclude from this that the 3′ portion of the recombinant βh3 gene is orthologous to the human δ gene. The correspondence of δ and βh3 in this region has been foreshortened by the incursion of a repetitive element that has displaced the majority of the βh3 3′ untranslated region. The evidence indicating two different ancestral adult genes prior to the mammalian radiation is not limited to δ and βh3. Comparisons of the entire 3′ untranslated regions of mouse, rabbit, and human adult genes revealed a grouping of rabbit ψβ2 with human δ and mouse βh3 as well as a distinct grouping of human β with rabbit β1 and mouse β1 and β2 (see Figure 1). In order for the β and δ equivalents to be distinguishable in the rodents, they must have duplicated at a time sufficiently prior to the mammalian radiation. Consequently, the homology observed today between δ and β in humans must be the result of a gene conversion event. Our interpretation drawn from the minimum distance trees, that βh3 is of ancient origin, is also supported by this 3′ sequence data insofar as βh3 and the adult mouse β genes must have separated prior to the primate–rodent radiation.

The fact that βh3 is a recombinant gene makes it clear that different portions of the gene have had different histories, and hence any analysis that pools the entire sequence is unlikely to be meaningful. If recombinant genes exist without obvious features defining the junction between the two parts, it would be less easy to decide how to segment the sequence for analysis. An example of a domain within a globin gene with a history different from the remainder is the gene-converted portion of the human γ gene (Slightom et al., 1980). Clearly, it will be important to determine whether each portion of a gene structure shows the same history.

```
(A)   Terminator
δ     TGA | GAT   CCT   GGA   CTG   TTT   CCT
          | |     | |   |||   ||    |||   |||   14/18 (78%)
βh3   AGA | GGC   CTT   GGA   CTA   TTT   CCT
          | |     |           ||      |         5/18 (28%)
β     TAA | GCT   CGC   TTT   CTT   GCT   GTC
          |  |    | |   |||   | |   |||    |    12/18 (67%)
β1    TAA | ACC   CCC   TTT   CCT   GCT   CCT

(B)   Terminator
δ     TGA | GAT   CCT   GGA   CTG   TTT   CCT
          | |     | |   |||   ||    |||   |||   14/18 (78%)
βh3   AGA | GGC   CTT   GGA   CCA   TTT   CCT
          | |     |                 |||   | |   7/12 (58%)
β     TAA | GCT   CGC   ---   ---   TTT   CTT
          |  |    | |               |||   | |   8/12 (67%)
β1    TAA | ACC   CCC   ---   ---   TTT   CCT
```

FIGURE 8. Alignments of the sequences just 3′ to the coding regions of several β-like globin genes. A. No gaps. B. Gaps to maximize homology.

CONCLUSION

What can we conclude from our foray into evolutionary analysis? Sequence divergence is clearly a complex process depending in a complicated way on the sequence interaction with the enzymology of mutation and repair as well as on the functions expressed by those sequences. A complete analysis will require that we continue to factor out and identify the various processes and rules that control divergence. Fortunately, the potential data base is large. However, unless we generate experimental approaches that test our models for such rules, we will need to rely solely on deductive methods, which, as the reader well knows, have limitations.

CHAPTER 2

EVOLUTION OF DUPLICATE

GENES AND PSEUDOGENES

Wen-Hsiung Li

Gene duplication is probably the most important mechanism for generating new genes and new biochemical processes that have facilitated the evolution of complex organisms from primitive ones. It is also important for generating many genes of the same function and thereby enables the production of a large quantity of RNAs or proteins. The fact that gene duplication has played a vital role in evolution can be seen from the comparison of the DNA contents of various organisms. The DNA content is the lowest in simple organisms such as viruses and bacteria and generally increases as the complexity of the organism increases (Britten and Davidson, 1969). For example, the DNA content of mammalian organisms is approximately 1000 times higher than that of *Escherichia coli,* which is in turn approximately 1000 times higher than that of phage φX174. The existence of numerous systems of homologous proteins or genes in higher organisms also indicates that gene duplication has occurred frequently in the evolutionary process (Dayhoff, 1972).

Gene duplication may occur in two different ways, that is, by tandem gene duplication and by genome duplication. Ohno (1970) has argued that genome duplication is generally more important than tandem duplication, because the latter may duplicate only parts of the genetic system of structural genes and regulatory genes and may disrupt the function of duplicated genes, whereas the former duplicates the entire genetic system with little harmful effect. Recent studies of genome organization in eukaryotes, however, suggest that tandem duplication has occurred quite frequently and is probably as important as genome duplication.

14

Gene duplication seems to have been important even in the refinement of genes in evolution. Many proteins of present-day organisms show internal repeats of amino acid sequences (for example, serum albumin), and these repeats often correspond to the functional or structural domains of the proteins (Barker et al., 1978). This observation suggests that the genes coding for these proteins were formed by internal gene duplication and that the function of the genes was improved by increasing the number of active sites or stability of the proteins produced. The recent finding that eukaryotic genes generally consist of several exons and introns (Figure 1) and that exons often correspond to functional or structural domains of proteins (Gilbert, 1979; Gō, 1981) also suggests that many genes in present-day organisms were formed by duplication of primordial genes that probably existed in the early stage of life and were presumably small in size and simple in function.

However, duplicate genes do not always lead to genes with new functions. A duplicate gene may accumulate deleterious mutations and become nonfunctional, as long as the other duplicate genes are functioning normally (Haldane, 1933). Considering the pattern of increase of genome size from bacteria to mammals, Nei (1969) predicted that the genome of higher organisms contains a large number of nonfunctional genes. Interestingly, recent DNA sequencing studies of eukaryotic genomes have revealed many nonfunctional genes (i.e., pseudogenes) in various genetic systems. Furthermore, comparisons of nucleotide sequences of pseudogenes and their functional counter-

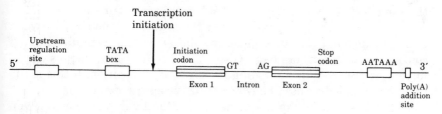

FIGURE 1. Schematic representation of an eukaryotic gene. The upstream regulation site controls the initiation of transcription. The TATA box is apparently involved in fixing the initiation of transcription within a narrow area. Exons are coding sequences and introns are intervening sequences. An intron starts with GT and ends with AG; there must be additional signals for splicing, but we still do not know what they are. The AATAAA (or a minor variation of it) near the poly(A) addition site might be important for poly(A) addition or RNA processing.

parts indicate that pseudogenes are subject to rapid nucleotide changes in the evolutionary process. This finding not only has an important implication for testing the neutral mutation hypothesis (see Chapter 11 by Kimura) but also provides useful information about the intrinsic rate of mutation (Li et al., 1981).

GENE ELONGATION

As mentioned earlier, most of the genes in eukaryotes are apparently products of the duplication of primordial genes. This increase in gene size, or gene elongation, was apparently one of the most important steps in the evolution of complex genes from primitive ones. This view is clearly supported by the nucleotide sequences of the ovomucoid genes in birds. Ovomucoid is a protein present in bird egg whites and responsible for the trypsin inhibitory activity of egg whites. The ovomucoid polypeptide can be divided into three functional domains (Figure 2). Each domain is capable of binding one molecule of trypsin or other serine proteases. The homologies at the amino acid level between domains I and II, I and III, and II and III are 46, 33, and 30%, respectively. These features suggest that the ovomucoid gene was derived from triplication of a primordial domain gene (Kato et al., 1978). Interestingly, recent DNA sequencing has shown that the DNA regions coding for the three functional domains are separated by introns (Stein et al., 1980). However, each of the three DNA regions consists of two exons interrupted by one intron rather than one continuous exon. Apparently, this pair of exons originally formed a primordial gene and was later triplicated.

The existence of two exons in each domain DNA region suggests that this region itself is a product of a more ancient gene duplication, the intron being a remnant of the ancient intergenic region. However, the homology between the two exons within the same domain DNA region is not high. It is possible that the sequence homology between them disappeared because of accumulation of many mutations. Of course, as some authors (e.g., Gilbert, 1979) proposed, introns might have been inserted in the coding region by transposition. If this is the case, no homology would be expected between the two exons.

A remarkable example of gene elongation by duplication is given by the $\alpha 2$ type I collagen gene [$\alpha 2$(I)]. Collagen is the main supportive protein of skin, bone, cartilage, and connective tissue in vertebrates. The $\alpha 2$(I) gene from chicken has been partially sequenced (Yamada et al., 1980; Wozney et al., 1981; Ohkubo et al., 1982); it has a length of approximately 38 kilobases and contains more than 50 exons, the largest number of exons ever observed in a gene. Twenty-one of the exons coding for the collagen helix have been sequenced. Two of these exons contain 45 base pairs (bp); twelve, 54 bp; four, 99 bp; and three,

DOMAIN I

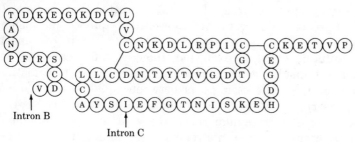

Intron B

Intron C

DOMAIN II

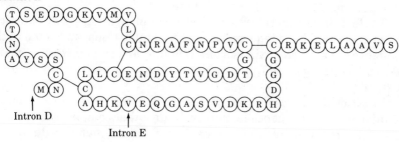

Intron D

Intron E

DOMAIN III

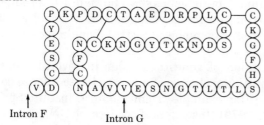

Intron F

Intron G

% HOMOLOGY

DOMAINS	A.A.	N.A.
I/II	46	66
II/III	30	42
I/III	33	50

FIGURE 2. The three functional domains of the secreted ovomucoid. The homologies between domains at the amino acid (A.A.) level and at the mRNA (N.A.) level are shown at the bottom of the figure. (From Stein et al., 1980.)

108 bp—all in multiples of the 9 bp that code for the triplet Gly-X-Y, where X and Y are often prolines (Figure 3). It is quite possible that all of these exons were derived from an ancestral exon of 54 bp by multiple duplications and recombinations (Figure 4). For example, a 99-bp exon could have arisen from an unequal crossing-over between two exons of 54 bp. Ohkubo et al. (1982) suggested that exons of 108 bp arose by precise deletion of the intervening sequence and fusion of two 54-bp exons. However, the precise deletion of an intron appears to be an unlikely event in evolution, though such a case has been found in a rat insulin gene (Lomedico et al., 1979). It seems more likely that an unequal crossing-over between a 99-bp exon and a 54-bp exon gave rise to a 108-bp exon and a 45-bp exon (Figure 4). At any rate, because of its regular repetitive structure, partial gene duplication can occur easily in this gene and would not disrupt the gene function as long as the duplicated part is a multiple of the 9-bp sequence coding for Gly-X-Y. This basic sequence of 9 bp seems to be necessary for forming a helical structure.

Table 1 shows more examples of gene elongation by duplication. All of them involve one or more domain duplications, and some of the sequences (e.g., ferredoxin, serum albumin, and tropomyosin α chain) were evidently derived from doubling or multiplication of a primordial sequence.

In all of the preceding examples the duplication event can easily be inferred from sequence homology. Many other complex genes might have also evolved by internal gene duplication, but the duplicated regions have become so diverged that sequence homology between them is no longer discernible. For example, the variable and constant region domains of immunoglobulin genes show no significant sequence homology, but their tertiary structures suggest that the two kinds of domains were derived from a common ancestral domain (see Hood et al., 1975).

Theoretically, elongation of genes can occur in several different ways. For example, mutational change of a stop codon to a sense codon may elongate the gene (for example, Hemoglobin Constant Spring; Weatherall and Clegg, 1979). Insertion of a piece of DNA sequence into one of the exons or mutation at a splicing site may also elongate the gene. However, these types of mutations would most probably

```
GGC CCT CCT GGG TTT CAA GGT GTT CCT GGT GAA CCT GGT GAA CCT GGT CAA ACA
Gly Pro Pro Gly Phe Gln Gly Val Pro Gly Glu Pro Gly Glu Pro Gly Gln Thr
19                                                                    36
```

FIGURE 3. DNA sequence of an α2(I) collagen exon. The number under the amino acids corresponds to the position of the residue in the polypeptide.

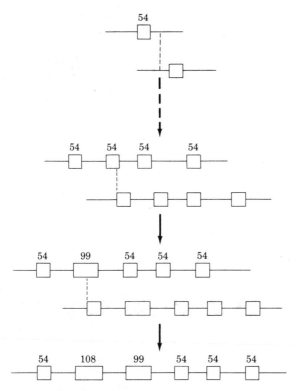

FIGURE 4. A hypothetical scheme for the evolution of the exons in the DNA region coding for the helical region in the α2(I) collagen polypeptide. The number on the top of a block denotes the number of base pairs (bp) in the exon. The dotted line denotes the place of unequal crossing-over. An unequal crossing-over between two exons of 54 bp can give rise to an exon of 99 bp and an exon of 9 bp, and an unequal crossing-over between an exon of 99 bp and an exon of 54 bp can give rise to an exon of 108 bp and an exon of 45 bp.

disrupt the function of the elongated gene because the elongated region would often include stop codons and frameshifts or coding frames that are not coordinated with the extant coding regions. Compared with these mechanisms, duplication of a domain is less likely to disrupt the gene function. Moreover, it can enhance the function of the protein produced by increasing the number of active sites or the stability of the protein. Because of these advantages, gene elongation seems to have depended largely on duplication of functional domains.

Occasionally, duplication of a domain may enable a gene to perform

19

TABLE 1. Proteins with domain duplications. (From Barker et al., 1978.)

Sequence	Length of protein*	Length of repeat	Number of repetitions	Percentage of repetition
Immunoglobulin ε chain C region	423	108	4	100
Immunoglobulin γ chain C region	329	108	3	98
Serum albumin	584	195	3	100
Parvalbumin	108	39	2	72
Protease inhibitor, Bowman-Birk (soybean)	71	28	2	79
Protease inhibitor, submandibular gland	115	54	2	94
Ferredoxin (*Clostridium pasteurianum*)	55	28	2	100
Plasminogen	790	79	5	50
Calcium-dependent regulator protein	148	74	2	100
Tropomyosin α chain	284	42	7	100

* Number of amino acid residues.

a new function. In fact, it seems that the coenzyme binding domain of several dehydrogenases, which binds nicotinamide adenine *dinucleotide* (NAD), was derived from duplication of a primordial *single-nucleotide* binding domain (Rossman et al., 1975). Moreover, two duplicated domains may diverge in function and enable the gene to perform a new or more complex function. For instance, the variable and constant region domains of immunoglobulin genes were probably derived from a primordial domain gene, but they now have distinct functions— antigen recognition by the variable regions and effector functions by the constant regions (Leder, 1982). Many complex genes might have arisen in this manner.

FORMATION OF GENE FAMILIES AND NEW GENES

Highly repetitive genes

Gene duplication has been important in increasing the number of copies of genes that are required for producing a large quantity of a specific kind of RNA or protein. Representative of these are the genes for rRNAs and tRNAs, which are required for translation of mRNAs, and for histones, which are the basic chromosomal proteins found in all eukaryotes and which must be synthesized in a large quantity during the S phase of the cell cycle (Elgin and Weintraub, 1975).

Table 2 shows the numbers of rRNA and tRNA genes for a variety of organisms. The mitochondrial genome of mammals contains only one 12S and one 16S rRNA gene. This is apparently sufficient for the mitochondrial translation system, because the genome has only 13 protein-coding genes (see Anderson et al., 1981). The mycoplasmas are the smallest self-replicating prokaryotes. Interestingly, the genome of *Mycoplasma capricolum,* the only species studied, contains only two sets of rRNA genes. The genome of *Escherichia coli* is 4 or 5 times larger than that of *M. capricolum* and contains 7 sets of rRNA

TABLE 2. Numbers of rRNA genes and tRNA genes per haploid genome in various organisms.

Gene	Organism	Number*	Genome size (base pairs)	References†
rRNA	Human mitochondrial genome	1	16,600	1
	Mycoplasma capricolum	2	1×10^6	2
	Escherichia coli	7	4×10^6	3
	Saccharomyces cerevisiae	140	5×10^7	4
	Drosophila melanogaster	130 ~ 250	2×10^8	4
	Xenopus laevis	400 ~ 600	8×10^9	4
	Human	300	3×10^9	4
tRNA	Human mitochondrial genome	22	16,600	1
	E. coli	100	4×10^6	5
	S. cerevisiae	320 ~ 400	5×10^7	4
	D. melanogaster	750	2×10^8	4
	X. laevis	7800	8×10^9	4
	Human	1300	3×10^9	4

* The numbers for eukaryotic species are rough estimates.
† References: 1. Anderson et al. (1981), 2. Sawada et al. (1981), 3. Kiss et al. (1977), 4. Tartof (1975), 5. Ozeki (1980).

genes. The number of rRNA genes in yeast is approximately 140, and the numbers in fruit flies and human are even larger. *Xenopus laevis* has a larger genome size and a larger number of rRNA genes than humans. Thus, there is a good correlation between the number of rRNA genes and genome size. This is also true for the number of tRNA genes (Table 2).

Highly repetitive genes such as the rRNA genes are generally very homogeneous, the sequence homology between member genes being very high. One factor responsible for the homogeneity may be purifying selection, because such genes may have very specific functional or structural requirements. Another factor is "concerted evolution," which occurs because of unequal crossing-over and gene conversion (Ohta, 1980; see Chapter 3 by Arnheim).

Moderately repetitive genes

In addition to highly repetitive genes, the genome of higher organisms contains many multigene families whose members have diverged to various extents in regulation and function. These include the collagen genes (Bornstein and Sage, 1980), the immunoglobulin genes (see below), and the hemoglobin genes. The hemoglobins in humans are encoded by two gene families: the α family and the β family. The α family consists of one embryonic (ζ) and two adult ($\alpha 1$, $\alpha 2$) genes, whereas the β family consists of one embryonic (ϵ), two fetal ($^G\gamma$, $^A\gamma$), and two adult (β,δ) genes (see Chapters 1 and 7 by Edgell et al. and Kazazian et al., respectively). Distinct globins are expressed at the embryonic ($\zeta_2\epsilon_2,\alpha_2\epsilon_2$), the fetal ($\alpha_2\gamma_2$), and the adult ($\alpha_2\beta_2,\alpha_2\delta_2$) stages of development. Some differences in oxygen-binding affinity have evolved among these globins. For example, the fetal hemoglobin (Hb F) has a higher oxygen affinity than the adult hemoglobin (Hb A), because it does not bind 2,3-diphosphoglycerate as strongly as Hb A and thus functions better in the fetus, which exists in a relatively hypoxic state (Wood et al., 1977). This example shows how gene duplication can permit refinements of a physiological system. As discussed by Dayhoff (1972), all these genes are evolutionarily related.

Immunoglobulin genes

One of the most elaborate systems of multigene families is that controlling the immune system in vertebrates. This system has a capacity to generate a virtually unlimited number of different antibodies (see Leder, 1982).

An antibody molecule consists of two identical light chains and two identical heavy chains (Figure 5). In most vertebrates, there are two types of light chains (κ and λ) and five types of heavy chains (μ,

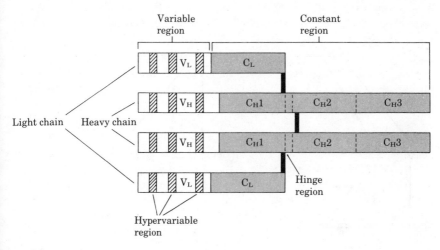

FIGURE 5. The basic structure of the immunoglobulin (IgG) molecule.

δ, γ, ε, and α). Both the light and the heavy chains can be divided into a variable (V) and a constant (C) region (Figure 5). The C region of a light chain has only a single functional domain (C_L), but that of the heavy chain consists of three domains: C_H1, C_H2, C_H3 (four domains if the chain is of the μ or ε type, and two domains if it is of the δ type). Each of the V and C domains is approximately 110 amino acid residues long. The V_L and V_H domains exhibit extensive sequence homology with each other, as do the C_L, C_H1, C_H2, and C_H3 domains. In spite of lack of homology at the amino acid level, the V and C domains have very similar tertiary configurations. These features suggest that all immunoglobulin genes were derived from a common precursor approximately the size of one domain (Hood et al., 1975). Figure 6 shows a hypothetical scheme for the evolution of the antibody gene families.

DNA sequencing has revealed that the variable region is not encoded by a single DNA segment but by two (V and J) or three (V, D, and J) separate segments, depending on whether it is a light or a heavy chain (Figure 7). There are usually many V, D, and J segments in the genome (Leder, 1982). During the process of development of the immune cells (B lymphocytes), one V and one J of the κ family are randomly joined to form a κ chain V region, probably by deletion of all intervening DNA. (The λ family has a somewhat different arrangement.) This process is known as V-J joining. Formation of a heavy chain requires V-D-J joining. In addition, there is the so-called heavy-

23

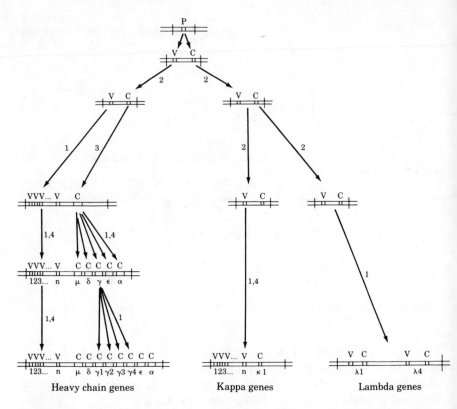

FIGURE 6. A hypothetical scheme for the evolution of the immunoglobulin gene families. The order of gene duplication events is unknown. A number of genetic mechanisms seem to have been employed in the evolution of these families, as indicated by the numbers adjacent to arrows. These are (1) gene duplication by intergenic, unequal crossing-over, (2) gene duplication by polyploidization or chromosomal translocation, (3) gene elongation by internal duplication, and (4) concerted evolution. Mechanisms 1 and 4 may be identical (see Chapter 3 by Arnheim). (Adapted from Gally and Edelman, 1972; Hunkapiller et al., 1982.)

chain class switching in which the C region is switched from μ to δ or one of the other types (Leder, 1982). The combinatorial joining and class switching mechanisms together with the existence of multiple DNA segments enable a mammalian organism to produce millions of different antibody molecules (Leder, 1982).

Multimeric proteins

Many multimeric proteins are composed of polypeptides encoded by duplicated genes. For example, the adult hemoglobin in mammals is

24

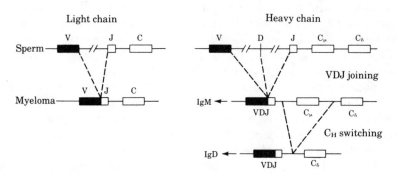

FIGURE 7. DNA rearrangements during the differentiation of antibody-producing cells. In light and heavy chain genes, V-J or V-D-J joining juxtaposes the gene segments encoding the V_L and V_H genes. Subsequently, in heavy chain genes, class (or C_H) switching may occur. The class switching leads to the expression of different immunoglobulin classes. Sperm indicates DNA undifferentiated with regard to antibody function, whereas myeloma denotes DNA in which one or more rearrangements have occurred. (From Hunkapiller et al., 1982.)

a tetramer consisting of two α chains and two β chains, which are encoded by two separate genes located on different chromosomes. The α and β chain genes were probably produced by genome duplication. Because myoglobin and some of the hemoglobins in jawless fish are monomers, polymerization of hemoglobin molecules in mammals probably occurred around the time of the α–β divergence. The mammalian hemoglobin has acquired the capabilities of (1) binding four oxygen molecules cooperatively, (2) responding to the acidity and carbon dioxide concentration of the red cell (the Bohr effect), and (3) regulating its oxygen affinity through the level of organic phosphate in the blood (see McLachlan, 1977). Apparently, the control of hemoglobins by two duplicate genes has facilitated the refinement of the function of the protein, because the two genes are subject to different mutations.

The collagens and the antibody molecules discussed earlier are also polymer proteins controlled by duplicated genes. More examples can be found in Edwards and Hopkinson (1977).

Genes with new functions

When a gene is duplicated, one copy may be free to change in structure and acquire a new function. In the long history of life, a large number of new genes seem to have been generated by this process. A good example is that of myoglobin and hemoglobin. The two proteins diverged about 800 million years (myr) ago (Table 3). Myoglobin remains a monomer and has a higher oxygen affinity than hemoglobin. As mentioned earlier, hemoglobin has become a tetramer and its function has become much more refined. The two proteins are functionally specialized: myoglobin is the oxygen carrier in muscle, whereas hemoglobin is the oxygen carrier in blood.

Another well-known example is that of trypsin and chymotrypsin, which diverged about 1500 myr ago by gene duplication. These two digestive enzymes have acquired distinct functions: trypsin cleaves polypeptide chains at arginine and lysine residues, whereas chymotrypsin cleaves polypeptide chains at phenylalanine, tryptophan, and tyrosine residues. Gene duplication also produced many other proteases related to trypsin (Barker and Dayhoff, 1980).

DIVERGENCE OF DUPLICATE GENES

Rates of regulatory and functional divergence

Table 3 shows the degrees of regulatory and functional divergence for a number of duplicate genes, together with estimates of divergence times. Although regulatory divergence tends to evolve faster than functional divergence, both types of divergence generally occur at a very slow rate. For example, the genes for chymotrypsin A and B probably diverged about 270 myr ago, but no discernible difference in regulation or function has occurred between these two genes. As another example, the genes encoding the H and M subunits of lactate dehydrogenase (LDH) were separated about 600 myr ago, but the two genes have not become completely independent in regulation and produce polypeptides that still have similarities in chemical reactivities and substrate specificities. The only example of rapid functional divergence is the two genes coding for the E and S chains of horse liver alcohol dehydrogenase (ADH). The divergence time between these two genes is approximately 10 myr, but they have already developed some differences in substrate specificities; the S chain has both steroid and ethanol activities, but the E chain has no steroid activity (Pietruszko, 1980). The E and S chains both have 374 residues in length but differ only by 6 amino acids. This suggests that in some cases a few amino acid differences can change the substrate specificity of an enzyme. Note, however, that in all of the cases with less than 40% sequence difference the duplicate genes show no marked functional difference.

There are several factors that can affect the rate of regulatory and functional divergence. One is the type of duplication. In the case of tandem duplication, two duplicate genes can begin to diverge immediately after the duplication; however, in the case of tetraploidization,

26

TABLE 3. Divergence in amino acid sequence (d), divergence in regulation (R), and divergence in function (chemical reactivities and specificities, aggregation properties, and places of action) between duplicate genes.[†]

Gene pair	d (%)	Time (myr)	R[‡]	Chem. diff.[§]	Aggreg. prop.	Places of action[‖]
Trypsin–chymotrypsin	64[¶]	1500[¶]	***	**	+	+
Hemoglobin–myoglobin	77[¶]	800	***	**	***	**
LDH M and H chains	26	600	**	**	+	*
Hemoglobin α and β chains	59[¶]	500	***	*	*	+
Immunoglobulin H and L chains	75[¶]	400[¶]	***	**	*	+
Lactalbumin–lysozyme	63[¶]	350[¶]	***	***	**	**
Growth hormone–prolactin	75	330	**	**	+	*
Chymotrypsins A and B	21[¶]	270[¶]	+	+	+	+
Carbonic anhydrases B and C	40	180	*	*	+	+
Hemoglobin β and δ chains	8[¶]	40[¶]	*	+	+	+
Insulins I and II, rat	4	30	*	+	+	+
Growth hormone–lactogen	15[¶]	23[¶]	**	*	+	**
ADH E and S chains, horse	2	10	?	*	+	+

[†] +, Similar; *, slightly different; **, moderately different; ***, markedly different; myr, million years.
[‡] Divergence in regulation refers to differential expression over tissues or developmental stages, or to differences in the rate of synthesis if the two genes are expressed only in one tissue.
[§] Chemical differences include differences in catalytic action and in binding to substrates, inhibitors, antigens, and so on.
[‖] Places of action refers to organs in the body or to types of differentiated cells.
[¶] Taken from Dayhoff and Barker (1972).

divergence cannot occur if tetrasomic segregation rather than disomic segregation occurs at meiosis (Ohno, 1970). In the case of autotetraploidization, it can take a very long time for disomic segregation to be regained; though in the case of allotetraploidization, the time can be very short (see Li, 1980). The salmonid fishes are thought to have arisen from an autotetraploidization event about 100 myr ago (Lim et al., 1975), whereas the catostomid fishes were apparently derived from an allotetraploidization event about 50 myr ago (Uyeno and Smith, 1972). In salmonids, some chromosomes still show residual tetrasomy, and the majority of duplicate loci do not show differential tissue expressions (May, 1980; Wright et al., 1980). By contrast, in catostomids all chromosomes show disomic segregation and the majority of duplicate loci show differential tissue expressions (Uyeno and Smith, 1972; Ferris and Whitt, 1979). This comparison shows that in the case of autotetraploidization the rate of divergence between duplicate genes can be extremely slow. A second factor is "gene correction"; by a mechanism such as gene conversion, two duplicate genes can correct each other so as to increase their homology (see Chapter 3 by Arnheim). This appears to have occurred in the two γ globin genes in man (Slightom et al., 1980). In the case of tandem duplication, the effect of gene correction can be important. The divergence times given in Table 3 were estimated by using amino acid and nucleotide sequences without considering this effect and assuming that divergence could occur immediately after duplication; thus, they may represent underestimates. Another important factor might be the breadth of substrate specificity. It would be more difficult for duplicate genes with a narrow specificity to develop differential specificities than for duplicate genes with a broad specificity. The slow divergence between the M and H subunits of LDH might be partly due to the narrow specificity of LDH. In Jensen's model for the emergence of a new gene (see below), broad substrate specificity plays an essential role.

Models for emergence of new genes

Several models have been proposed for the emergence of a new gene from a redundant duplicate. A model of gradual evolutionary modification proposes that a new enzyme could evolve from a duplicate copy of a gene to catalyze a similar type of reaction and that evolution would proceed by "stepwise lurches," involving only a minimal change in conformation (Waley, 1969). Thus, for example, a series of amino acid replacements in pyruvate kinase, affecting the region binding pyruvate and phosphoenol pyruvate but not much changing the nucleotide binding region, could give rise to, say, 3-phosphoglycerate kinase. Alternatively, the two kinases could have arisen from a common, less specific precursor. Similarly, a dehydrogenase could evolve from a duplicate copy of a gene for another dehydrogenase, a phosphatase from another phosphatase, and so on. Jensen (1976) and Jensen and Byng (1982) have proposed a similar model. They argue that primitive enzymes possessed very broad substrate specificities and that gene duplication provided the opportunity for specialization. Such

28

evolutionary opportunities may still exist because many contemporary enzymes exhibit considerable substrate ambiguity. The model of gradual modification appears to be reasonable, not only for enzymatic genes but also for nonenzymatic genes. As discussed earlier, functional divergence between duplicate genes generally occurs at a very slow rate.

Horowitz (1965) proposed that biosynthetic pathways could have arisen from "retrograde evolution." Consider an organism living in a primitive environment containing the end product Z of a biosynthetic pathway and also potential intermediates Y, X, and so forth. As Z is gradually depleted, the organism could respond by developing a new enzyme that catalyzes the conversion of Y into Z. As Z is the product of biosynthesis of the new enzyme as well as the substrate of the old enzyme, both enzymes must recognize Z. Thus, the two enzymes would be structurally similar, and modification of a duplicate copy of the old enzyme would be an easy way for producing the new enzyme. As the supply of Y becomes depleted, a modified duplicate of the second gene will be selected if it can catalyze the conversion of X to Y. In this manner, a biosynthetic pathway is created. Horowitz argued that the organization of genes in some operons in prokaryotes supported his model. For example, all of the genes involved in the pathway of tryptophan biosynthesis (but no other genes) are inside the tryptophan operon in *E. coli*, and a simple explanation is that this operon arose from a series of tandem duplications. However, recent DNA sequencing work shows no significant homology between genes in this operon (Yanofsky et al., 1981) and thus lends no support to the model. It is quite possible that some steps in a biosynthetic pathway have arisen in a retrograde manner, but whether all steps in a pathway have arisen in such a manner is doubtful. The main objections to this model are the uncertainty of the availability of many intermediates in the primitive environment, the extreme chemical lability of many intermediates, and the barriers that might be encountered in transporting intermediates into the cell (Hartman, 1975; Jensen, 1976).

Another model is reactivation of a silenced duplicate, that is, a pseudogene in the present terminology (Koch, 1972; Rigby et al., 1974). It is argued that during the silenced period the sequence can accumulate many drastic mutational changes so that it would have a very different nature if it is reactivated. Although a pseudogene would have a relatively much better chance of becoming a new gene than a totally random sequence, the probability of reactivation would still be very small. This is because a pseudogene often contains multiple major defects such as frameshifts (see later), which cannot be easily corrected. One might argue that gene conversion can reactivate a pseudogene. In this case, however, the resultant sequence would not be a

29

new gene, for it would be similar to the "template" sequence. Another difficulty is that the mutations accumulated during the silent period are random and have not been tested by natural selection so that the reactivated gene product is unlikely to be functional or useful to the organism (see also Zuckerkandl, 1975). For these reasons, this model would seem to be of limited applicability.

EVOLUTION OF PSEUDOGENES

Generation of pseudogenes

A pseudogene is a DNA segment that shows high homology to a functional gene but contains defects such as nonsense and frameshift mutations that prevent it from producing a functional product. The first pseudogene was described by Jacq et al. (1977) for the *Xenopus* 5S rRNA gene system. Subsequently, pseudogenes have been found in the α and β globin gene clusters (see Chapter 1 by Edgell et al. and Little, 1982, for a review), the immunoglobulin gene families, and many other multigene families that have been subject to extensive DNA sequencing. The existence of pseudogenes appears to be a widespread phenomenon.

The majority of pseudogenes studied to date have features suggesting that they were derived from nonfunctionalization of duplicate genes. A typical example is the rabbit globin pseudogene $\psi\beta2$, which is located between embryonic gene $\beta3$ and the adult gene $\beta1$ (see Figure 1 in Chapter 1), has two introns at positions identical to those in $\beta1$, and has retained the transcription initiation site and the signal AATAAA near the tail (Lacy and Maniatis, 1980). Apparently, gene duplication and subsequent loss of function of redundant duplicates is the major process of generating pseudogenes. A number of pseudogenes, however, have features characteristic of messenger RNAs, suggesting that they arose via an RNA intermediate (e.g., Hollis et al., 1982; Wilde et al., 1982). The mouse globin pseudogene $\psi\alpha3$, for instance, has cleanly lost its two introns (Nishioka et al., 1980) and is located on chromosome 11, instead of chromosome 15, where the α globin family exists (Leder et al., 1981). Possibly, it was picked up by a retrovirus, lost its introns when the virus passed through an RNA stage in its life cycle, and was carried back to the mouse genome as part of the DNA copy of retroviral RNA (Goff et al., 1980; Leder et al., 1981). This possibility is supported by the finding that the pseudogene is located between segments of retroviral DNA (Lueders et al., 1982). If this hypothesis is true, this gene should have been a pseudogene since the time of its occurrence.

Another way for generating pseudogenes is by duplication of existing pseudogenes. In fact, in the *Xenopus* 5S rRNA system, a DNA segment containing a pseudogene and a normal 5S gene has been duplicated tandemly many times (Jacq et al., 1977).

Table 4 shows the defects in some globin and immunoglobulin

30

TABLE 4. Defects in some structural pseudogenes.*

Pseudogene	TATA box	Init. codon	Frame shift	Premature stop	Essential amino acid	Splice GT/AG	Stop codon	AATAAA	V-J or V-D-J joining
GLOBIN PSEUDOGENES									
Human ψα1		+	+	+	+	+	+	+	NA
Human ψζ1				+					NA
Mouse ψα3	+		+	+		+			NA
Mouse ψα4			+		+				NA
Mouse βh3	?	+	+	+	+	+	?	?	NA
Goat ψβ^x	+		+	+	+	+	+	+	NA
Goat ψβ^z	+		+	+	+	+	+	+	NA
Rabbit ψβ2			+	+	+	+			
IMMUNOGLOBULIN PSEUDOGENES									
Human ψVκK100	?	NA	+	+	+		NA	NA	+
Human λψ1	NA		+	+	?	+		NA	
Mouse ψV_H3	?		+	+	+		NA	NA	
Mouse ψV_H6	?		+	+	+	+	NA		+
Mouse Jκ3	NA	NA				+	NA	NA	

*, +, Mutated; NA, not applicable.

31

pseudogenes. Human $\psi\zeta1$ and mouse $J_\kappa3$ contain only a single defect, which is probably the direct cause of gene silencing. All other pseudogenes have multiple defects, and it is difficult to know which defect was the primary cause of silencing. The majority of these pseudogenes show one or more frameshift mutations, each of which leads to one or more in-phase (premature) stop codons. Many of them also show defects in regions for essential amino acids.

Rate of nucleotide substitution

Pseudogenes are apparently subject to no functional constraint and thus would accumulate nucleotide substitutions at a rate equal to the mutation rate. It is therefore interesting to estimate the rate (b) of nucleotide substitution in pseudogenes. It is also interesting to know the time (T_d) of divergence between a pseudogene and its functional counterpart and the time (T_n) since it became nonfunctional (Figure 8).

These problems have been studied by several authors (Kimura, 1980; Miyata and Yasunaga, 1981; Li et al., 1981). Li et al. (1981) used the model shown in Figure 8. The time since divergence between mouse and human is known to be approximately 80 myr, so that the rate (a_i) of nucleotide substitution at the ith position of codons can readily be estimated from sequence comparison. Considering the number of nucleotide substitutions between each sequence pair at each of the three positions of codons, we can set up equations for obtaining T_d, T_n, and b. [We assume that once a gene becomes nonfunctional the substitution rate (b) is the same for all positions of codons.] Li et al. (1981) applied this model to data for mouse $\psi\alpha3$, human $\psi\alpha1$, and rabbit $\psi\beta2$. I have recalculated the rate for rabbit $\psi\beta2$ by using human β instead of both human β and mouse β major as a reference, because the latter appears to have evolved faster than the other mammalian β globin genes. The results are shown in Table 5. [The results for goat $\psi\beta^X$ and $\psi\beta^Z$ are obtained by W.-H. Li and T. Gojobori's (unpublished) method.]

According to the estimates given in Table 5, mouse $\psi\alpha3$ diverged from its functional counterpart $\alpha1$ about 27 myr ago and became a pseudogene 4 myr later. Actually, T_d and T_n are not significantly different, so the results are not incompatible with the hypothesis that mouse $\psi\alpha3$ was already a pseudogene at the time of its occurrence. Cleary et al. (1981) have proposed that goat $\psi\beta^X$ and $\psi\beta^Z$ were derived directly from an ancestral pseudogene by duplication. Li and Gojobori's (unpublished) sequence analysis supports this hypothesis and suggests that the duplication occurred about 10 myr ago. The ancestor of these two pseudogenes was duplicated from a normal β globin gene about 46 myr ago and became a pseudogene 10 myr later (Table 5). The

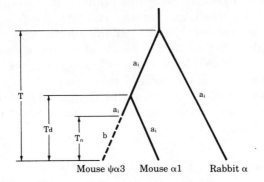

FIGURE 8. Plausible phylogenetic tree for mouse $\psi\alpha3$, mouse $\alpha1$, and rabbit α. T denotes the divergence time between mouse and rabbit, T_d the time since divergence of mouse $\psi\alpha3$ and $\alpha1$, and T_n the time since nonfunctionalization of mouse $\psi\alpha3$. a_i denotes the rate of nucleotide substitution per site per year at the ith position of codons in the normal globin genes, and b the rate of substitution for mouse pseudogene $\psi\alpha3$. (From Li et al., 1981.)

TABLE 5. Times since gene duplication (T_d), times since nonfunctionalization (T_n), and rates of nucleotide substitution per site per year (b) in pseudogenes.*

Gene	T_d (myr)	T_n (myr)	b ($\times 10^{-9}$)	a_1 ($\times 10^{-9}$)	a_2 ($\times 10^{-9}$)	a_3 ($\times 10^{-9}$)
Mouse $\psi\alpha3$	27 ± 6	23 ± 19	5.0 ± 3.2	0.75	0.68	2.65
Human $\psi\alpha1$	49 ± 8	45 ± 37	5.1 ± 3.3	0.75	0.68	2.65
Rabbit $\psi\beta2$	44 ± 10	43 ± 42	4.1 ± 3.4	0.94	0.71	2.02
Goat $\psi\beta^X$ & $\psi\beta^Z$	46	36	4.4	0.94	0.71	2.02
Average			4.7	0.85	0.70	2.34

* a_1, a_2, and a_3 denote the rates of nucleotide substitution for the first, second, and third positions of codons in functional genes, respectively. The a_i values in the first two cases are average values for mouse $\alpha1$, human α, and rabbit α, and those for the last two cases are average values for mouse β major, human β, and rabbit $\beta1$. myr, million years.

other two pseudogenes also appear to be considerably older than mouse $\psi\alpha3$; but, as in the case of mouse $\psi\alpha3$, the difference between T_d and T_n is small in each case. This suggests that a redundant duplicate can become a pseudogene in 1 or 2 million years.

The rate of nucleotide substitution is very high for all four pseudogenes. The average rate is approximately 5×10^{-9} per nucleotide per year. This is two times higher than the average rate at the third position of codons in functional genes, which is in turn approximately three times higher than the average rates at the first and second positions. It has also been found that the substitution rate in pseudogenes is slightly higher than the rate of synonymous substitution (Miyata and Hayashida, 1981). Thus, pseudogenes appear to evolve at the highest rate. The implication of this finding for the neutralist–selectionist controversy is discussed in Chapter 11 by Kimura. As noted earlier, the substitution rate in pseudogenes may reflect the rate of intrinsic mutation.

Pattern of nucleotide substitution

Because pseudogenes are apparently subject to no functional constraint, their pattern of nucleotide substitution would reflect the pattern of intrinsic mutation. With this aim, my colleagues and I have inferred the pattern of nucleotide substitution in pseudogenes by comparing pseudogene sequences with the sequence of their functional counterparts (Gojobori et al., 1982). In Table 6, the values in row A denote the relative frequencies of substitutions from A to T, C, and G, respectively. The other values are similarly defined. The result shows that mutation does not occur randomly among the four types of nucleotides. For example, when C mutates, it changes more often to T than to A or G. The mutations A → G, T → C, C → T, and G → A are called transitions, and the other eight types are called transversions. Obviously, if mutation occurs at random, transversions should be twice

TABLE 6. Relative frequencies (%) of different types of nucleotide substitutions in pseudogenes.* (From Gojobori et al., 1982.)

Original nucleotide	A	T	C	G
A	—	4.7 ± 1.9	5.2 ± 0.8	11.4 ± 1.6
T	4.5 ± 1.0	—	6.2 ± 1.8	4.6 ± 1.8
C	8.3 ± 1.4	22.0 ± 1.8	—	4.7 ± 1.0
G	16.0 ± 1.1	7.0 ± 1.5	5.5 ± 0.8	—

(Column header group: **Mutant nucleotide**)

* The relative frequency of the substitution from the ith to the jth type of nucleotide (i,j = A,T,C or G) in a sequence is equal to $p_{ij} \times (\Sigma_i \Sigma_{j \neq i} p_{ij})^{-1} \times 100\%$, where p_{ij} is equal to the number of $i \to j$ substitutions divided by the number of nucleotides of the ith type in the ancestral sequence.

as frequent as transitions. Yet the sum of the relative frequencies of the four transitions is approximately 56% (Table 6), that is, transitions have actually occurred more often than transversions. The substitution pattern suggests that a sequence under no functional constraint will become rich in A and T. Interestingly, this prediction appears to hold for noncoding sequences, the major parts of which are presumably subject to very weak functional constraints (Gojobori et al., 1982).

Rate of fixation of a pseudogene in a population

As mentioned earlier, a duplicate gene may lose its function and become a pseudogene, as long as the other duplicate genes are functioning normally. However, the rate of fixation of pseudogenes depends on population size as well as on the rate of null mutation. This is because the fixation of pseudogenes in a population occurs mainly by random genetic drift. In an extremely large population, the effect of genetic drift is negligible, so that pseudogenes may never be fixed in the population (Fisher, 1935; Nei and Roychoudhury, 1973). Many authors have studied the rate of fixation, assuming that there are two duplicate loci and that mutation occurs irreversibly from the normal to the nonfunctional or null state (Li, 1980; Watterson, 1982, and references therein). All authors except Watterson (1982) have relied on simulation or numerical solution of differential equations.

Table 7 shows some simulation results by Li (1980). In the table, v denotes the rate of mutation from the normal to the null state per gene per generation; N, the effective population size; r, the recombination value between the two loci; and T, the mean time until fixation of the null allele at one of the two loci. It is assumed that the population is initially free of null mutants and that double null homozygotes are lethal but all other genotypes are normal. In the case of $r = 0.5$ (i.e., no linkage), Watterson (1982) has shown that T is roughly given by

$$T = N \log N - N\psi(2Nv) + 2.53N$$

where $\psi(\cdot)$ denotes the digamma function.

We note from Table 7 several interesting properties. First, tight linkage ($r = 0.001$) has only a minor effect on T, as long as population size is not extremely large. Second, T is small in a small population but increases with population size. This is because as population size increases the effect of random drift becomes weaker and selection becomes effective. When N is 10^6, it takes approximately 10 million generations for a pseudogene to be fixed in the population. Third, in

TABLE 7. Mean time (T) until fixation of a null allele at one of two duplicate loci. (From Li, 1980.)

v^*	N^*	r^*	T (in generations)
10^{-5}	10^2	0.5 0.001	50,900 (50,900)† 50,700 (51,000)
	10^3	0.5 0.001	62,400 (57,300) 57,000 (51,200)
	10^4	0.5 0.001	163,000 (133,000) 134,000 (104,000)
	10^5	0.5 0.001	1,160,000 (934,000) 936,000 (685,000)
	10^6	0.5 0.001	10.8×10^6 (8.7×10^6) 9.3×10^6 (7.0×10^6)
10^{-4}	10^2	0.5	5,870 (5,500)
	10^3	0.5	14,000 (11,600)
	10^4	0.5	127,000 (82,900)
	10^5	0.5	970,000 (756,000)

* v, Null mutation rate; N, effective population size;
 r, recombination value.
† The values in parentheses are standard deviations.

a small population T is largely determined by v, the null mutation rate. Indeed, when N is of the order of 100, T is roughly given by $1/2v$. For example, when $v = 10^{-5}$, $N = 100$, and $r = 0.5$, T is approximately 50,900 generations. In a large population, however, an increase in v has only a minor effect on T. For instance, if $N = 10^5$ and $r = 0.5$, T decreases from 1.16×10^6 to 0.97×10^6 when v increases from 10^{-5} to 10^{-4}. Fourth, the standard deviation is as large as the mean. Thus, the fixation time is subject to large random errors.

The preceding computation shows that the rate of fixation of pseudogenes is highly dependent on population size and mutation rate. At the present time, however, there is still no data to test the details of these relationships. In practice, the rate of fixation would depend on several other factors such as the effect of null alleles on the fitness of heterozygotes (Takahata and Maruyama, 1979; Li, 1980), attainment of disomic segregation at meiosis (Li, 1980), and differentiation of the regulatory system of the duplicate genes (Ferris and Whitt, 1979; Li, 1982). All these factors would complicate the relationships among T, N, and v in Table 7. In general, however, these factors have an effect to prolong the fixation time (Li, 1982).

36

In a multigene family, the gene number may fluctuate from time to time as a result of unequal crossing-over. If the number becomes larger than the optimal number, some of the genes will be free to become pseudogenes. In this case, the mean time for the first pseudogene to become fixed in the population is expected to be considerably shorter than that for the case of two duplicate genes, because any of the multiple genes can become nonfunctional. This seems to be in agreement with actual data, because most of the multigene families include some pseudogenes at present.

As mentioned earlier, molecular data indicate that pseudogenes are widespread in the genome of higher organisms. Ohno (1972) has suggested that a large proportion of DNA in higher organisms is nonfunctional, that is, junk. If this is true, much of the junk DNA might have evolved by nonfunctionalization of duplicate genes (Nei, 1969). As long as the junk DNA does not impede the cellular physiology, they can stay in the genome without being eliminated. Gene duplication seems to have provided higher organisms with a sufficient amount of DNA for having such a "luxury" as well as for having many elaborate genetic systems discussed earlier.

CHAPTER 3

CONCERTED EVOLUTION

OF MULTIGENE FAMILIES

Norman Arnheim

The genomes of eukaryotic organisms are composed of highly repeated, moderately repeated, and single-copy DNA sequences (Britten and Kohne, 1968). The fact that multiple copies of DNA segments exist can pose special problems in understanding their evolution. Each individual member of a gene family does not necessarily evolve independently of the others. Through genetic interactions among its members, the family may evolve together in a concerted fashion, as a unit. This article explores some of the factors that influence the concerted evolution of multigene families, emphasizing in particular the consequences of this mode of evolutionary behavior.

Repeated sequence classes have been found to be organized in two different ways (see Davidson and Britten, 1973). The members of some repeated gene families are found on all of the chromosomes of a species; and even within a single chromosome, the multiple copies are interspersed throughout its length. The human *Alu*I DNA family, for instance, is composed of approximately 300,000 members, which are interspersed with single-copy DNA throughout the genome at intervals of approximately 2500 base pairs (bp) (Houck et al., 1979). Other repeated gene families often have a clustered arrangement. All the members of these multigene families may exist together in one region of a single chromosome or on a number of different chromosomes each having a tandem array of family members (see Long and Dawid, 1980). In the South African clawed toad *Xenopus laevis*, for example, the hundreds of copies of the genes that code for the 18S and 28S rRNAs are found on a single chromosome, whereas in human they are found on five pairs of chromosomes.

38

Repeated gene families were originally detected by DNA reannealing studies. The members of a gene family can be defined as those
DNA segments that retain nucleotide sequence similarity that is recognizable under a particular set of conditions used in DNA–DNA
hybridization experiments. The variation among the members of a
given family can be estimated by carrying out DNA melting studies
on the reannealed DNA. If the members of a family are very similar
to one another, then denaturation of genomic DNA followed by reas-
sociation results in the formation of well-paired, double-stranded molecules. Each strand of these duplexes constructed *in vitro* will by
random chance have originated from different family members. The
thermal stability of each new duplex, therefore, will be a function of
the sequence similarity between the two strands; the better paired
they are, the more thermal energy is needed to dissociate them. The
temperature range over which the *in vitro*-constructed DNA duplexes
dissociate can therefore be used to estimate the degree of variation in
sequence similarity among family members. Such studies have shown
that some repeated-sequence families are quite homogeneous. With
the advent of restriction enzyme analysis and DNA sequencing procedures, a large body of data that further confirms this conclusion has
been obtained.

REPEATED DNA AND CONCERTED EVOLUTION

The existence of highly homogeneous, repeated-sequence families can
be explained in several different ways. A homogeneous family may
have an important function that depends upon its specific nucleotide
sequence. If new variants arise, they would be expected to be edited
out of the family by natural selection (Figure 1). The high levels of
conservation in the structure of the 18S and 28S ribosomal RNAs
within and between eukaryotic species might be accounted for by
purifying selection. Of course, the greater the number of members in
the family, the more difficult it is to imagine how purifying selection
can act to keep the family homogeneous, especially if the number of
genes is in the range of tens- to hundreds-of-thousands, as is the case
in many multigene families. It can also be argued that a very recent
amplification of a sequence is the basis for the homogeneity of the
gene family within a species. However, if a gene family arose by a
sudden amplification event of a unique DNA segment, it would be
expected that the homogeneity of the family would slowly disappear.
Over evolutionary time, mutations would accumulate in the family

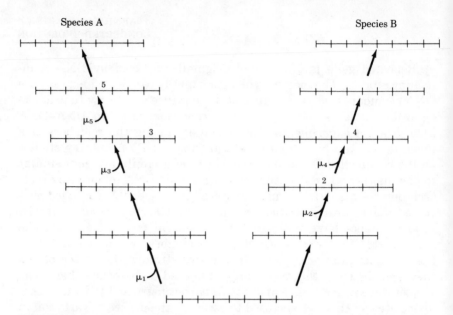

Species A   Species B

FIGURE 1. Purifying selection scheme for maintaining homogeneity among members of a tandemly arranged, repeated-gene family. The common ancestor of species A and B is shown as having a series of identical repetitive elements. Mutations (μ) would be expected to occur independently in both lineages. If the specific nucleotide sequence of the elements played an important role in its function, homogeneity would be expected to be observed within, as well as between, species.

members through genetic drift. These mutations would be expected to occur in those regions of the gene in which function does not depend upon a specific nucleotide sequence (Figure 2). Under the independent evolution model, comparison of the elements from the same multigene family between two species that diverged after the gene amplification would be expected to show that the degree of intraspecific variation among the repeated elements would be approximately equal to the degree of interspecific variation.

With either of the preceding models, multigene families would be expected to exhibit levels of interspecific variation no greater than the level of intraspecific heterogeneity. However, the results of one of the first evolutionary studies of a multigene family—the ribosomal genes (rDNA) in amphibians—did not conform to these expectations. Like most higher organisms, the genes coding for the 18S and 28S ribosomal RNAs in *Xenopus* are present in hundreds of copies and are organized in a tandemly arranged fashion (for a review, see Long and Dawid, 1980). Each gene or repeating unit consists of a transcribed and non-transcribed segment (Figure 3). The transcribed segment codes for a

40

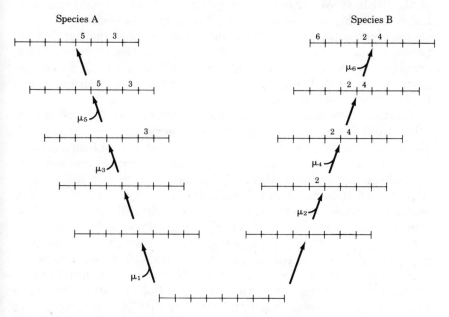

FIGURE 2. Independent evolution model for multigene families. The common ancestor of species A and B has a series of repeated elements, which arose by gene duplication. In the lineages leading to species A and B, independent mutations would be expected to occur and could, without having any selective advantage, be fixed by genetic drift.

large RNA precursor (45S) from which the 18S and 28S ribosomal RNAs are produced by means of enzymatic cleavage. These transcribed portions are separated from one another by the nontranscribed spacer (NTS) region, which can vary considerably in size among species. In *Xenopus*, the ribosomal gene repeating units are found on a single pair of chromosomes (Pardue, 1974).

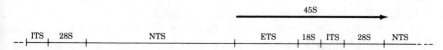

FIGURE 3. Diagrammatic representation of a typical eukaryotic ribosomal gene repeating unit. The transcribed portion that codes for the 45S precursor RNA consists of an external transcribed spacer (ETS), the 18S gene (18S), the internal transcribed spacer (ITS), and the 28S gene (28S). Transcription units are separated from each other by nontranscribed spacer (NTS).

41

Comparison of the structures of the ribosomal genes in two *Xenopus* species was carried out by nucleic acid reannealing techniques (Brown et al., 1972). In *Xenopus laevis*, the 18S and 28S gene segments were found to be highly homogeneous. NTS segments showed the same degree of intraspecific sequence homogeneity, suggesting that they too served a nucleotide sequence-specific function. The degree of homogeneity so closely approximated that of the 18S and 28S gene sequences that a highly important function for the nontranscribed spacer regions was to be expected.

Studies to confirm the evolutionary conservation of the NTS were carried out by analyzing the ribosomal gene repeating units in a closely related species, *Xenopus borealis* (previously misidentified as *Xenopus mulleri*). Like *Xenopus laevis*, the transcribed and nontranscribed segments were both highly homogeneous within this species. As expected from their intraspecific homogeneity, the 18S and 28S sequences of the two *Xenopus* species were virtually identical. However, whereas homogeneity of the NTS segments was observed within each species, little, if any, homology was detected *between* the NTS segments of these two species (Brown et al., 1972). These data represented a paradox. On the one hand, the NTS regions within a species are homogeneous, suggesting that the individual repeating units have not diverged from one another and implying that these segments have an important nucleotide sequence-specific function. On the other hand, the same reasoning would lead to the prediction that the NTS segments of closely related species should be highly homologous. This prediction was not confirmed, suggesting that the NTS segments evolved independently. However, the observation that the degree of interspecific variation among the NTS sequences was far greater than the intraspecific heterogeneity cannot be accounted for by either the purifying selection or independent evolution models.

Similar paradoxes have also arisen from the analyses of small multigene families (Zimmer et al., 1980; Slightom et al., 1980; Lauer et al., 1980; Leigh Brown and Ish-Horowicz, 1981). For example, duplicate α hemoglobin genes occur in all higher primates. Amino acid sequence data have shown that during the evolution of apes and men interspecific differences among the α globin proteins have accumulated; on the average they differ by 2.5 amino acid substitutions. By contrast, the duplicate genes within a species are on the average ten times more similar to one another. Because all of the higher primates have duplicate genes, it would be natural to assume a single α globin gene duplication event occurred prior to the divergence of these species from one another. If, subsequent to this event, the duplicate genes evolved independently of one another, the intra- and interspecific differences between the α globin proteins would be expected to be more or less the same. Because intraspecific divergences are ten times lower,

42

independent evolution of these duplicate loci within a species is unlikely (Zimmer et al., 1980); some mechanism must act to keep them homogeneous. An analogous observation was obtained in a study on the duplicate γ globin loci in humans. The duplicate γ globin gene loci on a single chromosome are more similar to one another than alleles at the same locus on homologous chromosomes (Slightom et al., 1980).

Finally, analysis of some very highly repeated sequences have revealed that many of the families that are highly homogeneous within a species are, surprisingly, very rapidly evolving and large differences exist between closely related species (for reviews, see Lewin, 1980; Brutlag, 1980). The satellite DNAs associated with the centromeric heterochromatin of chromosomes are examples of this class of DNA.

MECHANISMS OF CONCERTED EVOLUTION

How can large interspecific differences between the members of a gene family be compatible with intraspecific homogeneity? One possibility is that each species has evolved its own nucleotide sequence that permits optimal function of the DNA segment. Through positive selection, each species could have refined the structure of the DNA segment so as to optimize the function of the repeated element within the context of their own evolution. However, it is difficult to imagine how a repeated gene family within a species can attain homogeneity with selection occurring in parallel in each member of the gene family. During the evolution of a species, exactly the same selectively advantageous mutations must occur independently and be fixed in each family member (Figure 4). This parallel evolution would be extremely unlikely.

Parallel evolution by positive natural selection is not necessary to explain why intraspecific variability among multigene family members is much less than interspecific differentiation. Even repeated DNA segments that have no function (or have a function that does not depend upon a specific nucleotide sequence) are capable of remaining homogeneous within a species over evolutionary time (Edelman and Gally, 1968; Hood and Talmage, 1970; Brown et al., 1972; see also Hood et al., 1975; Tartof, 1975; Smith, 1973, 1976; Long and Dawid, 1980). This phenomenon is known as horizontal, coincidental, or (more recently) concerted evolution. In Figure 5, the common ancestor of species A and B has a set of identical repetitive elements. In the lineages leading to species A and B, independent mutational events would be expected to occur at random. Concerted evolution

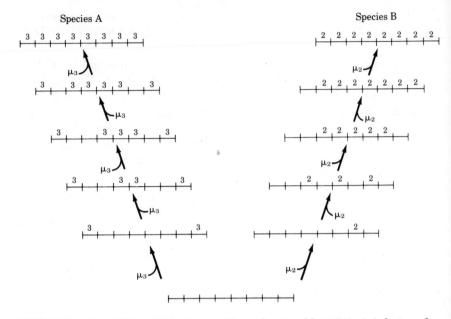

FIGURE 4. Parallel evolution by positive selection. Mutations (μ) that confer a selective advantage in each lineage must arise independently in each repeating unit. For the family to remain homogeneous within a species, exactly the same mutations must occur. However, different mutations could be selectively advantageous in other lineages. This would result in intraspecific homogeneity and interspecific heterogeneity.

events, labeled C, which involve genetic interactions among the family members, occur at random intervals and result in the elimination of most spontaneous mutations. Eventually, these genetic interactions result in the fixation of a particular mutation in all of the members of the array. Different mutations would be fixed in the two lineages and each species would thus become homogeneous for different mutations. Concerted evolution can thereby account for the greater degree of interspecific than intraspecific variation among the multigene family members.

Several genetic mechanisms that facilitate concerted evolution have been proposed; gene conversion and unequal crossing-over are especially important. Both mechanisms depend upon an inherent asymmetry such that the products of these events can be either enriched or depleted with respect to the parental types. In the case of unequal crossing-over, the duplication and deletion products of recombination among misaligned family members alter the frequency of individual members in the original array (Figure 6). Computer simulations involving repeated cycles of these events, in conjunction with genetic drift, result in the production of homogeneous tandem arrays

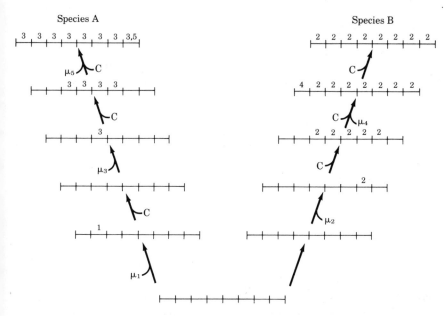

FIGURE 5. Concerted evolution model for multigene families. Spontaneous mutations (μ) that have no selective advantage and that occur in each lineage could be fixed or lost in every family member as a consequence of concerted evolution events (C) and genetic drift. For further details, see the text.

where all the members were derived from only one or a few of the original family members (Smith, 1973, 1976; Ohta, 1980). Gene conversion events (see Radding, 1978) have also been shown to be theoretically capable of producing concerted evolution (Birky and Skavaril, 1976; Ohta, 1977a; Nagylaki and Petes, 1982). All models of concerted evolution depend on an additional feature: for a gene family to evolve in a concerted fashion, the rate of mutational divergence among family members must be less than the rate of fixation. Experimental determination of the rates of gene conversion and unequal crossing-over in yeast are in accord with this requirement (Petes, 1980; Szostak and Wu, 1980; Klein and Petes, 1981).

REPEATED GENE FAMILIES DISTRIBUTED AMONG MANY CHROMOSOMES

The genetic mechanisms by which concerted evolution occurs have certain restrictions in the case of repeated-sequence families whose

45

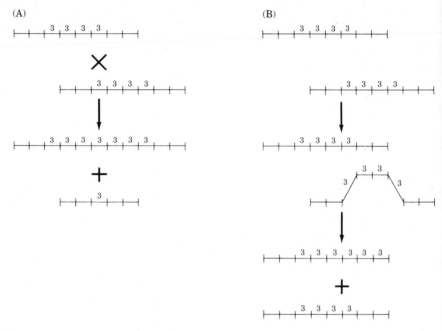

FIGURE 6. Models of unequal crossing-over (A) and gene conversion (B). Both of these events are shown to occur between two homologous chromosomes, each possessing 50% wild-type repeats and 50% spontaneous mutation 3-containing repeats. A. As a result of unequal crossing-over, both daughter chromosomes have an altered frequency of the two repeat types when compared to the parental frequencies. In addition, the number of copies of the repetitive element are also altered in the daughter chromosomes. B. Gene conversion changes the frequency of the two types of elements in one of the daughter chromosomes but does not alter the total number of repeats in either product.

members reside on nonhomologous chromosomes. In the case of the highly repeated and highly interspersed *Alu*I family, frequent interchromosomal recombination events would have disastrous cytogenetic consequences if DNA segments were shuffled among the whole chromosome set. Gene conversion events involving only the *Alu*I sequences would avoid this problem while leading to a family homogeneity. However, the larger the family, the more difficult it is for random gene conversion events to accomplish this (Birky and Skavaril, 1976; Ohta, 1977a; Nagylaki and Petes, 1982). On the other hand, if certain *Alu*I sequences were preferentially capable of initiating conversion events, more rapid fixation, sufficient for a very large family, could occur (Nagylaki and Petes, 1982). How might this preferential conversion be controlled? Data from several laboratories suggest the following possible explanation. Some (but not all) *Alu*I segments are capable of being transcribed into RNA molecules (Elder et al., 1981). Recent evidence strongly argues that certain classes of small RNA transcripts can be converted into DNA by a reverse transcription process and subsequently inserted back into the genome (van Arsdell et al., 1981; Jagadeeswaran et al., 1981). If a large number of DNA copies of a few *Alu*I transcripts could also act as donors in gene conversion events, the concerted evolution of the *Alu*I family might be explained. Preferential gene conversion, or molecular drive (Dover, 1982a,b), has also been proposed to account for concerted evolution in other large multigene families.

Over the last several years, we have been analyzing the human rDNA multigene family as a model system for studying the concerted evolution of a multichromosomally distributed, repeated-gene family. Among the advantages of using this system are the relatively low number of copies of these genes, their well-defined chromosomal location, and the known function of the family. We (Arnheim et al., 1980) examined the structure of the ribosomal genes in apes and men in order to examine the relationship between concerted evolution and the chromosomal organization of this multigene family. For example, man, chimpanzee *(Pan troglodytes)*, and pygmy chimpanzee *(Pan paniscus)* have approximately 400 ribosomal genes, which are distributed among nucleolus organizers located on five pairs of chromosomes (Henderson et al., 1972, 1976; Evans et al., 1974; Tantravahi et al., 1976). If ribosomal genes on nonhomologous chromosomes experience no genetic interaction, it might be expected that each nucleolus organizer region would evolve independently. Under these circumstances, we might also expect to detect extensive variation among the genes in certain portions of the repeating unit, within individuals of a species. On the other hand, the greater the degree of genetic exchange between ribosomal genes on nonhomologous chromosomes, the more likely it is that the whole gene family will evolve in a concerted fashion.

Our analysis of ape and human rDNA structure involved the digestion of total DNA with restriction endonucleases, agarose gel electrophoresis of the resultant DNA fragments, transfer of these fragments to nitrocellulose filter paper (Southern blotting technique), and detection of the fragments that contain any 18S or 28S genes. Seven different restriction enzymes were used to examine more than 20 different restriction enzyme sites. The restriction enzyme maps from five species of apes are shown in Figure 7.

Different regions of the ribosomal gene in human and the apes have different characteristic evolutionary patterns (Figure 7). The transcribed segments have undergone few changes; this is consistent with the general finding that the 18S and 28S RNA gene sequences have been highly conserved throughout evolution. Our data also show that differences can exist among species in the structure of the non-transcribed spacer, without any comparable variation existing within or among individuals of a species. This demonstrates the concerted evolution of the ribosomal gene family.

A summary of some of the data obtained from the analysis of

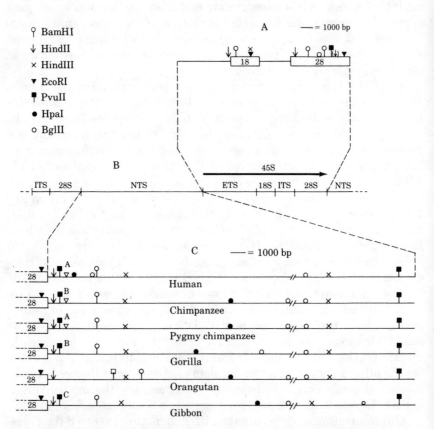

FIGURE 7. Restriction enzyme maps of human and ape ribosomal genes. A. Map of the transcribed region of the gene. B. Schematic representation of the repeating unit. The orientation of the gene with respect to the origin of transcription of the 45S precursor RNA is also given. C. Map of the nontranscribed spacer region. The break in the horizontal line indicates the spacer regions not mapped in our study. The human, chimpanzee, and pygmy chimpanzee show length heterogeneity (∇) in the region of the NTS between the *Pvu*II and *Bam*HI sites adjacent to the end of the 28 S gene. With the exception of the human *Bgl*II site near the *Bam*HI site in the NTS, none of the polymorphic restriction enzyme sites are shown.

48

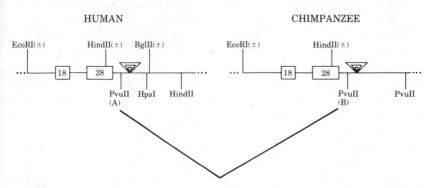

FIGURE 8. Summary map of key restriction enzyme sites in human and chimpanzee ribosomal genes. The restriction enzyme sites above the gene are polymorphic within these species. The inverted triangles denote length polymorphism in the NTS (see the legend to Figure 7). The sites below the gene are monomorphic at that position and unique to each species.

human and chimpanzee rDNA is shown in Figure 8. Consider, for example, the structure of a portion of the nontranscribed spacer in the human and chimpanzee. Each human ribosomal gene has a *Hpa*I site in the NTS region 3′ to the 28S gene. The chimpanzee lacks this site. The fact that none of the other great apes have a *Hpa*I site in this position suggests that the most recent common ancestor of human and chimpanzee also lacked this site. Therefore, the *Hpa*I site most probably originated in the human lineage after the chimpanzee–human evolutionary divergence and was eventually fixed in every human ribosomal gene. Other restriction enzyme sites in the nontranscribed spacer also demonstrate species-specific homogeneous sites (Figure 8).

We know that the ribosomal genes in human and chimpanzee are distributed among five pairs of chromosomes. It is likely that the common ancestor of these species had the same multichromosomal distribution. Thus, in spite of the fact that the ribosomal genes are found on nonhomologous chromosomes, *each chromosome pair has not evolved independently.* The whole gene family has undergone concerted evolution. It is likely that the mutation that gave rise to the *Hpa*I site in the human lineage arose only once and was propagated throughout the whole ribosomal gene family as a result of genetic interactions among rDNA sequences on nonhomologous as well as homologous chromosomes. It is extremely unlikely that the same *Hpa*I mutation arose independently on each of the five chromosome pairs and was subsequently fixed rather than being eliminated in each chromosome pair.

Our restriction enzyme analysis also revealed a second set of structural features in ape and human rDNA (Figure 8). There are restriction enzyme site polymorphisms within an individual, as well as between individuals in a species. These polymorphisms are of two types. One particular restriction enzyme site (*BglII*; P. Seperack and N. Arnheim, unpublished) is absent in all of the apes but is found in approximately 70% of the human ribosomal genes. It presumably arose since chimpanzees and humans last had a common ancestor and can be classified as a transient polymorphism. Such polymorphisms would be expected from the mechanisms of concerted evolution, as depicted in Figure 5. Two additional polymorphisms were detected. In contrast to the transient polymorphism, both of these polymorphisms were found in several different species, suggesting that they have been maintained during the concerted evolution of the rDNA family in a number of independent lineages. This raises the possibility that natural selection may be important in maintaining these polymorphisms in the face of the fixation events (the *HpaI* site) that led to the concerted evolution of the rDNA family as a whole.

We also examined the chromosomal distribution of these polymorphisms within humans. The model of genetic exchange between rDNA clusters on nonhomologous chromosomes would predict that, within sampling fluctuation, the polymorphisms should occur uniformly on all nucleolus organizer chromosomes. In order to determine the distribution of the human polymorphisms among the various chromosomes, we examined human–mouse somatic cell hybrids in which one or a few human nucleolus organizer chromosomes are isolated on a constant background of mouse genetic material. The results of this analysis (Krystal et al., 1981) are consistent with a general uniform distribution of all of the polymorphisms over the chromosomes (although a few exceptions were found) and a model of concerted evolution that involved exchanges between sister chromatids or homologous chromosomes and between nucleolus organizers on nonhomologous chromosomes. The first events, if relatively frequent, would tend to cause the homogenization of all members of a single cluster to the same form of any given polymorphism. The second event would, relatively infrequently, introduce blocks of rDNA genes from other chromosomes into the cluster, restoring heterogeneity. If these events involved unequal crossing-over, it would cause the number of rDNA genes per cluster to fluctuate randomly. In fact, estimates of the number of rDNA genes on the human nucleolus organizer containing chromosomes by *in situ* hybridization are consistent with this mechanism and reveal extensive copy number variation among the chromosomes within an individual and among the same chromosome when compared between different individuals (Warburton et al., 1976). Because of the placement of the ribosomal genes on human chromosomes, unequal crossing-over among rDNA sequences on nonhomologs will not have disasterous cytogenetic consequences (see the next section). Based on our analysis of individual chromosomes from two human individuals, we might expect that any given nucleolus organizer chromosome, in the human population as a whole, can carry all of the nucleotide sequence and length variants that are polymorphic in rDNA and will do so in the proportions that these forms exist in in the total human genome. No unique combination of rDNA polymorphisms would be expected to be characteristic of any one of the nucleolus organizer-containing chromosomes.

PHYSICAL BASIS FOR INTERCHROMOSOMAL rDNA INTERACTIONS

The model of concerted evolution involving genetic interactions between rDNA sequences on nonhomologous chromosomes in man is supported by data showing that ribosomal genes from nonhomologous chromosomes can associate with one another in the human germline. Both light microscopic and electron microscopic studies of the structure of nucleoli during meiotic prophase in human spermatocytes (Ferguson-Smith, 1964) and oocytes (Mirre et al., 1980) has shown that rDNA from two or three nonhomologous chromosome pairs can often contribute to the formation of a single nucleolus that is the cellular site of rRNA synthesis. The physical proximity of rDNA sequences from nonhomologous chromosomes in this organelle could provide an opportunity for genetic interactions. It is highly significant that these interactions were observed in human germ cells, since only germline genetic exchanges would contribute to the concerted evolution of a gene family. One cytological study (Mirre et al., 1980) also found that nucleoli in mouse oocytes do not contain rDNA sequences for more than one chromosome. This could limit the opportunity for nonhomologous exchanges.

The chromosomal distribution of mouse rDNA variants is in striking contrast to the distribution of human rDNA polymorphisms. The ribosomal genes in mouse inbred strains also have a multichromosomal distribution (Henderson et al., 1974; Elsevier and Ruddle, 1975; Dev et al., 1977). In mouse rDNA, a region (VrDNA) 5' to the origin of transcription varies in length among ribosomal genes because of differences in the number of copies of a small repetitive sequence (Arnheim and Kuehn, 1979). An inbred strain usually has four to six

51

major size classes of VrDNA regions. Using mouse recombinant inbred strains specifically constructed for genetic analysis, we determined the chromosomal linkage relationships among these polymorphic VrDNA regions (Arnheim et al., 1982). Our studies on five of the length variant classes showed that they can be divided into three discrete linkage groups. If genetic exchanges were common among rDNA clusters on nonhomologous chromosomes, each of the VrDNA size classes would have been expected to have representatives on every chromosome carrying rDNA. However, each nucleolus organizer appears to have ribosomal genes with one (or, at most, two) different VrDNA size classes that are unique to that chromosome pair and are apparently absent from nonhomologous rDNA-containing chromosomes.

In strains C57BL/6J and C3H/HeJ, one of the VrDNA classes appears to be on the same chromosome pair, although present in different amounts in each strain. Because these inbred mouse strains are not closely related, the conservation of the linkage relationships might considerably predate the origin of the inbred strains. The simplest explanation of our results is that genetic exchanges among nucleolus organizer regions on nonhomologous chromosomes occur relatively infrequently in mice when compared to fixation events occurring within a chromosome. Of course, the VrDNA regions on nonhomologous chromosomes could not have evolved totally independently of one another because they are homologous in nucleotide sequence and with respect to certain restriction enzyme sites (Arnheim and Kuehn, 1979). As Smith (1976) has pointed out, rare interchromosomal exchanges during mouse evolution could account for this. Thus, one interpretation of the different patterns of chromosomal distribution of rDNA variants in mice and men is that the relative rate of interchromosomal exchange among rDNA clusters is lower in mice.

The molecular basis is unknown for the fact that rDNA sequences on nonhomologous human chromosomes can cooperate in the formation of common nucleoli whereas mouse nucleoli are composed of rDNA sequences from only one chromosome pair. However, we find it interesting that the genetic events responsible for the evolution of a multigene family might be significantly influenced by factors related to their biochemical function in the cell.

Another possibly significant influence on the concerted evolution of the rDNA family is the position on the chromosome of these sequences. In humans and apes, it is found on the stalks (Goodpasture et al., 1976; Henderson et al., 1972, 1976; Evans et al., 1974; Tantravahi et al., 1976) between the short arm and satellite body of the D and G group chromosomes: 13, 14, 15, 21, 22. The recovery of the products of genetic exchanges among rDNA clusters on nonhomologous chromosomes could be affected by the chromosomal location of the nucleolus organizer. If a reciprocal exchange occurred between two rDNA clusters on nonhomologous human chromosomes, only rDNA and satellite genetic material would be translocated (Figure 9). That such events actually occur can be inferred from cytogenetic studies (Therman and Kuhn, 1976; Gimelli et al., 1976; Tomkins, 1981). One possible product of meiosis and fertilization, after a reciprocal exchange, is shown in Figure 10. This zygote would be duplicated for chromosome 14 rDNA and satellite sequences and deleted with respect to the comparable chromosome 21 segments. It is considered unlikely

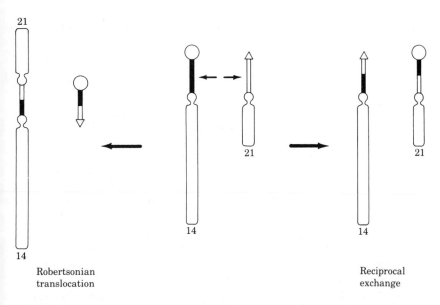

Robertsonian
translocation

Reciprocal
exchange

FIGURE 9. Cytogenetic consequences of genetic exchanges between human rDNA sequences on nonhomologous chromosomes. The satellite segments of chromosomes 14 and 21 are indicated by an open circle and an open triangle, respectively. The rDNA sequences for both chromosomes are immediately below the satellites and are represented as a filled (chromosome 14) or open (chromosome 21) rectangle. The centromeres of the chromosomes are shown as a constriction between the long and short chromosomal arms. In a reciprocal exchange, only the rDNA and satellite segments move to a nonhomologous chromosome. The breaks in the rDNA might also be resolved by the fusion of chromosomes 14 and 21, yielding a Robertsonian translocation that is dicentric and an acentric rDNA and satellite-containing fragment that would be lost. In order to simplify the figure, we have shown exchanges involving two chromosomes. In fact, these interactions would be expected to take place after DNA replication. Because homologous chromosomes are also paired, each bivalent involved contains four chromatids.

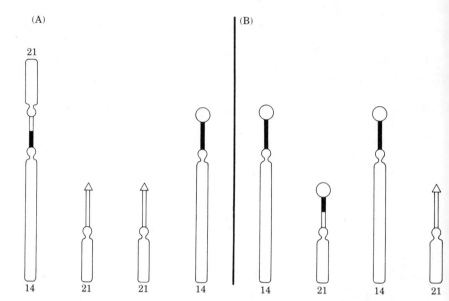

FIGURE 10. Possible zygotic products of rDNA exchanges between nonhomologous chromosomes. A. Following a Robertsonian translocation, duplication and deletion of rDNA and satellite sequences also result but, more significantly, trisomy for one of the chromosomes involved (in this case, chromosome 21) can also occur. B. One result of a reciprocal exchange followed by meiosis and fertilization is a zygote that is duplicated for chromosome 14 satellite and rDNA sequences and deleted for the comparable chromosome 21 DNAs.

that dosage changes involving these particular sequences would be physiologically disadvantageous. The survival of chromosomes carrying rDNA cross-over products would therefore not be affected. Repeated episodes of these events could account for the spread of rDNA sequences among the D and G group chromosomes, a consequence that is required for the concerted evolution of this gene family. In inbred mice, however, rDNA is located very close to the centromere (Henderson et al., 1974; Elsevier and Ruddle, 1975; Dev et al., 1977) on the long arm of chromosomes 12 and 15–19. The cytogenetic consequences of a reciprocal exchange between two rDNA clusters on nonhomologous chromosomes in mice is not obviously innocuous. Such an event would result in the exchange between two chromosomes of all the distal material, including the centromeres. If this led to genetically unbalanced gametes, the products of exchange could be lost. Many other mammalian species have in their genomes multiple sites that carry rDNA. Within any one species, the position of these sites within each chromosome is usually the same (see Hsu et al., 1975); they may be located in heterochromatic short arms (apes, human, *Peromyscus*),

54

in centromeric heterochromatin (mouse, field vole), or in telomeric regions (Chinese hamster). It would seem advantageous to have rDNA sequences located terminally in homologous positions on nonhomologous chromosomes as this would lessen the chance that genetic exchanges between these sequences would yield deleterious chromosomal rearrangements.

It seems likely that the interaction of rDNA sequences on nonhomologous chromosomes in the germ cells of apes and humans contributes significantly to the concerted evolution of the rDNA family. There are good reasons to believe that these same interactions can have severe clinical consequences to our species. It has been known for some time that the rDNA-containing chromosomes do not distribute themselves randomly with respect to the other chromosomes during mitotic cell divisions, but preferentially associate with one another (see Jacobs et al., 1976). This pattern of chromosome association has been related to chromosome trisomy, one of the major reproductive burdens in humans (Ohno et al., 1961; Ferguson-Smith and Handmaker, 1961, 1963). Trisomy 21 or Down's syndrome, for example, affects an average of 1 out of every 500 to 1000 births. The predominant cause of this condition is meiotic nondisjunction, and it has been suggested that the physical association of one chromosome 21 with another acrocentric chromosome might increase the chance that the two chromosomes 21 do not disjoin properly during meiosis. The cytological data mentioned previously indicate that chromosome association during meiosis and mitosis (Henderson et al., 1973) are undoubtedly due to the participation of rDNA sequences from more than one chromosome pair in the formation of a common nucleolus. This functional association could contribute to the disproportionately high frequency of trisomy conditions involving the human D and G group chromosomes in spontaneous abortions (Carr and Gedeon, 1977). Mirre et al. (1980) have pointed out that the association of rDNA sequences during oogenesis might also contribute significantly to the well-known positive correlation between the frequency of trisomy and advanced maternal age. The association of the human rDNA-containing chromosomes is observed up to the late diplotene stage, where oogenesis in the female embryo is blocked until ovulation occurs after puberty. Therefore, nonhomologous chromosome interactions mediated by rDNA may persist from before birth until ovulation, a period that can range from 12 to approximately 40 years. The longer the interaction exists, the more likely it may be that a nondisjunction event will occur.

Another kind of chromosomal abnormality might also arise out of the nucleolar association of rDNA sequences from nonhomologous chromosomes. The same breakage event that can result in a reciprocal exchange of rDNA sequences between nonhomologous chromosomes can also be resolved by the formation of a dicentric Robertsonian translocation chromosome and an acentric fragment (Figure 9). Trisomy for either of the two chromosomes involved could result following meiosis and fertilization (Figure 10). For example, approximately 3% of all Down's syndrome cases are the consequences of Robertsonian translocations. In fact, Robertsonian translocations are the most common structurally abnormal chromosome found in humans and are most frequently observed to involve the rDNA-containing chromosomes (see Therman, 1980). Although some cases of Robertsonian translocation involving the D and G group do appear to result from a break in the ribosomal genes themselves, most result from breaks in the short arm of the chromosome immediately below the rDNA (Brasch and Smyth, 1979; Mattei et al., 1979; Gosden et al., 1981; Mikkelsen et al., 1980). It is likely that the prerequisite for both events is the association of rDNA sequences from nonhomologous chromosomes in the nucleolus.

FUNCTIONAL CONSEQUENCES OF CONCERTED EVOLUTION

Genetic interactions between nonhomologous chromosomes containing rDNA are necessary for the concerted evolution of the rDNA family in apes and humans and other species with a multichromosomal distribution of these sequences. In the case of humans and apes, the physical bases for these genetic interactions most likely have a functional origin and lie in the association of rDNA sequences from nonhomologous chromosomes in the formation of common nucleoli. These interactions, however, also appear to have evolutionary consequences in the form of cytogenetic abnormalities. This disadvantageous aspect of the concerted evolution of the rDNA family might be outweighed by other features of this evolutionary process. The concept of concerted evolution can explain how sequences that are thought to have no nucleotide sequence-specific function can remain homogeneous within a species. It is also clear, however, that genetic interactions among members of a gene family allow for selectively advantageous mutations to be propagated throughout all of the copies of the repeated sequence. The absence of these interactions would necessitate parallel evolution, with the same selectively advantageous mutation arising independently in each gene (Figure 4). Clearly, the phenomenon of concerted evolution is a more efficient mechanism for fixing selectively advantageous mutations throughout all of the members of a multigene family.

Evidence from the study of the ribosomal gene systems in several

56

eukaryotes (see below) supports the idea that the evolution of sequences that are important for the initiation of ribosomal RNA synthesis have evolved very rapidly when compared to sequences that are critical for properly initiating the synthesis of messenger RNAs, tRNAs, and other gene products. This rapid evolution may have been the consequence of the fixation of adaptively significant mutations.

In higher organisms three different RNA polymerases function in transcription (see Lewin, 1980). RNA polymerase I transcribes only the rRNA genes. RNA polymerase II transcribes all of the messenger RNA coding genes and RNA polymerase III is used to synthesize tRNAs, 5S RNA, and some small-molecular-weight RNAs. The recent development of recombinant DNA technology and both *in vivo* and *in vitro* methods of assessing transcriptional activity has led to the finding of certain DNA sequences that are vital to the correct initiation of transcription (Figure 11). In the case of RNA polymerase II, these

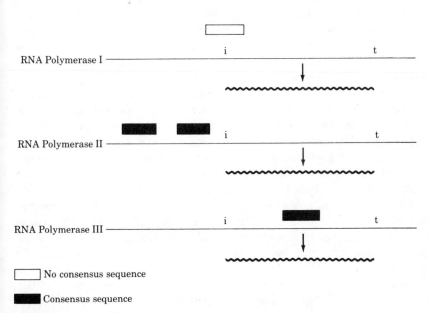

FIGURE 11. Transcriptional control signals for eukaryotic RNA polymerases. Regions that have been defined experimentally as being necessary for the proper initiation of transcription are indicated by solid and open rectangles depending upon whether a consensus sequence has been established. The wavy line represents the RNA product that has a specific initiation point (i) and termination site (t).

57

sequences are 5' to the origin of RNA synthesis (see Shenk, 1981). Several distinct blocks of DNA have been experimentally defined for several genes as serving this function. When the nucleotide sequences of these regions are compared among protein coding genes, including those from phylogenetically diverse organisms, a clear consensus sequence can be identified at each block. Thus, it appears that throughout eukaryotic evolution certain sequences in defined positions 5' to protein coding sequences have been maintained in order to insure proper initiation of transcription by RNA polymerase II. This idea is substantiated by a good deal of experimental evidence showing that cloned genes from a variety of species are capable of being transcribed in heterologous cell-free transcription systems. A dramatic example of this is the fact that the silk fibroin gene from *Bombyx mori* is faithfully transcribed in a human cell-free transcription system (Tsujimoto et al., 1981). In addition, when cloned gene fragments coding for proteins from sea urchin (Grosschedl and Birnstiel, 1980) or chicken (Wickens et al., 1980) are injected in *Xenopus* oocytes, faithful RNA transcripts are produced. Finally, interspecific cell hybrids between humans and rodents have been used for human gene mapping experiments over the last decade. These experiments depend upon the ability of human genes to be expressed in hybrid cells carrying all of the mouse chromosomes but only a random selection of human chromosomes (see Ruddle and Creagan, 1975), and their success demonstrates that the RNA polymerase II transcription machinery in mammals is highly conserved.

The data on RNA polymerase II contrast strongly with what is known about RNA polymerase I transcription. Nucleotide sequence comparisons of the regions around the origins of rDNA transcription in yeast (Valenzuela et al., 1977; Klemenz and Geiduschek, 1980), *Drosophila* (Long et al., 1981), *Xenopus* (Sollner-Webb and Reeder, 1979), mouse (Miller and Sollner-Webb 1981; Urano et al., 1980; Bach et al., 1981a), and human (Miesfeld and Arnheim, 1982) can be made without finding homologies common to most or all of them. Species that are relatively closely related have sequences around the origin of transcription which are highly homologous. Comparison of two *Xenopus* species, *Xenopus laevis* and *Xenopus borealis,* reveals a 13-bp segment (CAGGAAGGTAGGG) from −9 to +4, relative to the origin of transcription, that is identical for both species (Bach et al., 1981b). It is within this region that sequences have been shown by direct experimentation to be required for transcription by RNA polymerase I (B. Sollner-Webb and R. Reeder, pers. comm.). Comparison of mouse and human sequences reveals an identical sequence of 15 base pairs (CTGACACGCTGTCCT) that runs from +3 to +18 (Miesfeld and Arnheim, 1982). These highly conserved amphibian and mammalian sequences, however, show no significant homology to one another.

Furthermore, when 200 base pairs both 5′ and 3′ to the human and mouse origin of transcription are compared using a modification of the Needleman-Wunsch-Sellers algorithm (Goad and Kanehisa, 1982), these sequences are found to be statistically no more similar to each other than any two randomly selected sequences having the same base composition. In general, comparisons of the regions around the origin of transcription among all studied species show that there has been a rapid rate of nucleotide substitution and no consensus sequence has been established.

The nucleotide sequence differences in the regions essential for rDNA transcription among different animal groups might also be the consequence of the fixation of selectively neutral mutations. If this is true, then rDNA genes from diverse species might be expected to be functionally equivalent in tests of transcriptional potential. On the other hand, these sequence differences might have resulted from adaptive evolutionary change, and functional differences in transcriptional behavior among rDNA sequences in different species might be detected. Available evidence supports the latter hypothesis. Interspecific cell hybrids between human and rodent cells do not express the rDNA genes of both species (Eliceiri and Green, 1969; Croce et al., 1977; Miller et al., 1976; Marshall et al., 1975; Tantravahi et al., 1979). There is a clear "nucleolar dominance" (see Rieger et al., 1979) such that, depending upon the chromosomal constitution of the hybrid, the rDNA sequences of either one, but not both, of the species are expressed (Table 1). For example, if the hybrid cell contains all of the human chromosomes and a few mouse chromosomes that carry rDNA

TABLE 1. Nucleolar dominance in interspecific cell hybrids in rodents and humans.

Interspecific cell hybrids	rDNA expression	Source
Mouse–rat	Both species	Kuter and Rogers (1975)
Mouse–Syrian hamster	Both species	Eliceiri (1972)
Human > mouse	Human	Croce et al. (1977)
Mouse > human	Mouse	Eliceiri and Green (1969); Miller et al. (1976)
Human > rat	Human	Tantravahi et al. (1979)
Syrian hamster > human	Hamster	Marshall et al. (1975)

(human > mouse), only human rRNA is synthesized. On the other hand, rodent–rodent cell hybrids do not exhibit nucleolar dominance (Eliceiri, 1972; Kuter and Rogers, 1975), and rRNAs from both species are expressed in these lines (Table 1). Together these data support the idea that there are aspects of rDNA transcription that are species-specific.

Finally, studies using cell-free transcription systems have shown that a HeLa cell-free extract, which is capable of supporting transcription of a cloned human ribosomal gene fragment, is not capable of transcribing a mouse rDNA clone; nor is a mouse cell-free extract able to initiate specific transcription from a human rDNA fragment (Miesfeld and Arnheim, 1982; N. Arnheim and R. Miesfeld, unpublished). Other interspecific studies using cloned DNA segments in *in vitro* transcription systems have given identical results (Grummt et al., 1982).

The data summarized in Table 2 indicate that rDNA sequences from different species are not functionally equivalent with respect to their transcriptional control signals and suggest further that the basis for these differences might lie in their adaptive significance.

We can conclude that the evolution of RNA polymerase I transcriptional control signals might be under different constraints than the signals necessary for transcription initiation by RNA polymerase II. For RNA polymerase I transcription units, mutations that favorably affect transcription initiation could be propagated throughout the rDNA multigene family as a consequence of the genetic interactions that occur among them. The initial increase in frequency of the advantageous mutant might be solely a result of stochastic processes until a threshold frequency is attained, at which point selection could act on the phenotypic consequences of this advantageous mutation. Because the rDNA family evolves in concert, the rate at which selec-

TABLE 2. Evolution of transcription initiation signals.

RNA polymerase transcription units	Rodent–human cell hybrids	Cell-free transcription systems
II	Genes of both species usually active (basis of gene mapping)	Clones from astonishingly diverse species are transcribed in heterologous systems (silk worm gene in HeLa cell extract)
I	Nucleolar dominance	Species specificity (mouse rDNA clone does not work in HeLa cell extract)

tively advantageous mutations are fixed could be very rapid. For RNA polymerase II transcription units, advantageous mutations affecting transcription initiation that occur in any one gene would not be expected to be propagated throughout all RNA polymerase II transcription units in the genome by genetic interactions. This would severely limit the rate of evolution of RNA polymerase II transcription initiation signals, because to change these signals throughout the genome would seem to require extensive parallel evolution.

Our analysis of the rDNA multigene family suggests that selectively advantageous mutations affecting transcription initiation can be fixed in all of the ribosomal genes by concerted evolution. This phenomenon, however, is not limited to ribosomal genes, or to signal sequences involved in RNA polymerase I transcription. Any multigene family that evolves in a concerted fashion has the potential to fix advantageous mutations through genetic interactions among the family members. One can imagine, for example, that selectively advantageous mutations affecting the function of immunoglobulin variable regions could be fixed in the multiple copies of the V region genes by gene conversion (Baltimore, 1981). Gene correction mechanisms not only fix neutral mutations, thus maintaining homogeneity among family members, but also provide an efficient way of fixing selectively advantageous mutations in multigene systems.

CHAPTER 4

EVOLUTION OF ANIMAL MITOCHONDRIAL DNA

Wesley M. Brown

A major goal of current molecular genetic research is to achieve an accurate description of how DNA sequences and organization change with time. This requires both a description of the kinds of changes that occur and an estimation of their respective frequencies of occurrence. Knowledge of the structure–function relationships of the DNA sequences must also be acquired, so that the effects and importance of the changes can be assessed. Both qualitative and quantitative aspects of the mechanisms that lead to changes in the DNA and of the mechanisms that prevent such changes (e.g., DNA repair) must be understood. Finally, we must know how all of these parameters vary among organisms and how the parameters are influenced by the environmental conditions in which different organisms live. It should be clear, even from this brief list of requirements, that the attainment of this goal will require the completion of many technically and intellectually formidable tasks. It should also be apparent, given the complex nature of the genome, that even such limited tasks as cataloguing and measuring the relative frequencies of the kinds of changes seen in DNA will require a variety of approaches.

The mitochondrial DNA (mtDNA) of multicellular animals has many unique properties that make it useful for studies of molecular evolution. The mitochondrial genome is small and simple, in contrast to the large size and complex nature of the nuclear genome. Among multicellular animals, the diploid nuclear genome varies greatly in size, ranging from 4×10^8 to 4×10^{11} base pairs (bp), subdivided into from 4 to over 250 discrete chromosomes. The animal mitochondrial genome, by comparison, is 25,000 times smaller than the smallest animal nuclear genome, varies by only a factor of 1.3 between size extremes, and consists of a single, duplex, closed-circular DNA molecule. In contrast to the complex nature of nuclear DNA, which can contain sequences that vary in copy number from 1 to 10^6 per haploid genome, mtDNA appears to contain sequences that occur only once per mitochondrial genome. Both spacer sequences between genes and intervening sequences within transcribed genes are absent from animal mtDNA. Also, the animal mitochondrial genome appears to be much less susceptible to frequent sequence rearrangements, in marked contrast to the very complex and relatively fluid sequence organization and structure of the nuclear genome. Finally, all of the genetic variables that are a consequence of biparental (sexual) inheritance may be absent from the genetics of mtDNA, which appears to be maternally (clonally) inherited.

An important consequence of these nuclear-mitochondrial differences is that many of the complexities and ambiguities that make nuclear data difficult to interpret are reduced or absent from mitochondrial data, because of the much simpler, uniparentally-inherited mitochondrial genetic system. Because of the apparently great reduction in the number of mechanisms of variation available to mtDNA, its evolution appears to proceed in a much simplified and more straightforward manner than the evolution of nuclear DNA. Mitochondrial DNA may thus be a useful model for the study of certain general aspects of molecular evolution. In addition, because mtDNA is a cytoplasmically inherited, organellar genome that is physically and genetically distinct from, but functionally integrated with the nuclear genome, it also provides a system in which we can study the evolution of nuclear–cytoplasmic interactions. For example, over 90 proteins are required to form functional mitochondrial ribosomes, all or nearly all of which must be encoded by nuclear genes and supplied to the mitochondria in a coordinated manner (O'Brien et al., 1980). Finally, because mtDNA contains genes for ribosomal RNAs (rRNA), transfer RNAs (tRNA), proteins, and regulatory functions, it also provides a good system in which the evolution of these different gene categories may be studied.

In this chapter I will summarize information about how animal mtDNA evolves; that is, how it varies within and between individuals of the same species and between individuals of different species. Although the primary emphasis will be on description, rather than on interpretation, some of the molecular mechanisms that might affect mtDNA variation and evolution will be discussed. The aspects of an-

63

imal mtDNA evolution that will be treated include variations in genome size, gene content, gene arrangement, replication and base composition; types, relative frequencies, locations and rates of change; possible molecular mechanisms that promote and restrict mtDNA variation; present problems and future prospects. For a complementary discussion of mtDNA variation and evolution from a population genetic perspective, see Chapter 8 by Avise and Lansman.

GENETIC STRUCTURE AND REPLICATION OF MAMMALIAN MITOCHONDRIAL DNA

The genetic structure of mammalian mtDNA is shown in Figure 1, along with a brief summary of some of the mitochondrial molecular biology that is germane to this discussion (see legend to Figure 1). Contained in the mtDNA are genes for the small and large rRNAs, 22 tRNAs, and 13 messenger RNAs (mRNAs). Five of the mRNAs are translated into proteins that have been identified as subunits of the mitochondrial electron transport system. The remaining mRNAs are presumably also translated, but their protein products have not been identified.

The animal mitochondrial genetic code is different from the nuclear code (Barrell et al., 1980); because of this, only 22 tRNAs are required by the mitochondrial translation system. As shown in Figure 1, over 90% of the mitochondrial genome is transcribed. Detailed sequencing studies of the DNA, RNA, and some of the proteins have demonstrated that mtDNA lacks intervening sequences, that intergenic spacer sequences are either lacking or, when present, consist of only one to a few nucleotides, and that complete colinearity exists between genes, primary transcripts, and mature gene products (Anderson et al., 1981, 1982; Bibb et al., 1981; Montoya et al., 1981). The remaining few percent of the genome that is untranscribed consists of a discrete sequence block (Figure 1) that contains the origin of heavy-strand replication (O_H) and that lies (in mammalian mtDNA) between the genes for proline and phenylalanine tRNAs (Figure 1). This region often contains a prominent structure, the D-loop, that is formed by the synthesis of a short piece of DNA (the D-loop DNA) that is complementary to the L-strand and that displaces the H-strand in this region (Kasamatsu et al., 1971; Clayton, 1982). The D-loop DNA may serve as a primer for daughter H-strand synthesis and/or it may serve as a means to expose the displaced H-strand sequence so that it can be recognized by a sequence-specific molecule (e.g., an RNA polymerase, thus serving as an initiator sequence for transcription).

Finally, as indicated in Figure 1, the information encoded in mtDNA is asymmetrically distributed between the two strands. Of

64

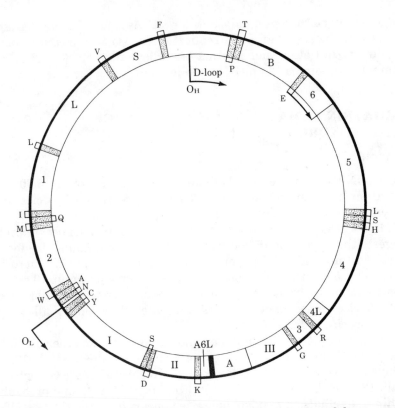

FIGURE 1. Structural, functional, and genetic organization of the mammalian mitochondrial genome. The thick (outer) and thin (inner) circular lines depict, respectively, the heavy (H) and light (L) strands of the mtDNA. O_H and O_L are the respective positions of the daughter H- and L-strand origins of replication; the directions of replication are shown by the arrows. S and L are, respectively, the genes for the small (12S) and large (16S) rRNAs. The genes for proteins are as follows: 1–6 (including A6L and 4L) are open reading frames for functionally unidentified proteins; I, II, and III are subunits of the cytochrome oxidase complex; A is subunit 6 of the mitochondrial ATPase complex and B is cytochrome *b*. The coding (i.e., template) sequences of all protein genes, except 6, and of both rRNA genes are on the H-strand. The tRNA genes, clockwise from O_H, are for the amino acids proline (*P*), threonine (*T*), glutamic acid (*E*), leucine (*L*), serine (*S*), histidine (*H*), arginine (*R*), glycine (*G*), lysine (*K*), aspartic acid (*D*), serine (*S*), tyrosine (*Y*), cysteine (*C*), asparagine (*N*), alanine (*A*), tryptophan (*W*), methionine (*M*), glutamine (*Q*), isoleucine (*I*), leucine (*L*), valine (*V*), and phenylalanine (*F*). The tRNA genes are labeled next to the strand on which their coding sequences occur. Most of the region between tRNA genes *F* and *P* is not transcribed; however, the transcriptional promotors for each strand are believed to be located in this region. This figure is based on the complete mtDNA sequences of human (Anderson et al., 1981), mouse (Bibb et al., 1981), and cow (Anderson et al., 1982).

the 37 transcribed genes, only 9 (8 for tRNAs and 1 for mRNA) are encoded by the L-strand. DNA replication is also asymmetrical, both in point of initiation and in time (see Clayton, 1982, for a superb summary of mammalian mtDNA replication).

VARIATION IN MAJOR FEATURES OF THE MITOCHONDRIAL GENOME

Genome size

The mtDNA of multicellular animals ranges in size from 15,700 to 19,500 bp. Size estimates for mtDNA from a variety of species are presented in Table 1. Given the broad taxonomic range represented in Table 1, the size variation is remarkably small. This observation suggests that a size of 15,700 bp is close to the minimum size for a functional mitochondrial genome in multicellular animals. There is no correlation between genome size and taxonomic group. This is illustrated most dramatically in the genus *Drosophila,* where the mtDNA size range among species is greater than among all other animal species so far examined.

Although microvariation in genome size (i.e., on the order of a few base pairs) occurs between conspecific individuals (Crews et al., 1979; Upholt and Dawid, 1977; Aquadro and Greenberg, 1983; Greenberg et al., 1982; R. L. Cann and A. C. Wilson, unpublished), and probably within individuals as well, genome size appears to be relatively stable within species and, in many instances, over even broader taxonomic groupings. Even in *Drosophila,* the mtDNAs of most species surveyed fall within a narrow size range (16,500 ± 500 bp), and the exceptions (Table 1) are confined to a few species in the *melanogaster* subgroup (Fauron and Wolstenholme, 1976). Among mtDNA from mammals, only that of the domestic rabbit (see Table 1) falls outside the extremely narrow range of 16,500 ± 200 bp. However, major mtDNA size variation occasionally occurs among closely related taxa and even between conspecific populations. Closely related species in the lizard genus *Cnemidophorus* show a mtDNA size range from 17,500 to 19,200 bp (Brown and Wright, 1979; Brown, 1981). Within some species of *Drosophila* (e.g., *D. melanogaster, D. simulans,* and *D. mauritiana),* mtDNAs from different populations vary in size by as much as 700 bp (Fauron and Wolstenholme, 1980a). Among vertebrates, individuals from one population of *Cnemidophorus sexlineatus* have a mtDNA that is approximately 1200 bp larger than that of individuals from six other populations (Brown, 1981; W. M. Brown and J. W. Wright, unpublished).

The occurrence of size variation constitutes unequivocal evidence that deletions and additions occur in animal mtDNA. How these occur

66

TABLE 1. Estimated sizes of mtDNA among animal species.

Species	Size (bp)	Source
VERTEBRATES		
Human	16,569	Anderson et al., 1981
Green monkey	16,400	Brown, 1981
Woolly monkey	16,300	Brown, 1981
Bush baby (*Galago*)	16,500	Brown, 1981
House mouse	16,295	Bibb et al., 1981
Golden hamster	16,300	Brown, 1981
Cow	16,338	Anderson et al., 1982
Domestic rabbit	17,300	Brown, 1981
Cnemidophorus lizards (11 species)	17,500	Brown and Wright, 1979
C. sexlineatus (Florida population)	18,000	W. Brown, unpubl. data
C. sexlineatus (outside Florida)	19,200	W. Brown, unpubl. data
Xenopus (2 species)	17,700	Brown, 1981; Dawid, 1972
SEA URCHINS		
Strongylocentrotus (3 species)	15,700	Brown, 1981
Lytechinus pictus	15,700	Brown, 1981
INSECTS		
Drosophila neohydei	15,700	Fauron and Wolstenholme, 1976
D. melanogaster (ORE-R)	19,500	Fauron and Wolstenholme, 1976
D. melanogaster (JAPAN)	18,700	Fauron and Wolstenholme, 1976

mechanistically and how they are distributed in the mitochondrial genome are two questions that arise immediately. There is, at present, no answer to the first question. A fairly detailed answer to the second, however, has been provided by data from many recent investigations. Microvariation in mtDNA size appears to result from deletions and additions of from one to a few base pairs (Crews et al., 1979; Brown et al., 1982; Aquadro and Greenberg, 1983; Greenberg et al., 1982; R. L. Cann and A. C. Wilson, unpublished). Although these events occur in all regions of the mtDNA, their frequency of occurrence varies greatly among different regions, being least in genes that code for proteins.

Macrovariation in mtDNA size appears to result from larger deletions and additions. These events appear to be confined exclusively to the D-loop and adjacent nontranscribed regions and to the homologue of this region in *Drosophila* mtDNA, the adenine + thymine-rich (A+T-rich) region. High resolution electron microscopy of heteroduplex molecules (formed by annealing H- and L-strands from different, closely related species that have D-loops of different sizes) demonstrated that large deletion/addition events are clustered near the ends of the D-loop (Upholt and Dawid, 1977). High resolution mapping of restriction endonuclease cleavage sites and sequence analysis of mtDNA from hominoid primates has revealed that a 95-bp deletion occurred in gorilla mtDNA in the region between the gene for phenylalanyl-tRNA and O_H (Figure 1) (Ferris et al., 1981a; J. Hixson and W. Brown, unpublished). Both electron microscopy and cleavage mapping studies have shown that an even more dramatic and rapid change occurs in the A+T-rich region of mtDNA from six closely related *Drosophila* species (Fauron and Wolstenholme, 1980a; Wolstenholme et al., 1979). Although mtDNA heteroduplexes for all *Drosophila* species combinations formed well-matched duplexes outside the A+T-rich region, no duplex formation could be observed within the region, except between the closest species where only 35% of the region was paired well enough to form a duplex (Fauron and Wolstenholme, 1980a). These results indicate that deletion/addition events in the A+T-rich region of *Drosophila* mtDNA are extensive and frequent and also that the primary base sequence changes with great rapidity. Similar conclusions may be drawn for the D-loop region of vertebrate mtDNAs, based on nucleotide sequence comparisons (Anderson et al., 1982; Bibb et al., 1981; Walberg and Clayton, 1981; Sekiya et al., 1980; J. Hixson and W. Brown, unpublished).

Gene content, arrangement, and replication

Data from a variety of studies suggest, albeit indirectly, that little or no variation in mtDNA gene content is likely to occur among multicellular animals. The major mitochondrial transcripts (i.e., rRNAs and mRNAs) appear to be similar in both number and size over a broad taxonomic range (insects to vertebrates: see, for example, Hirsch et al., 1974; Bonner et al., 1977; Rastl and Dawid, 1979a,b; and Ojala et al., 1981). The animal mtDNA genes whose products have been associated with known functions (i.e., the 22 tRNAs, 2 rRNAs, and 5 of the proteins; see Figure 1) appear to occur ubiquitously, not only in animal mtDNA, but also in the mtDNA of lower eukaryotes. The narrow size range for animal mtDNA is also consistent with a hypothesis of little or no variation in gene content, particularly because the events that cause differences in size appear to occur predominantly, if not exclusively, in the nontranscribed regions of the mtDNA.

The order in which the mitochondrial genes are arranged also appears to be relatively stable. Among mammals, mtDNAs from three species (human, house mouse, and cow) belonging to three different orders (primates, rodents, and artiodactyls) have been completely sequenced (Anderson et al., 1981, 1982; Bibb et al., 1981). The gene arrangement in these mtDNAs is identical, indicating that stability in this respect has been maintained for the 80–100 million years that have elapsed since the divergence of the three lineages. A less direct determination, based on the comparison of the mammalian mitochondrial gene arrangement with that of the clawed frog (*Xenopus*; Figure 2), suggests that the gene order may have been stable since the divergence of the lineages leading to mammals and contemporary amphibians, a period of approximately 350 million years. The unambiguously conserved features in this comparison are the adjacent arrangement of the D-loop and rRNA genes and the directions of DNA replication and RNA transcription (Attardi et al., 1976; Dawid et al., 1976; Ramirez and Dawid, 1978). A conclusion of probable conservation in the arrangement of the remaining genes is also warranted (Figure 2) but is based strictly on the sizes and relative order of the RNA transcripts of *Xenopus* (Ramirez and Dawid, 1978; Rastl and Dawid, 1979a,b) compared to the precisely known positions of the genes and transcripts of mammalian mtDNAs (Anderson et al., 1981, 1982; Bibb et al., 1981; Ojala et al., 1981; Montoya et al., 1981).

The first indication that the gene order was not a fixed feature of animal mtDNA came from a series of studies of *Drosophila* mtDNA (Wolstenholme et al., 1979; Goddard and Wolstenholme, 1980; Fauron and Wolstenholme, 1980a). Earlier studies had demonstrated the presence of an A+T-rich region in this mtDNA (Bultmann and Laird, 1973; Peacock et al., 1973; Polan et al., 1973; Wolstenholme, 1973), a feature that, so far, has not been found in other animal mtDNAs. The position of this region was mapped relative to the rRNA genes and several restriction endonuclease cleavage sites in *D. melanogaster* mtDNA (Klukas and Dawid, 1976) and in the mtDNA of several other *Drosophila* species (Fauron and Wolstenholme, 1980a). It was shown (Goddard and Wolstenholme, 1978) that the mtDNA of *D. melanogaster* replicates unidirectionally from an origin within the A+T-rich region, that the direction of replication is toward the rRNA genes (which lie immediately adjacent to one end of the A+T-rich region)

69

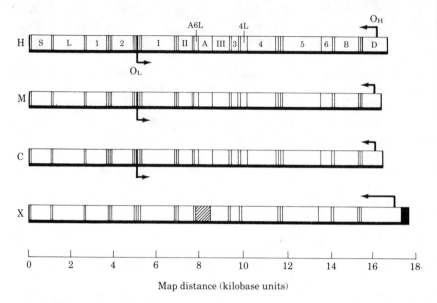

FIGURE 2. Comparison of the genetic maps of mtDNA from human (H), mouse (M), cow (C), and frog (X). Abbreviations are the same as for Figure 1, except for the nontranscribed region, which is here labeled D. The maps for H, M, and C are derived from sequence data (see Figure 1 legend). The map for X (*Xenopus laevis*) is from transcription mapping studies (Ramirez and Dawid, 1978; Rastl and Dawid, 1979a,b), and the differences in the number of tRNA genes at several positions in *Xenopus* are, thus, probably not real, but an artifact of the lower resolution achievable by this method. The darkened area at 8 kilobases in *Xenopus* indicates lack of resolution of the A6L and A transcripts in this species. The size shown for *Xenopus* mtDNA was determined by contour length measurements from electron micrographs (Dawid, 1972; Brown, 1976, 1981). The size estimates of the H, M, and X D-loop DNAs are from Gillum and Clayton (1978) and Brown et al. (1978).

and that the directions of DNA replication and rRNA transcription are identical. This is the reverse of the situation in mammalian mtDNA. These findings have been extended to other *Drosophila* species (Goddard and Wolstenholme, 1980).

One conclusion to be drawn from the preceding results is that the gene order in *Drosophila* is different from that in vertebrates. The data published as of this writing are insufficient for a determination of the specific kind or extent of the changes that have occurred. Unpublished sequencing studies indicate that such events have occurred, as manifested by significant differences in gene order between *Drosophila* and vertebrate mtDNAs (D. Wolstenholme, pers. comm.). The results of transfer–hybridization experiments also suggest that there may have been major rearrangements in the gene order of sea urchin mtDNA as compared with that of either *Drosophila* or vertebrates (R. Britten, pers. comm.).

A second conclusion from the *Drosophila* studies is that the mode of *Drosophila* mtDNA replication appears to differ in fundamental ways from that of the other animal species that have been examined. In addition to originating in a unique A+T-rich region, *Drosophila* mtDNA replication may not proceed via a D-loop mechanism. Few to no D-loop structures could be found in *Drosophila* mtDNA after extensive electron microscopic analysis conducted by three independent groups of investigators (Klukas and Dawid, 1976; Rubenstein et al., 1977; Wolstenholme et al., 1979; *but* see Zakian, 1976, for a possibly contradictory result). Also, using a different method of analysis, no D-loop DNA was detected in lysates of *Drosophila* mitochondria (Rubenstein et al., 1977).

Because the origins of replication can be so precisely defined, mtDNA provides an excellent system in which to examine the evolutionary genetics of this important class of DNA sequences. At present, exact sequence details are known for only a few mammalian species (Anderson et al., 1981, 1982; Sekiya et al., 1980; Walberg and Clayton, 1981; Clayton, 1982; J. Hixson and W. Brown, unpublished). Even relatively gross features of replication have been well characterized only for mammals, a few other vertebrates, *Drosophila,* and a sea urchin (see Clayton, 1982, for details and references). The mtDNA of the sea urchin appears to replicate via a D-loop mechanism that is similar (perhaps identical) to that of vertebrates (Matsumoto et al., 1974; Clayton, 1982). These observations could indicate a closer phylogenetic relationship between vertebrates and sea urchins than between either of these and insects. However, there is an obvious need for more information at all levels of resolution and from many additional taxa before valid evolutionary inferences about mtDNA replication can be drawn.

Nucleotide base composition

The base composition of animal mtDNA, expressed as percentage of guanine plus cytosine (G+C) content, ranges from a low of 21%, in *Drosophila melanogaster,* to a high of 50%, in the domestic duck (Table 2; see Borst and Flavell, 1976, and Brown, 1981, for more extensive species lists and for references to primary data). One possibly meaningful correlation between mtDNA base composition and evolution is that the mitochondrial G+C content among invertebrates (range 21 to 43%) appears to be generally lower than among vertebrates (range 37 to 50%).

A striking vertebrate–invertebrate division is seen (Table 2) when the base compositional bias of the complementary single DNA strands is investigated. This bias is indicated by the relative ability of the strands to separate in alkaline cesium chloride gradients and is due to a difference in the guanine plus thymine (G+T) content between the strands (Brown, 1981). Among vertebrates, the H- and L-strand density differences in alkaline cesium chloride gradients range from 13 to 43 mg/ml; among invertebrates, the range is 5 to 10 mg/ml (Table 2; also see Borst and Flavell, 1976; Brown, 1981). Of the ver-

TABLE 2. Percentage G+C of duplex mtDNA and strand separability in alkaline CsCl.*

Species	Percentage G+C	Strand Separation (mg)
VERTEBRATES		
Human	44	41
Green monkey	43	41
Woolly monkey	40	33
House mouse	37	31
Domestic duck	50	—
Chicken	46	43
Lizard (*Uma notata*)	38	—
Frog (*Rana pipiens*)†	40	13
Frog (*Xenopus laevis*)	39	23
Fish (*Salmo gairdneri*)	41	—
SEA URCHIN		
Strongylocentrotus purpuratus	39	≤5
MOLLUSK		
Mussel (*Mytilus californianus*)	43	≤10
INSECT		
Drosophila melanogaster‡	21	≤5
ECHIURID WORM		
Urechis caupo†	42	—
NEMATODE WORM		
Ascaris lumbricoides†	31	—
FLATWORM		
Hymenolepis diminuta†	32	—

* Unless otherwise noted, values are from Brown (1981).
† Borst and Flavell (1976).
‡ Polan et al. (1973).

72

tebrate mtDNAs characterized in this respect, those from "warm-blooded" species show a greater strand bias (range: 31 to 43 mg/ml) than do those from "cold-blooded" species (13 and 23 mg/ml). The bias in G content is particularly pronounced in both birds and mammals (Borst and Ruttenberg, 1969; Brown, 1976, 1981).

It is impossible, at present, to know if the preceding correlations are meaningful, because there are so few data for invertebrate and for "cold-blooded" vertebrate species. The observed strand bias could simply represent a historical accident, in which the common ancestor of the vertebrates possessed a "biased" mtDNA, due to chance alone, which has been "locked in" by virtue of common descent. Such a hypothesis is difficult to defend, however, because the high rate of sequence evolution of animal mtDNA (discussed later) is a potent randomizing force that opposes the maintenance of such a bias. A selectionist hypothesis, in which the strand bias takes on an unknown functional role, is much more tenable.

Thermal dissociation and sedimentation equilibrium analyses indicate that, with the notable exception of *Drosophila* mtDNA, neither G nor T shows a highly clustered distribution in animal mtDNA, and this has been confirmed by recent sequencing studies. Because the majority of the mitochondrial RNAs are transcribed from the H-strand, they are relatively depleted in G+T content as a function of the amount of strand bias present in the mtDNA from which they arise. Formation of secondary structures in single-stranded polynucleotides is not only sequence specific (i.e., a function of the degree of self-complementarity of the contained sequences) but is also a function of base composition. Guanine is especially capable of forming stable base pairs with itself, and thus an RNA with a high G-content would tend to have a large amount of intramolecular secondary structure as well as a strong tendency to form aggregates.

In the much-simplified mitochondrial system, the efficiency of RNA processing and protein synthesis may depend on the ability of mRNAs to remain unaggregated and relatively free of secondary structure. The energy demands placed on most invertebrates, and particularly on sessile invertebrates (e.g., sea urchins) may be low enough so that a high degree of efficiency in mitochondrial gene expression is not a strongly selected feature, and an "average" base composition with no strand bias can easily be tolerated. Perhaps the high energy demands placed on the mitochondrial system by "warm-bloodedness" are the chief selective factors that gave rise to and maintain the strong G-bias that is seen in avian and mammalian mtDNAs. The low G+C content of *Drosophila* mtDNA (only ~25%, even when the A+T-rich region is excluded) may represent an alternative solution taken by an insect whose rapid development places a similarly high demand on energy production.

NUCLEOTIDE SEQUENCE EVOLUTION

The variations seen in major features of the mitochondrial genome are manifestations of changes in the nucleotide sequence of the mtDNA, that is, of base substitutions and sequence rearrangements. In order to gain a more complete picture of how animal mtDNA evolves, we need answers to a number of questions about these sequence changes: At what rate and in what proportion do these events occur? Where do they occur in the mtDNA, and in what kinds of genes? What structural and functional roles are played by the sequences in which the events occur, and what functions can be assigned to the exact nucleotide positions at which they occur? Preliminary answers are available for most of these questions, based on data derived almost exclusively from studies of vertebrate mtDNAs. Although some data also exist for *Drosophila* mtDNA, the remaining major groups of animals are unstudied.

The nucleotide sequence of mtDNA evolves rapidly

Thermal dissociation analysis of heteroduplexes formed from H- and L-strands isolated from different *Xenopus* species gave the first indication that the rate of sequence evolution might be higher in mitochondrial than in nuclear DNA (Dawid, 1972). This same study also provided unambiguous evidence that (1) within the mtDNA itself the rate of change was much slower in the rRNA and tRNA genes than in the remainder of the genome and (2) the mitochondrial rRNA genes were less conserved than nuclear rRNA genes. Cleavage map comparisons of sheep and goat mtDNAs and electron microscopy of their heteroduplexes demonstrated that the D-loop region was the site of several rearrangements (most probably deletions and additions) and also the most rapidly changing portion of the mtDNA (Upholt and Dawid, 1977).

Studies employing a series of primate species provided the basis for an unambiguous demonstration that the nucleotide substitution rate is appreciably faster in mitochondrial than in single-copy nuclear DNA (Brown, 1976; Brown et al., 1979). Human and green monkey (*Cercopithecus aethiops*) mtDNAs are nearly identical in size (Table 1), base composition, and G+T strand bias (Table 2) (Brown, 1976, 1981), even though the lineages leading to these two species diverged at least 25 million years ago (Simons, 1972). Thermal dissociation analysis of heteroduplex mtDNA from these species (Figure 3) showed

74

that the nucleotide sequences differed at 21% of their positions (min-imally) despite the similarities in their physical properties (Brown, 1976; Brown et al., 1979). In contrast, similar analyses of single-copy nuclear DNA heteroduplexes from these species indicated only a 5 to 6% sequence difference (Kohne et al., 1972; Benveniste and Todaro, 1976), a factor that is 3.5 to 4 times less. The mtDNA data (Figure 3) also indicated (1) that the differences occur throughout the genome, because the heteroduplex curve does not become congruent with the homoduplex curves, but (2) that their distribution is uneven, that is, that some sequences have changed more than others, because the duplex to single-strand transition occurs over a much broader tem-perature range for the heteroduplex mtDNA than for the homodu-plexes.

Restriction endonuclease cleavage site mapping, combined with a knowledge of the origin and direction of mtDNA replication, provided a further means by which mtDNAs could be compared (Brown and Vinograd, 1974). From such cleavage map comparisons quantitative

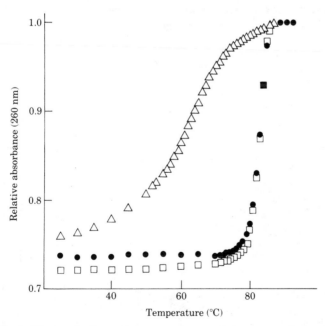

FIGURE 3. Thermal dissociation analysis of homo- and heteroduplex mtDNA. The dissociation of human (●) and green monkey (□) homoduplexes and a human H-strand × green monkey L-strand heteroduplex (△) are shown. From Brown et al. (1979).

estimates of the degree of sequence difference could be derived (Uphold, 1977; Nei and Li, 1979; Gotoh et al., 1979). Based on such comparisons among mtDNAs from a variety of mammalian (mostly primate) species (Brown et al., 1979), the estimated pairwise sequence differences, when plotted versus the estimated divergence times (Figure 4), revealed three important facts about mtDNA evolution. The first was that the rates of change of mtDNA and single-copy nuclear DNA are grossly different for closely related species pairs, i.e., those having times of divergence of 15 million years or less (Figure 4); the initial rate of change for mtDNA is approximately ten times greater than that for single-copy nuclear DNA, as shown by comparing the respective slopes in Figure 4. The second was that the rate of change of mtDNA is the same as or slower than the rate of change of single-copy nuclear DNA at greater times of divergence (say, > 25 million years). The third was that only 25–30% of the total mtDNA sequence is involved in the fast rate, as shown by the extrapolation of the slow rate curve for mtDNA back to zero time of divergence (Figure 4). It should be noted that these three conclusions are true even if the *absolute* divergence times used in Figure 4 are wrong. The conclusions depend only on maintenance of reasonable proportionality between the *relative* divergence times and on the assumption that estimates of sequence divergence from cleavage map comparisons are reasonably accurate. Direct sequence comparisons among the mtDNAs of some of

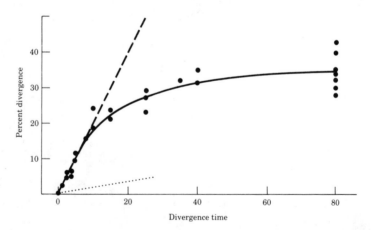

FIGURE 4. Dependence of percentage sequence divergence of mtDNA on divergence time. The points represent estimates from pairwise comparisons of restriction endonuclease cleavage maps (data from Brown et al., 1979; Brown, 1980; Ferris et al., 1981a). The initial rate of mtDNA sequence divergence is shown by the dashed line and the rate of divergence of single-copy nuclear DNA by the dotted line. Divergence times are in millions of years. Modified from Figure 3 of Brown et al. (1979).

the same primate species provide evidence that this latter provision has been met (Brown et al., 1982; W. Brown and D. Shumard, unpublished).

Substitutions predominate over deletions/additions

The predominance of nucleotide substitutions in the evolution of animal mtDNA has been assumed, on theoretical grounds, in most mtDNA comparisons. The frequent occurrence of deletions/additions of > 50 bp had been ruled out by data from analyses of heteroduplex molecules in the electron microscope (Upholt and Dawid, 1977; also earlier, unpublished studies by M. L. Simon and R. L. Hallberg), except in the D-loop region. Although smaller deletions/additions could have been observed in comparisons of mtDNA among conspecific individuals, none were seen using high-resolution gel electrophoresis of restriction endonuclease digests (Brown and Goodman, 1979; Brown, 1980; Ferris et al., 1982). However, empirical proof of the predominance of substitutions came only after DNA sequencing became possible.

   Comparison of the complete sequences of human (Anderson et al., 1981), house mouse (Bibb et al., 1981), and cow (Anderson et al., 1982) mtDNA confirmed not only the major role played by substitutions in the evolution of mammalian mtDNA, but also the absence of large deletions/additions in the transcribed portions of the genome and their presence in the nontranscribed portions (particularly in the D-loop region). In addition, these studies demonstrated that small deletions/additions (from 1 to 4 bp) were also present in the tRNA and rRNA genes, but not in the genes coding for proteins. The interspecific differences in the sizes of the proteins listed in Table 4 are due to differences in the positions of the postulated translation-initiation codons, and not to nucleotide deletions/additions.

Dynamics of nucleotide substitutions

The sequence comparisons among human, house mouse, and cow mtDNAs provide a wealth of information to both molecular biology and evolution. However, the picture they provide is static, because it represents a single divergence time that is well out on the "slow" region of the mtDNA curve in Figure 4. In order to develop a dynamic picture of the evolution of mtDNA (or of any other molecule), it is necessary to make comparisons among species that represent a variety of divergence times. For mammalian mtDNA, comparisons among species pairs that have diverged within the last 15 million years (i.e., during the period in which the events that give rise to the rapid evolutionary rate can most easily be discriminated) should be the most informative.

Among primates, the closely related species group composed of human, chimpanzee, gorilla, orangutan, and gibbon forms one such assemblage, with pairwise divergence times ranging from 5 to 10 million years (Figure 5) (Wilson et al., 1977). A homologous fragment of the mtDNA from each of these was cloned, sequenced, and compared (Brown et al., 1982). This 896-bp fragment contains three tRNA genes (*L, S,* and *H*) and portions of two adjacent protein genes (4 and 5 in Figure 1). Collectively, differences at 284 positions were observed among the five sequences; all of these were nucleotide substitutions save one, which was a 1-bp deletion. Although the tRNA genes constitute 22% of the 896-bp sequence, only 13% of the substituted positions were located in them, thus indicating (on a substitution per base pair basis) that their substitution rate is approximately half that of the protein genes. The 1-bp deletion that was seen occurred in a tRNA gene.

On a pairwise basis, the percentage sequence differences for five hominoid species agreed with the evolutionary relationships as shown in Figure 5. The sequence data also fell on the mtDNA curve of Figure

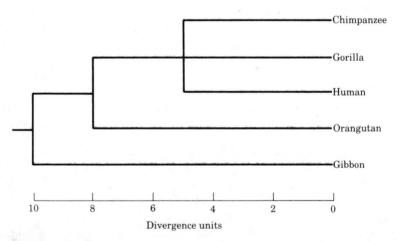

FIGURE 5. Evolutionary tree for the hominoid primates. Because of the uncertainties and variation in divergence time estimates, one divergence unit is estimated to equal from 1 to 1.5 million years. Although parsimony analysis of mtDNA data suggests a closer association between chimpanzee and gorilla than between either of these and human (Ferris et al., 1981a,b,; Brown et al., 1982), the difference is so small that the branching of these three species is depicted as an unresolved trichotomy.

4, thus directly confirming the reliability of the conclusions based on estimates from cleavage map comparisons as regards both quantitation of sequence divergence and rapid initial rate of sequence evolution. A most unexpected result, however, was the predominance of transitions (A ↔ G and C ↔ T substitutions) over transversions (A or G ↔ C or T) at lesser divergence times (Brown et al., 1982). The ratio (S/V) of transitions (S) to transversions (V) appears to change as a function of divergence time, as illustrated in Table 3. This apparent change is due to the time-dependent accumulation of multiple substitutions at many of the nucleotide positions. The highest S/V so far observed, from a mtDNA sequence comparison among seven conspecific individuals (humans) indicated that transitions occur 24 times more frequently than transversions (Aquadro and Greenberg, 1983; Greenberg et al., 1982). Available evidence suggests that this skewed ratio is due to a higher rate of transition-causing mutations, rather than to a higher probability of fixation for transitions. Based strictly on the number of substitutional pathways leading to transitions and transversions (two and four, respectively) and assuming an equal probability for each, the expected S/V is 0.5. This value is approached only in comparisons among species pairs whose estimated times of divergence are ≥ 80 million years (Table 3). The picture that has emerged from these data is one in which the kinetics of nucleotide substitution are biased toward transitions by a large factor (say, ≅25) but in which

TABLE 3. Dependence of the ratio of transitions (S) to transversions (V) on divergence time.

Divergence time, millions of years*	Percentage S	S/V	Source
<0.3	96	24.0	Aquadro and Greenberg (1983)
5	92	11.5	Brown et al. (1982)
8	77	3.3	Brown et al. (1982)
10	72	2.6	Brown et al. (1982)
25	58	1.4	W. Brown and D. Shumard, unpublished
80	45	0.8	Brown et al. (1982)

* The hominoid divergence time estimates are from Figure 5, assuming one unit = 10^6 years. Estimates for the hominoid–green monkey and the primate–rodent–artiodactyl divergences are, respectively, 25 and 80 million years.

the equilibrium condition approximates an unbiased condition, with $S/V \cong 0.5$. This equilibrium is attained only after a considerable divergence time has elapsed (see Brown et al., 1982, for further discussion).

Dynamics of sequence rearrangements

The dynamics of sequence rearrangements in mtDNA are less well studied than those of nucleotide substitutions. Based on high resolution mtDNA cleavage map comparisons among 135 individuals (humans), R. L. Cann and A. C. Wilson (unpublished) estimate that oligonucleotide deletion/addition events occur approximately 0.5 times as frequently as do substitutions in those mtDNA regions that do not give rise to functional RNA transcripts. Available sequence comparison data indicate that rearrangements are even less common, by a factor of at least 100, in regions that express functional RNA transcripts.

We have obtained data (J. Hixson and W. Brown, unpublished) that provide support for the hypothesis that the large regions of non-homology that occur in portions of the nontranscribed, O_H-containing region of animal mtDNA (Anderson et al., 1982; Walberg and Clayton, 1981) do not represent single deletion/addition events but rather are the cumulative result of many, closely adjacent, oligonucleotide deletion events. Other investigators have compared a 900-bp sequence in the D-loop region from the mtDNA of seven individuals (humans) (Aquadro and Greenberg, 1983; Greenberg et al., 1982). Their investigation revealed that both substitutions and small (1 or 2 bp) deletion/addition events were frequent. Restriction endonuclease cleavage site map comparisons among primate mtDNAs revealed that gorilla mtDNA contained a deletion of ~95 bp in this region (Ferris et al., 1981a). The sequence of this region of gorilla mtDNA, when compared with the human sequence, does not show a single large deletion, but rather shows a complex series of smaller deletions that add up to a total of 96 bp (J. Hixson and W. Brown, unpublished). Even though this species pair is quite recently diverged, it is already impossible to confidently discern individual deletion events. Examination of even more closely related species pairs will be necessary if a better understanding of the dynamics of change in this region is to be obtained.

Sequence changes in specific genes

Both thermal stability data and cleavage map comparisons showed that changes in sequence were distributed throughout the animal mitochondrial genome (Dawid, 1972; Brown, 1976; Brown et al., 1979; Ferris et al., 1981a,b). Within the limits of their resolving power,

80

these studies also indicated that the frequency distribution of the changes was different in different parts of the genome, but that parts representing the extremes in this respect probably differed by no more than a factor of 10, and possibly less. By employing nucleotide sequence data, we are now able to make such comparisons at the maximum level of resolution. Comparisons have been made for the entire mtDNA sequence of two species pairs (human and cow, by Anderson et al., 1982; and human and mouse, by Bibb et al., 1981) and for the sequence of a smaller region among five primate species (human and apes, by Brown et al., 1982). The reader is referred to these articles for detailed, gene-by-gene comparisons and discussion of specific evolutionary implications. Only the more general features of these studies are summarized here.

In comparisons among species (human, mouse, cow) with divergence times of 80–100 million years, mitochondrial protein genes exhibited a range in nucleotide conservation from 61 to 79% and in amino acid conservation from 46 to 91% (Table 4). Both within and between species-pairs, the relative degrees of nucleotide and amino acid conservation among the genes correlated well, as may be seen in Table 4. This suggests that the relative functional constraints/selection pressures on the genes have been similar since the divergence of the three taxa. Although it is likely that the constraints have operated primarily at the amino acid rather than at the nucleotide level, the data are insufficient to exclude the possibility that some constraints may also operate directly on the nucleotide sequence.

Nucleotide substitutions in the protein genes occur most frequently in the third positions of codons. Most of these changes are silent, that is, they do not lead to concomitant amino acid substitutions. Although silent substitutions constitute approximately one-half of the observed pairwise sequence differences among species with divergence times of 80–100 million years, the actual rate of accumulation of these substitutions is several times higher. This has been shown in mtDNA comparisons among species pairs with shorter divergence times. The reason for the dependence of frequency on divergence time is that the positions at which nucleotide substitutions are silent are under few, if any, functional constraints and therefore can change at a maximum rate. (If no constraints exist, this rate equals the mutation rate.) A much slower rate occurs in the remaining positions (replacement sites) in which nucleotide substitution causes a replacement in the encoded protein, of one amino acid with another. This slower rate results from the fact that only a few of the many possible amino acid replacements in the protein are also compatible with maintenance of function. Thus,

81

TABLE 4. Comparison of mitochondrial protein genes of human (H), mouse (M) and cow (C). (From Anderson et al., 1981, 1982, and Bibb et al. 1981.)

Gene*	Protein size (amino acids)†			Percentage amino acid conservation†		Percentage nucleotide conservation	
	H	M	C	H/M	H/C	H/M	H/C
I	513	514	514	90	91	77	79
III	261	261	261	86	87	76	79
B	380	381	379	78	79	74	73
1	318	315	318	78	78	71	73
A	226	226	226	75	78	72	73
II	227	227	227	71	73	71	70
4L	98	97	98	67	74	69	75
4	459	459	459	67	74	68	72
3	115	114	115	65	74	66	70
5	603	607	606	63	70	65	71
2	347	345	347	57	63	61	66
6	174	172	175	52	63	63	65
A6L	68	67	66	46	52	65	63

* Genes listed in order of decreasing amino acid conservation (averaged for H/M and H/C).
† Amino acid sequences were inferred from the nucleotide sequences.

after a short divergence time, only the least constrained nucleotide positions are likely to have undergone substitutions. With increasing divergence time, more of these sites undergo substitution, until finally, at saturation, all of these sites have been changed at least once, and many of them more than once. Thus, in comparing taxa with divergence times sufficiently great to allow saturation, nearly all of the silent *sites* will be counted as different, but this number will represent considerably less than the actual number of silent *substitutions* that have occurred. Also, because substitutions will continue to accumulate at replacement sites long after the silent sites have reached saturation, the apparent (i.e., observed) ratio of silent/replacement substitutions will decrease. The "true" ratio can be estimated using empirical data from comparisons among species whose divergence times are considerably less than the time needed to saturate the silent sites. This interval corresponds, in primates, to pairwise divergence times of less than 15 million years (Figure 4). For the protein-coding sequences examined among species whose divergence times fall within this interval, the average silent/replacement substitution ratio (incorporating multiple hit corrections) is approximately 5. As predicted in the preceding discussion, the apparent (i.e., observed) ratio decreases with greater divergence times (e.g., it is approximately 3 for pairwise comparisons between human, cow, and mouse).

Our estimates of the silent/replacement substitution ratio for mitochondrial protein genes (Brown et al., 1982) are slightly lower, but similar to the estimates made by Perler et al. (1980) for nuclear genes. This similarity in ratio suggests that these mitochondrial and nuclear proteins are under fairly similar selective constraints, despite the more rapid rate of evolution for the mitochondrial genes. This relatively low ratio indicates quite clearly that the rapid evolution of animal mtDNA is not due solely to silent substitutions, as has been proposed by some investigators (G. Brown and M. Simpson, 1982). Because the evolutionary rate is proportional to the product of the mutation and fixation rates and because only the fixation rate is affected by selective constraints, these results suggest that an elevated rate of mutation contributes significantly to the higher rate of mtDNA evolution.

Nucleotide substitutions are the predominant type of change seen in mitochondrial tRNA genes, although 1- to 2-bp deletion/additions also occur. Changes have been observed in all structural regions of the tRNAs, except in the anticodons themselves. Many structural features that are invariant among nuclear tRNAs show extensive variation among mitochondrial tRNAs. The substitution rate for the tRNA genes is approximately the same as the replacement rate for protein genes (Brown et al., 1982), but this is higher by a factor of at least 100 than the rate for nuclear tRNA genes. Although a higher mutation rate in mtDNA can explain part of this increase, the remainder must be due to other factors, such as the possibility of drastically lowered functional constraints on the mitochondrial tRNAs. Because of the simplified mitochondrial genetic code, there are 30% fewer tRNAs in the mitochondrial than in the cytoplasmic (i.e., nuclear) system, a factor that, by itself, might result in a lowering of constraints on the interactions between the tRNAs and their charging enzymes (aminoacyl-tRNA synthetases) and between the tRNAs and their ribosomal binding sites. A second factor is the strong possibility that regulation of the expression of 13 mitochondrial protein genes necessitates many fewer translational-level controls than are required for the coordinated regulation of the expression of thousands of nuclear

83

genes. If this is the case, then those aspects of tRNA structure that are involved in such regulatory interactions would also be under greatly reduced constraints.

A predominance of nucleotide substitutions with some oligonucleotide deletion/additions also characterizes the changes seen in the genes for the small (12S) and large (16S) rRNAs. The direct sequence comparisons that are presently available all involve species pairs with long divergence times (e.g., human, cow, and house mouse). Changes have occurred in rRNA structural regions that are postulated to be single-stranded in the ribosome, and also in regions postulated to be duplex. Although extensive changes are observed, their distribution is markedly nonrandom, and numerous short (i.e., 10–20 bp) sequences of high conservation have been observed. Many of these conserved sequences show complete, or nearly complete, homology, not only with the mitochondrial rRNA genes of other species, but also with the cytoplasmic rRNA genes of both eukaryotes and prokaryotes (Küntzel and Köchel, 1981). This indicates that there is a direct relationship between the primary nucleotide sequence and ribosomal function for some parts of the rRNAs.

Because sequencing data are lacking for species pairs with short and intermediate divergence times, the evolutionary rate for the mitochondrial rRNA genes is not well characterized. However, less direct comparisons (thermal stability measurements and cleavage map comparisons) already provide convincing evidence that their evolutionary rate relative to nuclear rRNA genes is at least tenfold greater, but that it is somewhat less than the rate of mitochondrial protein genes (Dawid, 1972; Brown et al., 1979; Ferris et al., 1981a,b). Many of the reasons for the high evolutionary rate relative to nuclear rRNA genes are the same as discussed earlier for the tRNA genes, involving the simpler mitochondrial genetic code as well as relaxation, or absence, of translational-level controls in the mitochondrial system.

MECHANISMS OF MITOCHONDRIAL DNA EVOLUTION

Clues to at least a few possible mechanisms underlying the evolutionary behavior of animal mtDNA come from structural and biochemical studies of the mitochondrial system (see Brown, 1981, for a recent review). However, most of these studies have been confined to analysis of the mtDNA itself. Consequently, little is known about the enzymes and enzymatic processes that may be involved in the replication and repair of mtDNA. Knowledge of the transmission and population genetics of animal mtDNA, both of which strongly influence how it evolves, is also poorly developed. The following discussion must, therefore, be regarded as quite limited in scope and both preliminary and speculative in its conclusions. It may, however, serve to point out some interesting areas for future investigation.

Close packing of mitochondrial genes lowers the frequency of genome rearrangements

The almost total lack of noncoding sequences (exclusive of the D-loop region) in animal mtDNA poses a formidable barrier to fixation of sequence rearrangements. Thus, transpositions, in order to "succeed," would have to occur by sequence excisions at exact gene boundaries followed by insertions elsewhere in the genome at the exact boundary between adjacent genes. Furthermore, any transposition, even if exact, might fail because of the manner in which mtDNA is transcribed and the transcript processed. Most mitochondrial genes lie between directly adjacent tRNA genes (Figure 1). Attardi and co-workers have hypothesized that this direct interspersion of tRNA genes is a structural feature that is needed for correct processing of the mitochondrial transcripts (Ojala et al., 1980, 1981). If this hypothesis is true, then RNA processing requirements would add to the constraints on rearrangement already imposed by the close packing of the mitochondrial genes. Such processing requirements would also explain why the coding region of animal mtDNA has been maintained at a constant and compact size for so long and among such a diverse array of taxa. The small size and compact organization of animal mtDNA have heretofore been unexplained, because the size of the mitochondrion itself placed no obvious constraints on the size of the mtDNA.

Faulty replication and repair may enhance the mutation rate

The fully mature mtDNA of vertebrates contains approximately one to two ribonucleotides per thousand deoxyribonucleotide pairs. The distribution of these ribonucleotides is random (although preferred sites for ribosubstitution exist) and each nucleotide position in the mtDNA can be ribosubstituted (Brennicke and Clayton, 1981). The existence of randomly distributed ribonucleotides in mature mtDNA is *prima facie* evidence for both faulty (i.e., error prone) replication and inefficient (or nonexistent) repair.

Mammalian mtDNA is replicated by γ-polymerase, an enzyme that both lacks the ability to edit its newly synthesized product (Ciarrocchi et al., 1979) and is prone to a higher rate of nucleotide misincorporation than, for example, the major nuclear DNA replication enzyme, α-

85

polymerase (Kunkel and Loeb, 1981). Additionally, some kinds of repair activities have been exhaustively sought, but never found in mammalian mitochondria (Clayton et al., 1974; Lansman and Clayton, 1975). At present, there is no convincing evidence that any DNA repair activity exists in animal mitochondria. It is conceivable that animal cells rely on organellar redundancy (there are an estimated 1000 mitochondria per somatic cell), rather than DNA repair, to maintain the function and viability of mtDNA.

mtDNA has a short turnover time

Unlike nuclear DNA, which has a generation time identical to the cell type in which it occurs, the generation time of mtDNA is usually much shorter (Rabinowitz and Swift, 1970). This probably results from the continual replacement of mitochondria that are defective or damaged, perhaps as a result of metabolic "wear-and-tear." Thus, mtDNA undergoes many more rounds of replication than nuclear DNA, with a consequent increase in the number of errors produced per cell generation.

Exposure to mutagenic agents may occur

Endogenous DNA-damaging agents, such as free radicals, are present in high amounts in mitochondria as normal metabolic intermediates in oxidative phosphorylation. Exogenous chemical mutagens (e.g., some alkylating agents) are also known to concentrate preferentially in the mitochondrial membrane. Given the proximity of these agents to mtDNA, frequent (and reasonable) speculations have been made that they contribute to an elevated mutation rate in mtDNA. Although this may be true, it should be pointed out that the endogenous agents are normal, enzymatically bound intermediates in a highly efficient and controlled process (oxidative phosphorylation) and that the probability of their being free to undergo random reactions must be quite small. Also, enzymes that inactivate these endogenous agents are found in high amounts in mitochondria, presumably as a specific means of protecting them from this kind of damage.

No data presently exist that permit reasonable speculation regarding the role of exogenous chemical mutagens in mtDNA mutation. Comparison of rates of mtDNA evolution among groups or taxa that differ greatly in their exposure, at the mitochondrial level, to such agents might provide such data.

Organellar redundancy and small genome size may enhance fixation of mutations

Because many copies of mtDNA are present in each cell, deleterious mutations in one or a few of these copies might have little effect on the cell itself. Thus, providing the mutations do not affect the ability of the mitochondria to replicate, such mutations could persist at low frequencies for a considerable period of time. During this time, full restoration of function could result from a second (compensatory) mutation in a defective mtDNA or from complementation by recombination between two differently mutated mtDNAs.

Evidence that the preceding speculation is reasonable comes from the results of several recent studies. We have obtained direct evidence of polymorphism within the mtDNA from germ-line cells of individual sea urchins (W. Brown and S. Hechtel, unpublished). Pedigree analysis of cow mtDNA has provided indirect, but equally compelling evidence for germ-line polymorphism in mammals (Hauswirth and Laipis, 1982). Biochemical genetic evidence has been obtained (Oliver and Wallace, 1982) that considerably strengthens the hypothesis, based originally on evidence from light microscopy, that mammalian mitochondria within the same cell are able to fuse with one another. Thus, compartmentalization of mtDNA molecules in different mitochondria of the same cell does not constitute a barrier to the potential for recombination. The hypothesis that mammalian mtDNAs recombine is less well supported. Horak et al. (1974) have provided suggestive evidence for recombination of mtDNA in an RNA/DNA hybridization analysis of interspecific somatic cell hybrids. Data that are consistent with, but do not themselves confirm, mtDNA recombination (or gene conversion) have been obtained in a sequence analysis of D-loop DNA among cows from the same maternal lineage (P. Olivo, W. Hauswirth, and P. Laipis, pers. comm.). However, the inability of mouse cells to recover from light damage ($<$ 1 lesion per molecule) to their mtDNA (Lansman and Clayton, 1975) may be interpreted to mean that recombination, even between mtDNA molecules within the same mitochondrion, either does not occur or is an extremely infrequent event.

The small size of and small number of genes encoded by the mitochondrial genome may be a significant factor in elevating the probability of fixation of mutations. For example, a mutation in a tRNA or rRNA gene affecting protein gene expression would be easier to compensate for (e.g., by a subsequent mutation) in the mitochondrial system, where all 13 protein genes are probably regulated as a single unit, than in the nuclear system, where a myriad of protein genes must be differentially, yet coordinately regulated. Even a drastic mutation, such as one leading to a change in the anticodon of a tRNA gene (and thus to a change in the genetic code), could be tolerated in a small genetic system if the concomitant amino acid changes were minimally compatible with protein function. It is conceivable that the differences in the nuclear and mitochondrial genetic codes could have arisen in this manner, as could the differences in mitochondrial genetic codes that have been observed among multicellular and various unicellular organisms (Jukes, 1981).

PRESENT PROBLEMS AND FUTURE PROSPECTS

Although the most general aspects of animal mtDNA evolution have been studied and described, many details, in particular those that deal with the dynamics of change, remain to be investigated. Knowledge of these details will be useful in erecting and testing hypotheses about mechanisms of evolution. Also, nearly all studies have dealt with either mammalian or *Drosophila* mtDNA, and even the most general information about mtDNA in other animal groups is nonexistent. Such information is necessary to assess the generality of the features of mtDNA evolution that are presently known and may be useful in suggesting phylogenetic affinities among these groups.

An even more pressing problem is the lack of detailed information about the transmission genetics of animal mtDNA. Assessments of the degree of intracellular mtDNA heterogeneity, the possible transmission of paternal mtDNA, and the potential for mtDNA recombination are needed. Most of our knowledge of mtDNA evolution and genetics has been derived using somatic tissues and cultured cells as DNA sources. Studies in which mtDNA from germline tissue is used are essential for addressing these questions.

A third problem is the paucity of information about the enzymes that carry out mtDNA replication and repair. Although there is strong evidence that vertebrate mtDNA is replicated by γ-polymerase, little is known about the properties of this enzyme and nothing is known about the replication complex of which it is part. There is evidence (see Brown, 1981) that the mtDNA polymerase may be different in *Drosophila*. If this is true, then the rapid evolution that appears to be characteristic of mammalian mtDNA would be less than universal, and rates of mtDNA evolution could vary considerably among various taxonomic groups.

The recent determinations of the complete nucleotide sequences for the mtDNA of human, house mouse, and cow, combined with the wealth of data from other evolutionary and molecular biological studies, have made this the most thoroughly characterized DNA. Functional identification of the remaining eight protein genes and of the sequences in the nontranscribed region that are important for gene expression and mtDNA H-strand replication are being intensely studied and should be completed in the next few years. With this information, it should be possible to begin to make precise assessments of such evolutionarily important parameters as functional constraints.

88

CHAPTER 5

PROTEIN POLYMORPHISM AND THE GENETIC STRUCTURE OF POPULATIONS

Robert K. Selander
and
Thomas S. Whittam

The study of the genetic structure of populations through the analysis of molecular polymorphism began at the turn of the century with the pioneering work of Landsteiner (1900) on blood types in humans, but it was not until protein electrophoresis was introduced to population genetics—in 1966—that data for other organisms began to rapidly accumulate. In conjunction with this growth of information, there has been extensive interaction between theoretical and empirical work in an effort to determine the biological meaning of polymorphic variation in the primary structure of proteins. The results of this interaction have been summarized at various stages in reviews by LeCam et al. (1972), Lewontin (1974), Nei (1975), Ayala (1976), Wright (1978), and Kimura (1982b).

One particularly active area of both theoretical and empirical research has centered on the neutral theory of molecular evolution (Kimura, 1968a,b, 1982a; King and Jukes, 1969; Kimura and Ohta, 1971b; Nei, 1975). Based on specific simplifying assumptions about the mutation process (e.g., infinitely many alleles), this theory has generated several testable predictions about levels of diversity in structural genes in natural populations. Notwithstanding the inability of the original version of the neutral theory to account for all aspects of molecular evolution, it has served well as a null hypothesis for those interested in the action of both natural selection and genetic drift (see Chapter 9 by Nei); and it has stimulated the development and application of a wealth of statistical techniques for testing specific hypotheses (Ewens, 1972; Yamazaki and Maruyama, 1972; Johnson and Feldman, 1973; Borowsky, 1977; Fuerst et al., 1977; Watterson, 1978; Yardley et al., 1977; Chakraborty et al., 1978, 1980; Gaines and Whittam, 1980; Latter, 1981; Skibinski and Ward, 1982; Slatkin, 1982; and others).

In this review, we will not attempt to catalog the extensive literature on genetic diversity within and among populations revealed by protein electrophoresis. Rather, we shall first consider how the recent demonstration of "hidden" allelic variation is likely to affect our understanding of population structure, and then we shall review several studies illustrating some of the types of structure that have been observed in diverse organisms, ranging from man to bacteria. These examples demonstrate some of the major evolutionary processes involved in the origin and maintenance of population structure and also show how dynamic processes operating in the past can sometimes be inferred from contemporary patterns of variation. An underlying theme of this chapter is the proposition that the study of the genetic structure of populations should involve a sequence of hypothesis testing in which models incorporating the fewest evolutionary forces are first applied.

THE PROBLEM OF HIDDEN VARIATION

It is common knowledge that electrophoresis cannot be expected to detect all allelic variation at a locus. For purposes of theoretical work, it has been assumed that the proportion of detectable variation is approximately 30% for the average protein, an estimate based simply on the frequency of random substitutions involving charged and non-charged amino acids (Nei, 1975; Wilson et al., 1977). The actual proportion for any specific protein could be drastically different. Moreover, under the neutral hypothesis, the expected proportion of alleles detectable by electrophoresis is a function of population size, and common electromorphs should be allelically more heterogeneous than uncommon or rare ones (Nei and Chakraborty, 1976; Chakraborty and Nei, 1976; Chakraborty, 1977).

In the past decade, population geneticists have made a major effort to measure amounts of cryptic variation within electromorphs (see reviews by Ayala, 1982; Coyne, 1982; Shumaker et al., 1982), using two basic experimental approaches (Ramshaw et al., 1979). The more common one is the "forward experiment," in which electrophoretic conditions are varied (sequential gel electrophoresis) or other biochemical techniques are used to discriminate allelic states within electromorphs. An early example of this approach is Bonhomme and Selander's (1978) survey of electrophoretic mobility, thermostability, and response to thiol reagents of 14 enzymes in laboratory strains of the house mouse. Only four new allelic variants were detected by thermostability studies, and thiol reagents revealed no additional variation. These results led Bonhomme and Selander (1978) to conclude that standard electrophoresis detects, on the average, approximately 50% of the alleles at structural gene loci in the house mouse.

A second approach, applied by Ramshaw et al. (1979) to human hemoglobins, is the "backward experiment," in which protein variants of known primary and tertiary structure are used to calibrate standard electrophoretic techniques. The consequence of this work is that sequential gel electrophoresis was able to detect 85 to 90% of the substitutions represented in the sample of 20 hemoglobin variants studied. The notions that only charge-changing substitutions are detected by electrophoresis and that each electromorph is homogeneous in nominal charge are clearly incorrect, at least for hemoglobin.

How does hidden variation affect the study of population structure through the analysis of protein polymorphism? To answer this question one needs to know both the numbers and the relative frequencies of hidden variants in natural populations. Rare alleles (whether detected or not) have little effect on the statistical measures of genetic diversity commonly used in analyzing population structure. One measure that reflects both the number and frequency distribution of alleles is the effective number of alleles, n_e (Crow and Kimura, 1970), which is the reciprocal of the sum of squares of the allele frequencies. Table 1 presents the effective number of alleles in seven species for which the resolving power of standard electrophoresis can be directly compared to those of other methods of detecting variation. As shown, urea denaturation increases n_e by approximately 25%, whereas two-dimensional electrophoresis generally yields lower estimates of n_e. Whether the results from two-dimensional electrophoresis are due to poorer resolving power or the sampling of a different, less polymorphic class of proteins is problematical.

The extent of increase in n_e revealed when electrophoresis is either sequential or is supplemented by a denaturing technique may depend partially on the molecular weight of the enzyme (Singh, 1979). Taking

91

TABLE 1. Hidden variation uncovered in natural populations of animal species.

Species	n_e*	n_e'*	n_e'/n_e	Technique†	Reference
D. pseudoobscura	1.38 (13)	1.65 (13)	1.12	Sequential	Ayala (1982)
D. subobscura	1.83 (8)	2.42 (8)	1.25	+ Urea denaturation	Loukas et al. (1981)
D. melanogaster	1.73 (4)	2.06 (4)	1.18	+ Heat denaturation	Ayala (1982)
	1.20 (21)	1.04 (54)	0.87	2D	Leigh Brown and Langley (1979)
D. simulans	1.06 (21)	1.00 (70)	0.94	2D	Ohnishi et al. (1982)
Mus musculus	1.06 (22)	—			Berry and Peters (1977)
	—	1.02 (72)	0.96	2D	Racine and Langley (1980)
Homo sapiens	1.06 (87)	—			Harris et al. (1977)
	—	1.00 (83)	0.94	2D	Smith et al. (1980)
	—	1.01 (61)	0.95	2D	Comings (1979)

* n_e, Effective number of alleles revealed by standard allozyme electrophoresis; n_e', effective number of alleles revealed by alternative techniques. The number of loci examined is given in parentheses.
† +, Standard electrophoresis plus denaturing; 2D, two-dimensional electrophoresis.

the ratio n_e'/n_e (Table 1) as a dependent variable and the estimated molecular weight of enzymes for *Drosophila melanogaster* (Koehn and Eanes, 1978, Table IV) as an independent variable, we have found a significant positive linear relationship for seven enzymes in four species of *Drosophila* ($y = 0.004x + 0.891$, $r = 0.85$, $p < 0.01$). Because our sample lacks enzymes with molecular weights between 60,000 and 120,000, the relationship reflects primarily a low ratio for a group of smaller (20,000–30,000) enzymes (*Adh, Mdh, Odh, αGpd*) and a relatively high ratio for the larger (130,000–140,000) enzymes (*Ao, Xdh*). Nonetheless, this result suggests that estimates of n_e are more precise for smaller proteins than for larger ones.

The Bogata population of *Drosophila pseudoobscura*

The data summarized in Table 1 suggest that hidden variation is not extensive enough to grossly alter present estimates of levels of variation within and between populations (Ayala, 1982; Shumaker et al., 1982). This may be wishful thinking, however, for the data are limited, and we already have one dramatic example of a major reinterpretation of genetic structure necessitated by the uncovering of hidden variation. This is provided by Singh's (1983) work on the relationship of the Bogata, Colombia, population of *Drosophila pseudoobscura* to its "mainland" counterpart in Central America.

Beginning with the early allozyme research on the *virilis* group by Hubby and Throckmorton (1965), there developed a body of data suggesting that relatively little genetic differentiation accompanies the process of speciation in *Drosophila* (Ayala and Powell, 1972; Richmond, 1972; Ayala, 1975; Throckmorton, 1977). (Inferences regarding genetic differentiation in relation to the process of speciation are based on the ergodic argument, which, in this application, is questionably valid; see Paterson, 1981.) Support for this hypothesis was provided from two sources. First, Prakash (1972) discovered that crosses between Bogata females and mainland males of *D. pseudoobscura* produce fertile females but sterile males. (Both male and female progeny of the reciprocal cross are fertile.) Second, electrophoretic comparisons of the Bogata and mainland populations at 21 protein loci revealed differentiation at only one locus (*Pt-8*), for which Bogata has a high frequency (87%) of an allele that is rare in mainland populations (Prakash, 1972). Subsequently, studies of 46 loci showed only two loci with an overlap in allele frequencies of less than 5% and only 14 loci with an overlap of less than 15% (Ayala and Powell, 1972; Prakash, 1977).

When Prakash (1972) showed that partial reproductive isolation has evolved, apparently without much genetic differentiation, low average heterozygosity in the Bogata population ($H = 0.068$ versus 0.140 on the mainland) was attributed to a postulated recent founding of the Bogata population (perhaps around 1960) and accompanying genetic drift (Prakash, 1972; Nei et al., 1975). [And Bryant (1974) interpreted the low heterozygosity of the Bogata population as evidence of lesser environmental variance in the tropics.] But Dobzhansky (1974) continued to maintain that the Bogata and mainland populations have been separated for thousands or millions of years.

Recent work has drastically altered our understanding of the extent of genetic differentiation between the Bogata and mainland populations. Singh (1983; see also Singh et al., 1976) used sequential electrophoresis to search for hidden variation at ten loci. [And Coyne and Felton (1977) also reexamined an eleventh locus, *Odh.*] For the reexamined loci, the total number of alleles increased by 178% (due mainly to *Est-5, Xdh,* and *Ao*); mean *H* increased by 27% on the mainland and by 114% in Bogata (mostly due to *Ao, Xdh, Pt-12, Pt-13,* and *Est-6*); and average genetic identity per locus between mainland and Bogata decreased from 87 to 24% for the 11 loci. Additionally, breeding experiments have detected genetic divergence between these populations in many fitness characters, including mating speed, developmental rate, viability, and fertility (Singh, 1983).

Singh's work on differentiation in molecular, physiological, and behavioral characters supports Dobzhansky's view that the Bogata population has been isolated from the main body of the species for a considerable period. But the evidence is not compelling, because rapid genetic differentiation could have occurred if there was a severe bottleneck. Thus, neither the allozyme data nor the information on chromosomal polymorphism (Ayala and Dobzhansky, 1974) provides direct evidence about the age of the Bogata population.

This example demonstrates that, in the absence of historical information, the interpretation of differentiation in terms of evolutionary factors is a difficult undertaking at best. We shall now turn to research on the genetic structure of two species of land snails for which some useful historical information is available. This work also illustrates the approach to the empirical analysis of population structure that we favor.

GENETIC STRUCTURE OF LAND SNAILS

Hypothesis testing

Our work begins with a test of the null hypothesis that stochastic variation in allele frequencies generated by random genetic drift is the primary cause of molecular evolutionary change. This is the logical point of departure, because alleles at every locus are potentially subject to the action of genetic drift in equal degree (Cavalli-Sforza, 1966; Lewontin and Krakauer, 1973). The advantage of this basically demographic approach over those based on more complex models of population structure in which selection is the predominant factor is that the stochastic model makes a number of specific, quantitative predictions, whereas no predictions are generated by more complex models that cannot be explained away if expectations are not met by the data. We shall invoke other evolutionary forces, such as a particular mode of natural selection, only when a stochastic model (sensu Selander, 1975) is rejected. In adopting this approach, we do not subscribe wholeheartedly to the neutral theory of protein polymorphism or to recent modifications of it—the theory of effectively neutral mutations (Kimura, 1979a). Nor do we in any way exclude natural selection as a factor determining population structure and evolutionary change at the molecular level. There is no need for an either–or attitude. It is simply that we prefer to begin with the simplest model, determining (as a baseline for further analysis and interpretation) the degree to which existing geographic (or in the case of bacteria, clonal) structure matches expectations of a random model. Deviations from the random model are a source of new hypotheses for further analysis.

Helix and Cepaea

In the following example, we can directly compare genetic structure in relatively young and relatively ancient populations of two related species of pulmonate land snails, *Helix aspersa* and *Cepaea nemoralis* (family Helicidae). The young population, *H. aspersa*, was introduced to California from France in 1859 and subsequently spread over much of the southwestern United States (Stearns, 1881; Selander, 1975). In contrast, *Cepaea nemoralis*, which is native to Europe, has inhabited river valleys in and around the Spanish Pyrenees at least since the most recent Pleistocene glaciation, and probably for millions of years. In both situations, the structure of the environment determines the pattern of subdivision of the population. For *Helix*, which occurs primarily in residential gardens on city blocks, the environment is subdivided by the way in which human habitations are organized (Selander and Kaufman, 1975). For *Cepaea*, the topography determined by the mountain slopes and river valleys of the Pyrenees fragments the total population into a series of linear patches of habitat (Jones et al., 1980; Caugant et al., 1982).

To quantify the genetic structure of these subdivided populations, we have employed Wright's (1943, 1965, 1978) hierarchical F-statistics to analyze allozyme data for eight polymorphic enzymes. F-statistics are particularly useful because they are directly related to the statistical test of genetic heterogeneity based on contingency tables (e.g., Workman and Niswander, 1970) and can easily be interpreted as ratios of genetic diversity within and among subdivisions (Nei, 1977; Chakraborty, 1980).

In preparation for analysis, we partitioned each population into three hierarchical levels. The basic sampling unit in *H. aspersa* in California is a city block, and the 140 blocks sampled were grouped

95

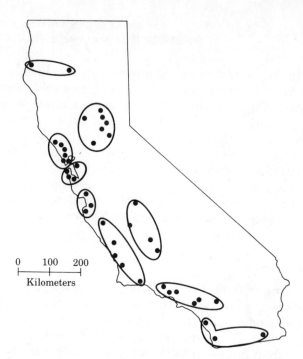

FIGURE 1. Grouping of 40 cities into 9 geographical regions of California for the purpose of analyzing population structure in the land snail *Helix aspersa*. Several samples, representing blocks, were available for each city.

into 40 cities and 9 geographic regions (Figure 1). For *Cepaea* in the Pyrenees, we similarly grouped 197 localities (hereafter called demes) into 13 regions and 2 major subdivisions. Each region is a river valley and the two major subdivisions are the north-facing and south-facing slopes of the Pyrenees (see Figure 2). Results of the hierarchical F analysis are presented in Table 2; and components of total genetic diversity, expressed in terms of heterozygosity (H), within each species are illustrated in Figure 3. Approximately 71 and 78%, for *Cepaea* and *Helix*, respectively, of the total diversity over the eight polymorphic loci studied is within demes (or blocks). Both species show similarly large components of diversity between demes within regions (rivers for *Cepaea*, blocks within cities for *Helix*); but in *Cepaea* there also is a large component of diversity between river valleys within

FIGURE 2. Locations of 197 demes of *Cepaea nemoralis* sampled in 13 river valleys in the Pyrenees of southern France. River valleys were assigned to two major subdivisions, representing the north-facing and south-facing slopes of the Pyrenees.

96

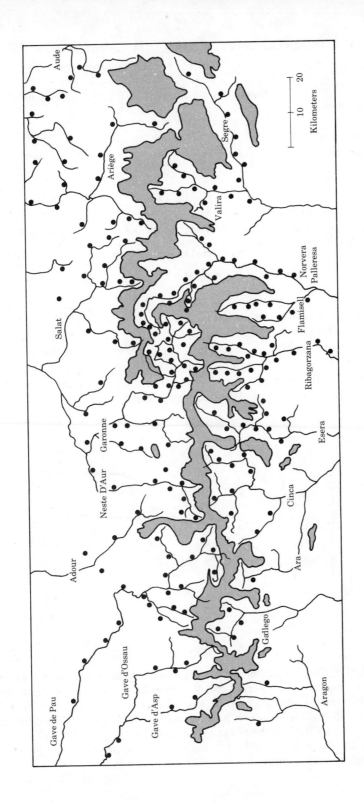

TABLE 2. Hierarchical genetic diversity in two species of snails.

Species	Number of loci	Number of samples	F-statistics*		
Helix aspersa (California)	7	140	F_{BC} 0.168	F_{CR} 0.030	F_{RT} 0.017
Cepaea nemoralis (Pyrenees)	6	197	F_{DR} 0.177	F_{RS} 0.118	F_{ST} 0.000

* F-statistics for three levels of subdivision for the two species of snails: *Helix aspersa* (B, blocks; C, cities; R, regions) and *Cepaea nemoralis* (D, demes; R, river valleys; S, subdivisions). Each statistic was corrected for sampling variation as described in Wright (1978, pp. 86–89).

each mountain slope, whereas *Helix* shows virtually no differentiation between cities or geographic regions in California. In both species, the standardized variances for alleles over different loci at the structural level of demes within regions (*Cepaea*) or blocks within cities (*Helix*) are essentially homogeneous. This circumstance is more easily explained by a random model of differentiation than by a complex "de-

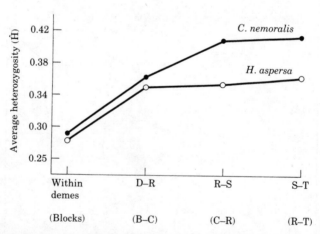

FIGURE 3. Components of genetic diversity in populations of two species of snails, *H. aspersa* in California and *C. nemoralis* in the Pyrenees. Four levels of subdivision are shown along the abscissa: within demes for *Cepaea* and on blocks for *Helix*; D-R, demes within river valleys for *Cepaea*; B-C, blocks within cities for *Helix*; R-S, river valleys within slopes for *Cepaea*; C-R, cities within regions for *Helix*; S-T, north and south slopes within total area for *Cepaea*; R-T, regions within total area for *Helix*.

terministic" model, which would require similar spatial variances in selective factors among blocks or demes (see discussion in Selander, 1975).

One can test the null hypothesis that genetic differentiation results from random drift of isolated demes by comparing observed correlations in allele frequencies with those predicted by random differentiation (Nei, 1965; Nei and Imaizumi, 1966a,b; Jacquard, 1974). Under this simplest of hypotheses relating to population structure, isolates diverge entirely by random genetic drift (i.e., no mutation, selection, or migration), and the expected correlations in allele frequencies are simply functions of average allele frequencies among the populations.

For each locus, we calculated both the observed and the expected correlations between the common allele and each alternate allele occurring at a frequency greater than 0.05. (Both variances and covariances were corrected for sampling variance within populations.) In Figure 4, values for allele pairs falling above the diagonal indicate less differentiation than expected under the random model, whereas values below the line indicate greater negative correlation than expected, and, hence, more differentiation than predicted by the random model. For *Helix*, observed correlations among demes for all but one locus (*Mdh-1*) are in good agreement with expected correlations (Figure 4A). For one pair of the three alleles at the *Mdh-1* locus (alleles *120* and *100*), the observed correlation is nearly zero, but the expected correlation is −0.391. Hence, there is less differentiation among blocks than expected, reflecting the circumstance that both alleles increase in frequency from north to south in California. This pattern of geo-

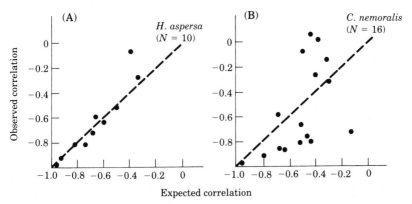

FIGURE 4. Comparison of observed correlations in allele frequencies with those expected under a model of random differentiation.

graphical variation indicates either the action of natural selection or an historical sampling event that altered allele frequencies at the *Mdh-1* locus (see discussion in Selander, 1975). In contrast with *Helix*, most of the observed correlations for *Cepaea* differ strongly from predicted correlations (Figure 4B). For six allele pairs at *Mdh*, *Pgm-2*, *Pgm-3*, *Ipo*, and *Lap*, the observed correlations within regions are positive, rather than negative as expected under the model of random differentiation; and for ten allele pairs at *Pgi*, *Mdh*, *Lap*, *Pgm-2*, *Pgm-3*, and *Ipo*, the observed correlations are more strongly negative than expected. Here we have evidence of a large historical component and/or the strong influence of natural selection in determining allele frequency distributions.

To further investigate the genetic structure of *Cepaea* in the Pyrenees, we employed a multivariate analysis of allele frequencies at all loci and extracted a new set of genetic variables by principal components analysis. Three groups of demes were revealed in a plot of the first two principal axes (Figure 5). These groups represent three major geographic subdivisions of the total population: western, central, and eastern groups of subpopulations, *each of which extends across the crest of the Pyrenees*. Because this pattern is evident over all the loci and because the physical and biotic aspects of the environments on the north and south slopes of the Pyrenees are drastically different, this pattern of population structure cannot be accounted for by an hypothesis invoking natural selection for adaptation to local conditions. Rather, we interpret the pattern as a result of the expansion and contraction of *Cepaea* populations in and around the Pyrenees in a past period of climatic fluctuation, such that the central population was isolated and today remains largely intact as a genetic relict. The differentiated populations on either side of the central population presumably achieved their present distributions in the Pyrenees by invading from the east and the west along both the north and south slopes.

Whereas selection apparently has had little effect on allele frequencies at enzyme loci in *Cepaea* in the Pyrenees, it has operated strongly on shell characteristics of banding and coloration, with two consequences: first, stabilizing selection has erased any historical record comparable to that revealed by the allozymes, so that shell color and banding show none of the population structure evident at the protein level; second, there is a tendency for banding patterns to change with elevation in river valleys (Jones et al., 1980). This pattern was observed in 7 of 13 river valleys.

In sum, random drift among blocks within cities accounts for most of the genetic diversity in the young population of *Helix* in California; and there has been virtually no genetic differentiation on a broader geographic scale. Perhaps in time a more complex population structure

100

will develop by fission and fusion of local populations and/or through natural selection moving allele frequencies toward equilibria determined by climatic or other environmental gradients. But, in any event, we now at least have a clear picture of the extent to which differentiation can be rapidly generated by random processes—a baseline against which to evaluate the degree of adaptive response, if and when it develops.

In the *Cepaea* population in the Pyrenees, there are large components of variation both within and between river valleys. With the much greater amount of time involved, a more complex population structure than that in *Helix* has evolved. The extensive differentiation

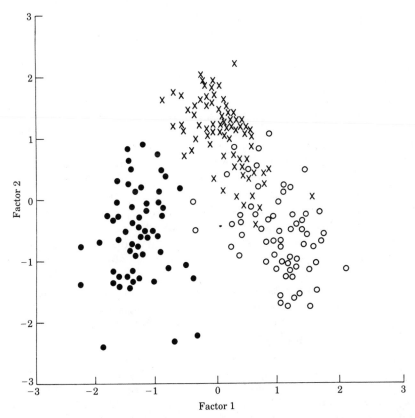

FIGURE 5. Plot of samples of *Cepaea nemoralis* with respect to factor scores along the first two principal axes. ○ indicates relic central group of subpopulations; × and ● indicate eastern and western groups, respectively.

101

that has occurred at the molecular level emphasizes the strength of selection pressures maintaining the polymorphic shell characters, the frequencies of which remain more or less uniform over the whole of the Pyrenees.

ANTHROPOLOGICAL GENETICS

The serious study of human genetic population structure began with Cavalli-Sforza and Edwards' (1964) attempt to measure genetic distance between geographic populations by using allele frequency data obtained by the typing of blood groups. In recent years, very extensive data on blood groups and protein-encoding loci have been used to attack several types of problems: (1) estimation of the genetic structure of aboriginal populations at various geographic levels and its relationship to aspects of demographic and social structure, as exemplified by the work of Neel and his associates on the Yanomama and other South American tribes (e.g., Neel et al., 1977; Smouse et al., 1981) and by the research of Kirk (1979), Birdsell et al. (1979), and Blake (1979) on the Australian aborigines; (2) estimation of the genetic relationships among human races and other geographic populations, as exemplified by the work of Nei and Roychoudhury (1982) on a worldwide level and by the studies of Rychkov and Sheremetyeva (1977) on North Asian ethnic groups; and (3) the reconstruction of historical patterns of migration of human populations.

Anthropological geneticists have several advantages over students of natural populations of plants and animals (Harrison, 1977). (1) Extensive genetic data have been gathered for humans on a worldwide basis. (2) Information on a wide range of morphological, metrical, linguistic, and cultural characters are available for comparison with the genetic data. Indeed, studies of human populations have interdisciplinary and international dimensions not approached in research on other organisms. (3) Extensive contemporary and historical demographic data are available—census numbers, effective population sizes, and historical records that can be used to reconstruct patterns and intensities of migration and sequences of historical events leading to subdivision and fusion of populations. (4) Age structure, fertility rates, age-specific mortality rates, and distributions of progeny sizes also are available and ultimately will be required for a thorough understanding of genetic population structure and evolution.

Much of the research in anthropological population genetics begins with the testing of simple models of structure incorporating the demographic parameters of effective population size and migration rate. And to a surprising degree these models have been successful in accounting for microgeographic and macrogeographic patterns of variation in human populations.

102

An excellent example of the way in which genetic data can play a major role in testing hypotheses regarding historical demographic processes is provided by the analysis of Menozzi et al. (1978) of the genetic consequences of the neolithic spread of farming into Europe from the Middle East. This was a slow and regular process, extending from 9000 to 5000 B.P. (before present), at a radial rate of advance of approximately 1 kilometer per year (Figure 6). Two hypotheses as to the nature of this process have been proposed from archaeological data:

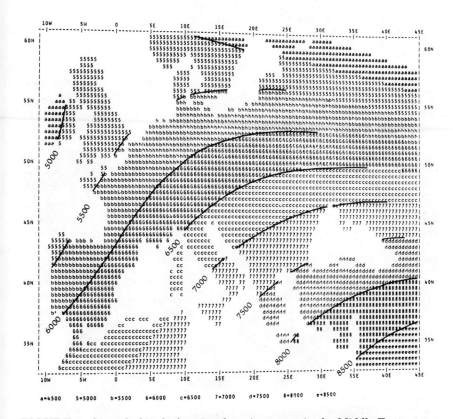

FIGURE 6. Spread of early farming from its center in the Middle East across western Europe, based on radiocarbon datings of remains of earliest farming culture in each region. (After Bodmer and Cavalli-Sforza, 1979.)

103

1. *Cultural diffusion of farming technology* into preexisting populations of hunters and gatherers living in Europe, which would have no direct effect on the underlying genetic structure.
2. *Demic spread,* or the diffusion of farmers themselves. According to this hypothesis, one might expect to find clines in allele frequencies at polymorphic loci in Europe extending radially from the Middle East—a SE–NW gradient across Europe.

To test these hypotheses, Menozzi et al. (1978) used principal components analysis of frequencies for 38 alleles at ten loci, including HLA-A and HLA-B (21 alleles), phosphoglucomutase-1, acid phosphatase, and a number of blood type loci. [See Piazza et al. (1981a) for methods of constructing synthetic gene-frequency maps.] As shown in Figure 7, the first principal component for the 38 alleles (which explained one-third of the total variance) shows a remarkable similarity to an archaeologically based map of the advance of early farming (Figure 6)—a general SE-NW gradient in the form of a series of concentric circles centered on the Middle East. This pattern is shown most clearly by the HLA alleles but is evident for other loci as well.

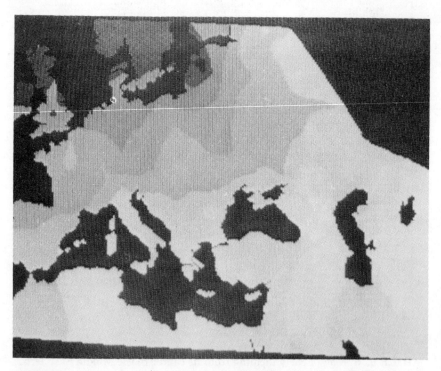

FIGURE 7. First principal component of frequencies of 38 independent alleles at nine loci. Shading indicates intensity of first principal component, which accounts for 27% of the total variation. (After Menozzi et al., 1978.)

These results clearly support the hypothesis of demic spread. Note that, in the absence of the historical information, all sorts of complex models involving selective environmental gradients might have been entertained.

A particularly interesting conclusion emerging from this study is that, among the first three principal components, there was no evidence of north–south clines, which could have been expected under an hypothesis of selective response to climate. [But see Piazza et al. (1981b) for evidence suggesting selective responses on a broader geographic scale.]

Sokal and Menozzi (1982) have recently extended the analysis of neolithic migration routes in Europe by applying techniques of spatial autocorrelation (adapted for population genetics by Sokal and Oden, 1978) to more firmly establish that demic spread was in fact the process underlying the neolithic revolution. Using data for the 21 HLA-A and HLA-B alleles from 58 localities, they measured spatial autocorrelation of allele frequencies at five distance classes along a "directed connectivity network" derived from archaeological information on paths of the spread of agriculture in Europe. Most of the allele frequencies show gradients along the routes of spread; and for 15 of the 21 alleles, the gradients are monotonic over the five distance classes. Figure 8 shows Moran's I correlograms of allele frequencies that load highly on principal axis I.

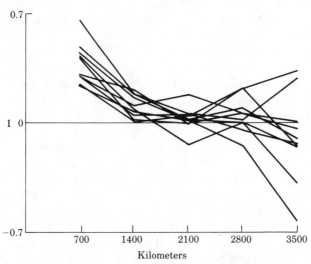

FIGURE 8. Spatial correlograms based on a directed network along hypothesized spread of agriculture in Europe. Correlograms of allele frequencies loading highly on principal axis I are shown. (After Sokal and Menozzi, 1982.)

105

Having discussed one of the more revealing analyses of population structure, in which historical information played a crucial part, we now turn to a group of organisms for which population structure has barely been explored and for which we are unlikely ever to have useful historical information.

GENETIC STRUCTURE IN ENTERIC BACTERIA

Although *Escherichia coli* has been the workhorse of molecular genetics, population geneticists have only recently begun to study the structure of its natural populations. The first attempt to assess overall genetic diversity by protein electrophoresis was made by Milkman (1973, 1975), who isolated clones from a wide variety of mammalian hosts and estimated an average genetic diversity of 0.23 at five enzyme loci. Subsequently, this estimate was revised upward by Selander and Levin (1980), who assayed 20 loci and found an average genetic diversity of 0.47. But, more importantly, their results contradicted Milkman's view of the population structure of *E. coli*. Milkman assumed that populations are essentially panmictic and have high rates of recombination that generate most of the extensive genetic diversity of strains occurring in individual hosts. Selander and Levin (1980), however, repeatedly recovered clones of identical allozymic profile from unassociated hosts—a highly improbable event if recombination rates are high. Their findings suggested that recombination is severely restricted and that genetic diversity within individual hosts is largely a result of the immigration of new clones.

Selander and Levin's (1980) interpretation derives support from three sources. First, low rates of phage-mediated and conjugative-plasmid transfer of genes have been observed in experimental populations in chemostats (Levin et al., 1977, 1979). Second, a high rate of turnover of unrelated strains within a single human host was observed by Caugant et al. (1981). Third, these workers found significant associations between many pairs of alleles at different loci, a result not readily compatible with the hypothesis of high rates of recombination.

The implication of restricted rates of recombination in *E. coli* is that natural populations consist of mixtures of numerous, more or less independently evolving clones. If so, the evolution of the "species " *E. coli* may, in many respects, be dominated by random sampling of lines (strains), which occurs frequently through either stochastic extinction or periodic selection of fitness mutations. As noted by Maruyama and Kimura (1980) and by Levin (1981), the extinction of lines has a profound effect on effective population size and, hence, on the amount of genetic diversity carried by the species. [Wright (1940) recognized early the general effect of extinction on population structure, as did

106

Nei (1976) in specific application to *E. coli*.] Levin (1981) has demonstrated, by computer simulation, that periodic natural selection of fitness mutations purges variation in populations with recombination rates that are sufficiently low relative to the rates of increase of mutant clones, because few if any of the alleles at other loci have a chance to become associated with the fitness mutations. Thus, the amount of genetic diversity carried by a species having an astronomically large standing crop—like *E. coli*—may be expected to be only moderate, perhaps not much in excess of populations of higher organisms, as has been observed (Selander and Levin, 1980). Another evolutionary consequence of the extinction and periodic selection models is that strong linkage disequilibrium should be generated (see Ohta, 1982a,b, for an analysis of the extinction model).

In an attempt to understand the overall genetic structure of *E. coli*, we have combined allozymic data from several sources into a single analysis. Our collection of clones includes those from the original studies of Milkman (1973), Selander and Levin (1980), Caugant et al. (1981), and a new study of urinary tract infections in human hosts in Sweden (Caugant et al., 1983). For comparative purposes, we have also included electrophoretic profiles of four species of *Shigella*, a group of pathogenic bacteria known to be closely related to *E. coli* on the basis of total DNA hybridization (Brenner et al., 1972), nucleotide sequencing of various genes, and other evidence. In tabulating allele frequencies for 12 loci common to all studies, we found 113 allelic states and 302 unique electrophoretic types, of which 279 were *E. coli* and 23 were *Shigella*.

Figure 9 is a projection of the 302 unique electrophoretic types onto the first two axes derived in a principal components analysis. In this analysis, each clone was represented by a binary code indicating the presence or absence of each allele. The first two axes account for only 10% of the total variance among all clones.

This study demonstrated that the observed combinations of allelic states in strains is highly nonrandom and that *E. coli* is subdivided into three groups of strains. (In Figure 9, symbols indicate assignments of strains to groups on the basis of a discriminant function analysis.)

By examining those alleles that have high loadings on the first two principal axes and thus determine the groups of *E. coli* clones, we can further test for deviation from random associations of alleles. Table 3 presents three tests of partial association (Bishop et al., 1975) for combinations of alleles at four different loci. The first test (comparison I) involves four alleles that have high negative loadings on

107

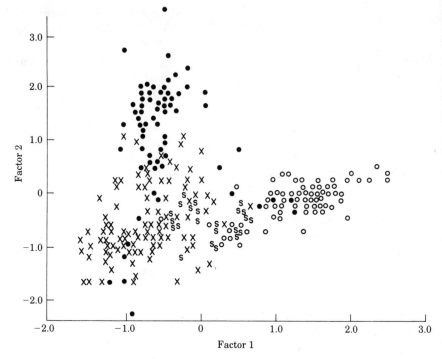

FIGURE 9. Plot of factor scores for electrophoretic types (ETs) of *Escherichia coli* ($N = 279$) and *Shigella* spp. ($N = 23$) for the first two principal axes. Groups of *E. coli* are designated by different symbols; S indicates *Shigella*.

factor 1 and determine the group in the lower left-hand corner of Figure 9. Three pairs of alleles showed strong positive association over electrophoretic types, and there was a significant three-way association between Idh^5, Pgi^7, and Adh^6. Comparison II includes alleles with high positive loadings on factor 1, which determines the second cluster of clones, of Figure 9. The third comparison (III) involves alleles with high loadings on factor 2, which determines the group at the upper-left of Figure 9. Thus, the groups represent highly nonrandom associations of alleles at different electrophoretic loci; and in each case both two-way and three-way associations among alleles are evident.

We have also compared degrees of differentiation among the groups of *E. coli* and *Shigella* by constructing dendrograms of the Euclidean distances between centroids (i.e., average locations), based on the first four principal axes. For comparison, we employed two methods of coding data: the BINARY method gives equal weight to each allele; and the RANGED method weights each locus equally. As shown in Figure 10, *Shigella* and *E. coli* are more similar to one another than

108

TABLE 3. Associations among alleles at different loci within electrophoretic types of *Escherichia coli.*†

Effect§	I	II	III
AB	0.28	1.25	52.07**
AC	48.23**	6.25*	24.88**
AD	0.22	2.55	4.79*
BC	9.87**	24.86**	0.00
BD	14.84**	20.42**	21.22**
CD	36.71**	25.43**	11.04**
ABC	0.41	0.03	2.30
ABD	0.96	8.19**	1.41
ACD	1.94	2.98	0.01
BCD	10.56**	0.88	23.06**

* $p < 0.05$.
** $p < 0.01$.
† This table summarizes associations in three comparisons (I,II,III) of alleles (see Table 4) at four different loci. For each comparison, the likelihood ratio statistic resulting from a test of partial association is given for two-way and three-way effects (see text).
§ Symbols defined in Table 4.

TABLE 4. Alleles used in comparisons presented in Table 3.*

Comparison	Symbol			
	A	B	C	D
I	Ak^5	Idh^5	Pgi^7	Adh^6
	(0.18)	(0.70)	(0.45)	(0.43)
II	Pep^7	Idh^2	Aco^7	Adh^1
	(0.14)	(0.24)	(0.33)	(0.22)
III	Ak^5	Pep^5	Idh^2	Mpi^3
	(0.18)	(0.62)	(0.24)	(0.41)

* The proportion of electrophoretic types including each allele is given in parentheses.

109

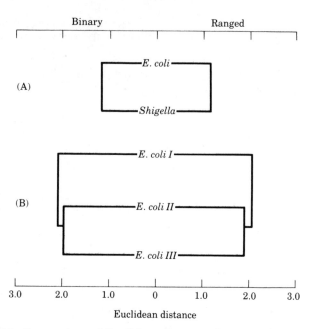

FIGURE 10. Comparison of Euclidean distances between the centroids of *E. coli* and *Shigella* (A) and of three groups of *E. coli* (B), determined from discriminate function analyses of factor scores for two methods of data coding (see text for explanation).

are the three groups of *E. coli*. Because the species of *Shigella* fall into two of the *E. coli* groups, a multiple origin of these pathogenic bacteria is suggested.

In sum, our exploratory studies of the structure of natural populations of *E. coli* indicate that restricted rates of recombination between clones has permitted the development of strong associations among alleles at different loci. *E. coli* also appears to be subdivided into three major groups of strains that can be distinguished by allozyme analysis. As yet, we have identified no biological correlates of this subdivision. Thus, none of the groups of strains is uniquely associated with urinary tract infection, and it does not seem that "resident" strains (versus "transient" strains; see Caugant et al., 1981) belong to any one group. Nor are there any apparent host (animals versus man) associations. But at this stage, our knowledge of the distribution and ecology of strains of *E. coli* is fragmentary.

DISCUSSION

The study of evolution relies heavily on statistical inference from observed patterns of variation to gain insight into the past action of

110

dynamic processes. In this chapter, we have demonstrated how the study of protein polymorphism by electrophoresis has advanced our understanding of the genetic structure and evolution of natural populations. Particular emphasis has been given to the importance of historical demographic information in interpreting patterns of population structure, and the examples discussed were selected to demonstrate the following four major points:

1. *Extensive demic differentiation can occur rapidly through genetic drift in subdivided populations.* Populations of *Helix*, which was introduced to California only about 125 years ago, have differentiated to the extent that almost 20% of the total genetic diversity is apportioned between demes (blocks) within cities. Because a simple model of random differentiation adequately accounts for most aspects of the population structure of *Helix*, we conclude that genetic drift has been the primary evolutionary factor responsible for this differentiation.

2. *Natural selection for local adaptation can conceal the genetic history of subdivision.* Our analysis of *Cepaea* populations in the Pyrenees revealed strong nonrandom differentiation in allozyme frequencies within and between river valleys and identified a relict group of subpopulations that presumably was isolated during a past period of climatic fluctuation. However, the existence of this relict subpopulation, which is strongly differentiated at the molecular level, is undetectable by examination of two polymorphic shell characters: color patterns are virtually uniform in frequency throughout the Pyrenees, and the frequency of banding varies with elevation within river valleys. Both of these conspicuous phenotypic characters are known to be subject to local selection pressures (reviewed by Jones et al., 1977).

3. *Adaptation can spread by interdemic selection.* The analysis of molecular polymorphisms in contemporary European human populations has revealed gradients in allele frequencies that agree remarkably well with those expected on the basis of archaeological evidence of the spread of early farming. Hence, there is good reason to believe that the neolithic revolution involved the asymmetrical diffusion of genes from the Middle East, caused by the movement of early farmers into the hunter–gatherer population of Europe.

4. *Clonal mixtures can exhibit complex multilocus structure despite recombination.* We have demonstrated that natural populations of *E. coli* exhibit a complex multilocus structure involving strong associations between alleles at different loci. This degree of asso-

111

ciation is maintained despite the occurrence of recombination (albeit apparently at low frequency) mediated by various mechanisms available to prokaryotes (e.g., Bodmer, 1970), and a generation time sufficiently short to allow for the rapid decay of disequilibrium of even tightly linked groups of genes. Our analysis further revealed that *E. coli* consists of three major groups of strains.

Contrary to the view of Ewens (1979), who concludes from an analytical study of isotropic migration that "the effect of subdivision is not important and that for many purposes the population can be taken as one large randomly mating population," studies of protein polymorphisms indicate that a great variety of organisms, ranging from bacteria to humans (see summary in Wright, 1978, Table 7.31), are strongly structured genetically and that their evolution cannot be understood without reference to this structure. We will conclude our discussion by considering one particularly important consequence of population structure—its effect on the rate of adaptive evolution.

Population structure and the rate of adaptive evolution

For half a century, Sewall Wright (1931, 1977, 1982) has advocated an evolutionary process in which population structure is a major determinant of the rate of adaptive evolution. Wright based his "shifting balance process" on three basic premises, two of which concern the phenotypic effects of different alleles. First, the quantitative variability of a phenotypic character is determined by a large number of polymorphic loci; and the alleles at the loci differ only slightly, if at all, in their selective value. Second, there are pleiotropic effects of most allele differences; there are manifold phenotypic effects of a particular mutant allele. The third premise concerns the relationship between the organism and the environment, as measured by the ultimate quantitative character—fitness. Wright pictured the combinations of different genes as a multidimensional field of genotypic fitnesses with multiple adaptive peaks and valleys.

The critical evolutionary problem in Wright's view (as well as in that of many others, e.g., Simpson, 1944; Rosenzweig, 1978) is how a population leaves one adaptive peak, against the action of natural selection, and reaches another peak. In a panmictic, homogeneous population, the rate of adaptive evolution is limited by the additive genetic variance in fitness; when it is exhausted, the rate of evolution depends solely on the rate of favorable mutations. Furthermore, mass selection can strand a population on an adaptive peak, because there is no mechanism to move the population across valleys of the adaptive topography against the action of natural selection. In a subdivided population, however, the random differentiation of subpopulations al-

112

lows a trial-and-error process in which different combinations of genes arise. If a local subpopulation hits upon a particularly favorable genetic combination (i.e., interaction system) under the control of a higher adaptive peak, the subpopulation will produce a surplus of individuals that can disperse and systematically shift equilibrium gene frequencies toward a higher adaptive peak. Wright called this process of differential migration from local populations "intergroup selection" (Wright, 1977).

The rate of adaptive evolution (i.e., the expected number of generations required to move to successively higher adaptive peaks) is accelerated in a subdivided species because immigration, rather than mutation, is the rate-limiting source of new genetic combinations, and the immigration rate is likely to be orders of magnitude greater than the rate of mutation.

Wright (1978) sees the shifting balance process as a third alternative view of evolution to the neutral and balance hypotheses (Lewontin, 1974). Like the neutral view, it depends on there being very large numbers of nearly neutral polymorphic loci, and it emphasizes the role of genetic drift in changing gene frequencies. It differs from the neutral view in that genetic drift operates primarily in local populations, which evolve somewhat independently. However, genetic drift is not the primary mechanism for fixation or near-fixation of an allele at a single locus in the species as a whole. The role of drift is to generate random combinations of alleles at many loci so that a favorable interaction system may emerge at a single locality.

Fixation or near-fixation of alleles in the shifting balance process is caused by selection, as hypothesized in the balance view. However, individual selection operates primarily in the local population to establish favorable genetic combinations, which then spread through the species via the asymmetrical diffusion of genes resulting from interdemic selection (Wright, 1977).

A consideration of Wright's theory makes it clear that the nature of the genetic structure of populations must be understood in considerable detail if we are to assess the action and interaction of various evolutionary forces. His model depends on a special type of structure in which subpopulations are sufficiently isolated to allow different allele frequencies to develop but not so isolated that genes in adapted complexes cannot spread with reasonable speed. Allozyme studies have disclosed appreciable amounts of geographic variation in allele frequencies within many species, but it is as yet unclear which structures observed in nature are sufficient for the operation of the shifting balance process (Harpending, 1974; Jorde, 1980; Charlesworth et al.,

113

1982). (For a critique of Wright's theory in relation to molecular evolution, see Nei, 1980.)

Even if it can be established that the "shifting balance" type of structure is widespread in natural populations, it will be necessary to determine whether the structure has *actually* accelerated the rate of evolution of favorable genetic combinations. In principle, one could compare species with various types of population structure and examine rates of adaptive evolution. But this is a complex and difficult problem, for several reasons. First, evolutionary biologists presently have no empirical measure of the degree of adaptation of a species to a particular environment and thus have no way of comparing levels of adaptation among species. Second, because fitness is an environment-specific attribute, environmental changes can alter adaptive topographies. Thus, adaptive changes could result from a population moving to a new peak or a shift in a peak caused by an environmental change. Third, the ecological components of population structure (i.e., local population densities and individual dispersal capabilities) could change through time. That is, population structure itself could evolve with changes in the ecological characteristics of an organism. And, as Wright (1977) noted, modification in population structure caused by changes in the breeding system (e.g., amphimixis to automixis) can profoundly affect the rate of adaptive change.

Finally, we should mention that genetic structure can affect the propensity of a population to undergo particular modes of speciation. For instance, Templeton (1980) has argued that speciation via the founder effect is less likely to occur if the founders are derived from a strongly subdivided population than from a large, panmictic one. In all, then, the study of the genetic structure of natural populations is of central importance in our continuing effort to understand the evolutionary process.

114

CHAPTER 6

ENZYME POLYMORPHISM

AND NATURAL SELECTION

Richard K. Koehn, Anthony J. Zera, and John G. Hall

The evolution of enzymes begins with the origin of mutational variants within local populations. The fate of these mutants will be determined by the magnitude and interaction of various microevolutionary forces, including population structure, population size, the forces of natural selection, and interactions with other loci. The magnitudes and patterns of variation and the evolutionary differentiation of enzymes must reflect the varying importance of these different processes.

Enzyme variants (allozymes) offer many advantages for the investigation of microevolutionary processes because they generally have a simple genetic basis and their phenotypes (i.e., catalytic properties) can be objectively and accurately quantified. Despite these advantages, the processes that direct (or have directed) the course of enzyme evolution in populations have proved enormously difficult to identify. Although there are several reasons for this, the most important has been the difficulty in attributing a role for natural selection in enzyme evolution. Early descriptive studies focused exclusively upon levels of enzyme polymorphism and patterns of differentiation of enzyme variants and have had only limited success in resolving this issue. Numerous arguments that enzyme evolution has occurred primarily in response to natural selection have neglected the role of structurally or functionally associated genes (i.e., linkage and epistasis). Conversely, population parameters, which rely exclusively on allele frequency distributions, are often consistent with theoretical predictions involving several evolutionary processes and do not necessarily distinguish among various competing hypotheses.

More recently, direct attempts have been made to study the role of natural selection in determining the microevolution of enzymes.

115

Enzymes have identifiable functions, and knowledge of these functions provides information on the performance of specific enzymes in particular environmental and physiological circumstances. This is not a totally new approach to the study of molecular adaptation. The comparative biochemistry of enzymes consists mainly of a description of catalytic properties of homologous enzymes from different species: differing properties of catalysis in each species are often considered to reflect the consequences of adaptive evolution of enzyme function and structure. This comparative biochemical approach, when applied to the study of enzyme variation within species, can be used to investigate the potential importance of natural selection in the initial stages of enzyme evolution. Unlike comparisons between species, we can experimentally test the effect that alternative biochemical phenotypes might have on physiological function and fitness.

Population geneticists describe evolutionary changes, or their expectation of changes, in terms of relative genotypic fitnesses (e.g., survival and reproduction). The pattern and magnitude of fitness differences will determine the direction and rate of evolution. If there are no differences in fitness among individual genotypes at a locus, evolution of that locus will be determined exclusively by mutation and genetic drift. The question of the adaptive significance of enzyme polymorphism can be restated: to what degree is it possible to describe a molecular mechanism for fitness differences among enzyme variants? Does phenotypic diversity exist among genotypes at an enzyme synthesizing gene such that natural selection determines the course of evolution of that gene, or alternatively, is its evolution a simple reflection of genetic drift or the adaptive evolution by its genetic neighbors?

MOLECULAR MECHANISMS OF ADAPTATION: HOW MIGHT THEY WORK?

Several premises underlie studies of adaptation that characterize the biochemical properties and the physiological consequences of enzyme variants at a particular locus. First, this approach presumes that the genetic diversity of enzymes, vis-à-vis enzyme polymorphism, represents phenotypic diversity and that natural selection can potentially discriminate among alternate phenotypes and thereby affect systematic genetic change of populations. For enzymes, this phenotypic diversity must occur specifically as different catalytic properties (i.e., biochemical phenotypes) of enzyme variants; these constitute the ultimate molecular bases of fitness differences. Second, this biochemical diversity is presumed to have a significant effect upon higher levels of biological organization. That is, different catalytic properties of allozymes must result in different physiological and metabolic consequences (i.e., physiological phenotypes). Although these relationships

116

between genotype and phenotype are fairly straightforward, the "mapping" of genetic variation onto fitness variation, via physiological function, further presumes that the functional structure of the genome is organized into fairly simple selective units.

We may illustrate these points by a few simple contrasting examples. The simplest of such examples is in Figure 1. Different alleles at locus A have different catalytic activities (i.e., biochemical phenotypes) with consequent differences in physiological performance that could ultimately produce fitness differences. Genetic variation at other loci, such as locus B, does not alter the expression of gene A. Variations of biochemical and physiological properties of an organism can be directly and totally attributed to different genotypes at locus A. The *evolutionary consequence* of such a scheme will be determined not by the molecular mechanism itself, which is also true for the other examples below, but by the relative values of the fitness differences. Although the relationship in Figure 1 is heuristically important because it simply portrays the concept of mapping phenotypes on genotypes, it is probably oversimplified; it implies that the function of an enzyme (and therefore its evolution) occurs without influence from other genes, either metabolically related enzyme genes or modifier genes.

More realistic, and therefore complex, relationships can be envisioned (Figure 2), but in this case it is still possible to understand the molecular basis of fitness variation. However, the molecular basis involves multiple loci. For example, the biochemical phenotypes re-

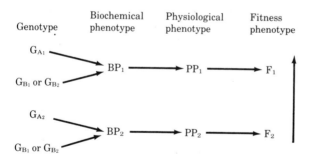

FIGURE 1. The simplest molecular basis of fitness variation where two alleles at one enzyme-synthesizing locus (A) result in biochemically different phenotypes (BP); these lead to different fitness phenotypes (F) through their different physiological phenotypes (PP). There are no effects of alleles at other loci (e.g., locus B). Fitness values may vary with the environment. Biochemical polymorphisms in *Tigriopus, Mytilus,* and *Fundulus* that are discussed in the text might seem to correspond to this simple scheme, but the correspondence is likely to be more apparent than real.

sulting from genetic variation at a structural enzyme locus (A) are affected by variation at a second locus (B). Locus B might represent a specific regulator of A or simply another enzyme locus that functionally interacts with A in metabolism. Many other fitness values can occur, depending on the number and combination of alleles at the A and B loci. In this scheme, the number of interacting genes in the phenotypic unit must be small—say, less than three. Both Figures 1 and 2 have the common property that allele substitutions at the A locus have a significant effect upon the biochemical phenotype, such that whatever the detailed molecular interactions might be, there is a unique fitness phenotype for each genotype.

Phenotypic distributions will become continuous when many loci functionally participate in the phenotypic unit and allele substitutions have individually small effects (Figure 3); in this case it would not be possible to "map" phenotypes at any level of biological organization to the individual genotypic determinants. If the phenotypic (and therefore selective) units of the genome are more like those illustrated in Figure 3, it will not be possible to investigate the molecular basis of fitness. Although our intuitive impression of how individual enzymes function in metabolism may be more closely represented by Figure 3, we will see that biochemical studies of enzyme polymorphism do not support this view.

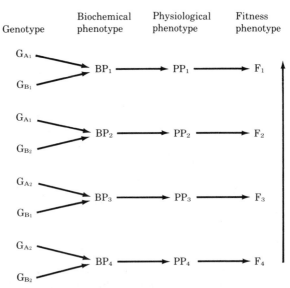

FIGURE 2. The same as Figure 1, but the biochemical phenotypes (and therefore fitness phenotypes) of locus A are not independent of alleles at other genes such as locus B. Allelic substitutions at each locus have relatively large effects. Several biochemical polymorphisms in *Drosophila* discussed in the text might correspond to this scheme.

118

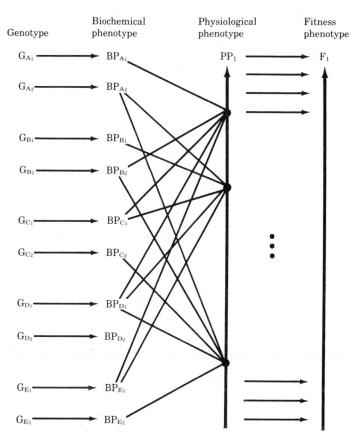

FIGURE 3. Same as Figures 1 and 2, but five loci contribute to variations in the physiological (and therefore fitness) phenotypes. Different combinations of alleles at loci A–E each result in a different physiological phenotype (only three combinations are illustrated). However, the phenotypic effect of a specific allele substitution is small so that the physiological phenotypes are continuously distributed rather than being discrete units that correspond to a simple genetic basis. Biochemical polymorphisms, because enzymes generally function within a complex metabolic network, might be presumed to be represented by this illustration, but examples discussed in the text do not support this.

BIOCHEMISTRY OF POLYMORPHISM: WHAT DO WE KNOW?

The study of the adaptive potential of allozyme variants in natural populations requires a demonstration that any observed structural differences, detectable by electrophoresis, also reflect functional differences. Most workers have generally concentrated on the most obvious possible mode of functional differentiation among allozymes, namely, "catalytic efficiency" divergence with respect to a particular substrate. For the purposes of this discussion, catalytic efficiency is defined as the relative rate v at which an enzyme catalyzes the conversion of its substrate to product under a specified set of conditions. The initial rate of an enzyme exhibiting Michaelis-Menten kinetics is determined by the following:

$$v = \frac{k_{cat}[E_0][S]}{K_m + [S]}$$

where k_{cat} equals turnover number (i.e., the maximum number of substrate molecules converted to product per active site per unit time; Fersht, 1977); K_m equals the Michaelis constant ($[S]$ at which v equals $\frac{1}{2} V_{max}$); $[E_0]$ equals total enzyme concentration; $[S]$ is the substrate concentration; and V_{max} equals $k_{cat}[E_0]$.

The reaction velocity v is determined by the kinetic parameters K_m and k_{cat}, and the contribution of these two parameters to v is determined by the substrate concentration at which the reaction proceeds. As $[S]$ becomes very high, this becomes approximately $v = k_{cat}[E_0]$; whereas at very low substrate concentrations, $v = k_{cat}[E_0][S]/K_m$. At intermediate concentrations of $[S]$, both k_{cat} and k_{cat}/K_m contribute to the reaction velocity.

From this discussion, it can be seen that there are two fundamental rate parameters that must be estimated in order to determine the catalytic efficiency of an allozyme: k_{cat} and k_{cat}/K_m. This has not been appreciated by allozyme workers; only studies of esterase-6 in *Drosophila melanogaster* (R. C. Richmond, pers. comm.) and lactate dehydrogenase in *Fundulus heteroclitus* (Place and Powers, 1979) have directly determined these two parameters. This is probably due in part to the necessity of obtaining pure enzyme in order to estimate k_{cat}. A conclusion regarding the relative catalytic efficiencies of two allozymes therefore depends upon the particular substrate concentrations chosen for assessing catalytic efficiency. Furthermore, the values of both k_{cat} and K_m are valid only for the experimental conditions under which they were measured. This turns out to be a major problem in evaluating kinetic data for allozymes because, in many cases, characterizations of allozymes have been done under nonphysiological conditions. A final problem has been the tendency of many workers to interpret K_m as a measure of affinity of the enzyme for substrate, but this is not necessarily true (Fersht, 1977; Greaney and Somero, 1980).

Until there is greater understanding of k_{cat} and k_{cat}/K_m variations for enzyme polymorphisms, a complete picture of the role of variations in catalytic rate in adaptation cannot be fully appreciated. A more detailed discussion of this point can be found in Hall and Koehn (1983).

120

We have emphasized here and elsewhere (Clarke, 1975; Koehn, 1978) the importance of identifying the physiological consequences of enzyme polymorphism. The ability to successfully relate biochemical diversity of alleles at a locus to physiological variation depends upon knowing the biochemical function of an enzyme and its role in metabolism.

A handful of polymorphic enzymes have emerged as principal foci of these multilevel investigations, many as a consequence of their patterns of allele frequency variation among natural populations. To date none of these studies fully describes the adaptive molecular mechanism of an enzyme polymorphism, although some come very close and each contains information that is consistent with an adaptive mechanism. Obtaining such information involves methodologies that are alien to traditional evolutionary genetics, many of which are technically difficult. Nevertheless, several studies deserve special comment, and a few will be discussed in detail.

ENZYMES OF *DROSOPHILA MELANOGASTER*

Drosophila melanogaster has long served as a model species in population genetic research. A number of enzyme polymorphisms have been intensively studied in this species, and we will present only the highlights of these studies. More detailed accounts of these polymorphisms can be found in Zera et al. (1983). In this chapter, notations are consistent with the following example: *Adh* for the alcohol dehydrogenase gene (locus); ADH for the enzyme product of the gene. Superscripts denote alleles, for example, Adh^F; ADH^F is the allozyme. A notation of $Adh^{F/F}$ represents the genotype, whereas ADH-FF is its enzyme product.

Alcohol dehydrogenase

The polymorphic alcohol dehydrogenase enzymes in *D. melanogaster* have received extraordinary attention, probably more than any other single polymorphic gene. There is a very large body of information on this enzyme; it is reviewed in detail by van Delden (1982).

Several alcohol dehydrogenase alleles have been found to segregate in natural populations, but the Adh^F and Adh^S alleles have combined frequencies of more than 90%. The structural difference between these two common alleles involves a single amino acid substitution (Fletcher et al., 1978): threonine to lysine at position 192.

121

There is geographic variation in the frequencies of the Adh^F and Adh^S alleles in natural populations, principally in the form of geographic clines. Oakeshott et al. (1982) have provided evidence that the patterns of geographic variation in the frequency of Adh alleles are maintained by selective environmental gradients. Latitudinal clines occur in Asia, North America, and Australia; and in all three continental regions, the frequency of Adh^S is negatively correlated with latitude (i.e., frequency decreases as latitude increases). In North America, the frequency of this allele also decreases significantly with more westerly longitudes, which because of the topography, is similar to increasing latitudes. In all three continental regions (and it is these parallel observations that make the data unique), the frequency of Adh^S is positively correlated with total rainfall for the wettest calendar month among sites but not with maximum monthly temperatures (except in North America). Although the absence of a correlation with maximum temperature suggested to Oakeshott et al. (1982) that the relative thermostabilities of ADH allozymes would seem to have little importance in the relationship of allele frequencies with temperature, other experimental evidence has demonstrated that differences in thermostability appear to have a direct effect on fitness (Sampsell and Simms, 1982).

A number of authors have demonstrated differences in *in vitro* ADH activity among ADH-FF, ADH-FS, and ADH-SS. Virtually without exception, the ADH-FF exhibits greater activity than ADH-SS (see Zera et al., 1983). Several authors have demonstrated that the concentration of enzyme is greater in $Adh^{F/F}$ individuals than in $Adh^{S/S}$ (Gibson, 1972; Lewis and Gibson, 1978). S. M. Anderson and J. F. McDonald (pers. comm.) have determined that this difference in allozyme concentration is due to differential rates of synthesis (i.e., genetic regulation); $Adh^{F/F}$ individuals contain a higher concentration of cytoplasmic ADH-mRNA than $Adh^{S/S}$ individuals. Other differences between the genotypes have been described, including thermostability (Anderson et al., 1980) and catalytic properties (McDonald et al., 1980). McDonald et al. (1980) reported genotype-dependent differences in K_m for ethanol, propanol, and butanol, with the K_m differences between genotypes greater for propanol and butanol than for ethanol. In addition to K_m differences between ADH allozymes, McDonald et al. (1980) described differences between the allozymes in the interactions for both cofactor and substrate—though these interpretations have been challenged (Hall and Koehn, 1981).

At present, it is impossible to know if activity differences among ADH genotypes are due to kinetic properties of the allozymes or to differences in enzyme concentration. All K_m determinations have been done near the optimum pH of the enzyme, rather than at physiological pH's, and this severely limits the inferences that can be made from

122

these *in vitro* characterizations. Fundamental kinetic parameters, such as k_{cat} and k_{cat}/K_m have yet to be determined. Although there is considerable evidence indicating that temperature is an important selective agent or that it modulates the role of natural selection, experimenters have not investigated the influence of temperature on the kinetic parameters of the allozymes. Alcohol dehydrogenase is thought to play a critical role in the ability of *D. melanogaster* to exploit habitats that contain high levels of alcohols (McDonald et al., 1977; Heinstra et al., 1982), but the physiological role of the enzyme is still unclear (van Delden, 1982); potentially it involves both detoxification and utilization of alcohol as a carbon source.

There is a large body of literature that collectively constitutes a forceful argument that natural selection is *directing* the evolution of the alcohol dehydrogenase polymorphism in *D. melanogaster*. However, it is still impossible to specify whether this is primarily a consequence of the functional differences between ADH allozymes or differences in regulatory variants that are closely linked to the *Adh* locus.

α-Amylase

α-Amylase catalyzes the hydrolysis of internal glucosidic linkages in polysaccharides composed of three or more glucose units, such as starch and glycogen. The *Amy* locus exhibits extensive polymorphism in *D. melanogaster*, and there are substantial activity differences among *Amy* genotypes (summarized in Doane et al., 1983; Treat-Clemons and Doane, 1982). There are activity differences between amylase genotypes (Hoorn and Scharloo, 1978); and when starch is added to the culture media of *Drosophila*, the frequency of *Amy* alleles can be experimentally manipulated. De Jong and Scharloo (1976) reported a selective advantage for the high-activity AMY[4,6] allozyme (the amylase "gene" is actually a duplicated locus, and the separate enzyme products of the two loci are designated as 4 and 6; see Dickinson and Sullivan, 1975) when flies were raised on high-starch, low-yeast diets. Similar results for flies raised on a high-starch diet was reported by Hickey (1977, 1979). Although selection for amylase genotypes can be accomplished under these specific culture conditions, it is unclear whether or not selection is directed specifically to the structural *Amy* locus. At least two polymorphic regulatory genes are linked (<5 map units) to the *Amy* locus (Doane, 1980). In addition, another locus, *adipose* (*adp*), located approximately 6 map units from the *Amy* locus, may have influenced some of the results reported in studies on

123

the structural gene (Doane, 1980). However, the effect of *adp* on these experiments, especially those of De Jong and Scharloo (1976), is now thought to be of less importance (see Zera et al., 1983). The limited biochemical data on AMY allozymes also precludes an assessment of their influence on this polymorphism in natural populations. At present, the *Amy* polymorphism can only be considered an attractive experimental system for future studies of molecular interaction and does not provide unequivocal evidence bearing on questions concerning the adaptive significance of enzyme polymorphism.

Esterase-6

The *Esterase-6* locus in *D. melanogaster* appears to have importance in reproduction (Richmond et al., 1980). EST-6 activity was found to be highly concentrated in the anterior ejaculatory duct of males. In a series of experiments using individuals bearing $Est-6^S$, and $Est-6^0$ (null) alleles, it was observed that EST-6 activity was depleted in males and transferred to females during the early stages of copulation. Two other observations suggested the role of EST-6 in reproduction: (1) a strain selected for decreased time to remating exhibited higher EST-6 activity, and (2) females would remate sooner if they had been inseminated by an $EST-6^{S/S}$ male than by and $EST-6^{0/0}$ male, at 25°C.

The allozymes have been purified and their kinetic and physiochemical properties determined (R. C. Richmond, pers. comm.). Although the allozymes possess similar K_m and k_{cat} values for four synthetic acetate or propionate esters, the allozymes do differ in k_{cat}/K_m for one of the substrates, β-naphthylpropionate. The physiological significance of this difference is not yet known.

Although these studies represent a significant advance in our understanding of the role of EST-6 in reproduction and the possible adaptive significance of this biochemical polymorphism, several basic questions still remain unanswered. For example, the specific substrate(s) hydrolyzed by EST-6 in the female reproductive tract are unknown. R. C. Richmond (pers. comm.) has observed *in vitro* hydrolysis by EST-6 of radiolabeled *cis*-vaccenyl acetate. This substance is stored in the ejaculatory bulb, is transferred during mating, and acts as an antiaphrodisiac in mating experiments (S. Maine and R. C. Richmond, unpublished).

α-Glycerophosphate dehydrogenase

α-Glycerophosphate dehydrogenase plays several important metabolic roles in insects, especially in the specialized metabolism of insect flight (Sackter, 1975; O'Brien and MacIntyre, 1972a,b). α-*Gpdh* polymorphisms are among the most intensively studied insect polymorphisms. There are several reasons for this: the enzyme plays an indispensible

124

role in an important biological process (flight) directly related to fitness; there is extensive background information on the metabolic role of this enzyme in flight metabolism; and the physiological consequences of differences in enzyme function may be quantified by measuring various flight efficiency parameters (Curtsinger and Laurie-Ahlberg, 1981). The α-$Gpdh$ polymorphism in $D.$ $melanogaster$ has been especially well studied.

The polymorphism consists of two common alleles, and the frequency of the α-$Gpdh^S$ allele is positively correlated with latitude in North America, Asia, and Australia (Oakeshott et al., 1982). These parallel clines on three different continents suggest that the polymorphism is influenced by latitudinally varying selection gradients. Although four different researchers have investigated the kinetic properties of α-GPDH allozymes, the adaptive significance (if any) of this polymorphism is unclear. Miller et al. (1976) and McKechnie et al. (1981), using crude adult and larval preparations, respectively, reported temperature-dependent differences in apparent K_m for the substrate dihydroxyacetone phosphate (DHAP) among the three α-$Gpdh$ genotypes. Miller et al. (1975) also reported temperature-dependent differences in specific activities among genotypes. However, the results of Miller et al. (1975) are complicated by the fact that they used crude adult homogenates; these contained two α-GPDH isozymes, which are known to exhibit a twentyfold difference in K_m for DHAP. Genotype-dependent differences in specific activity could be due to differences in enzyme concentration, as was observed for Adh genotypes, rather than to differences in kinetic properties of α-GPDH allozymes. Other studies of genotype-dependent differences in apparent K_m for DHAP using homogeneously purified α-GPDH allozyme have given contradictory results. G. Bewley (pers. comm.) found no difference in K_m for DHAP between purified adult allozymes at several temperatures, whereas in a preliminary study G. E. Collier (pers. comm.) obtained results similar to Miller et al. (1975). It is important to bear in mind that all of these studies have focused on genotype-dependent differences in K_m which, as discussed earlier, does not necessarily reflect differences in catalytic efficiency. k_{cat}/K_m has not yet been measured for α-GPDH allozymes in $D.$ $melanogaster$.

Glucose-6-phosphate dehydrogenase/6-phosphogluconate dehydrogenase polymorphisms and pentose shunt flux

The works of Kacser and Burns (1973; 1979; 1981) have emphasized that the effect of changes in catalytic activity of mutant enzymes usually results in only a negligible change in metabolic flux. This is because metabolic enzymes do not function in isolation but are embedded in complex metabolic pathways wherein a number of enzymes participate in flux. Even a large change in enzyme activity, according to Kacser, leads to only negligible changes in flux. These interpretations, based upon mathematical models derived from experimental results in yeast, suggest that significant physiological effects of enzyme polymorphism should not occur, simply because catalytic differences between allozymes are small. Yet, the study of Cavener and Clegg (1981) suggests a significant effect of biochemical polymorphism on pentose shunt flux.

The pentose shunt (Figure 4) provides an alternate pathway for the catabolism of glucose. Although usually less important than the glycolytic pathway in terms of the percentage of glucose catabolized, the pentose shunt serves the important role of generating reduced

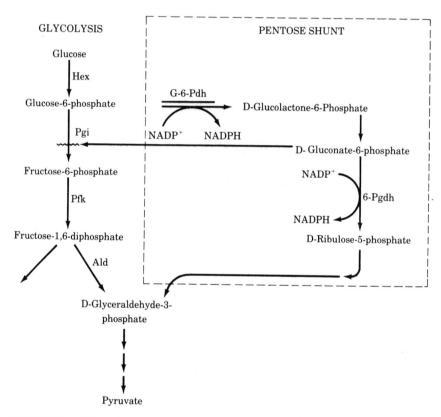

FIGURE 4. The pentose shunt and portions of glycolysis showing the functions of glucose-6-phosphate dehydrogenase (G-6-Pdh) and 6-phosphogluconate dehydrogenase (6-Pgdh) and the inhibition of phosphoglucose isomerase (Pgi) by 6-phosphogluconate (D-gluconate-6-phosphate).

nicotinamide adenine dinucleotide phosphate (NADPH), primarily for lipid synthesis, and ribose phosphates for nucleic acid synthesis. Production of NADPH by the pentose shunt is the responsibility of two enzymes: glucose-6-phosphate dehydrogenase and 6-phosphogluconate dehydrogenase. Both enzymes are polymorphic in *Drosophila melanogaster* (Powell, 1975). In addition, both enzymes catalyze metabolically related reactions in the pentose shunt (Figure 4), use the same cofactor, are coordinately regulated (Geer et al., 1978; 1979; Lucchesi et al., 1979; Laurie-Ahlberg et al., 1980), and are inhibited by similar compounds. The two polymorphisms are thus ideal candidates for investigating the possibility of molecular interaction and have become the subjects of considerable research activity.

Cavener and Clegg (1981) have performed an imaginative experiment in which pentose shunt flux was measured with respect to two di-locus genotypes, G-6-$pdh^{A/A}/6$-$Pgdh^{F/F}$ and G-6-$pdh^{B/B}/6$-$Pgdh^{S/S}$. This experiment was based on the fact that the C-1 carbon of glucose is always lost as CO_2 when glucose is oxidized in the pentose shunt, whereas neither C-1 nor C-6 is lost when glucose is oxidized via glycolysis. Thus, the relative flow of glucose through the pentose shunt can be determined by measuring the relative incorporation of $[1\text{-}^{14}C]$ glucose versus $[6\text{-}^{14}C]$ glucose into lipids and protein. The G-6-$pdh^{B/B}/$ 6-$Pgdh^{S/S}$ genotype had a significantly greater pentose shunt activity than the G-6-$pdh^{A/A}/6$-$Pgdh^{F/F}$ in flies on media containing either 9 or 290 mM sucrose. The difference between the two di-locus genotypes was magnified at the higher sucrose concentration, because this causes a greater total pentose shunt activity. The authors also found genotype-specific differences in relative ^{14}C incorporation into lipid versus protein. These results are consistent with the reported activity differences between the G-6-pdh and 6-$Pgdh$ genotypes (Bijlsma, 1978) and suggest that genotypes of differing activity affect flux. These differences could be due to either enzyme concentration or catalytic differences. Not only did pentose shunt fluxes differ between the genotypes, but there were "higher order" effects of these differences, measurable as different allocations of carbon to protein and lipid (but only at 9 mM sucrose).

Hughes and Lucchesi (1977) have shown marked differences in fitness between individuals hemi- and homozygous for a common (A or B) versus a null G-6-pdh allele, where the difference was contingent upon the allelic state of 6-$Pgdh$. W. Eanes (pers. comm.) has carried this a step further by demonstrating fitness differences between individuals hemi- or homozygous for either the G-6-pdh^A or G-6-pdh^B alleles; these differences were also contingent upon the allelic state of

127

6-Pgdh. No differences in viability fitness could be demonstrated with a normal activity *6-Pgdh* allele; however, in the presence of a "leaky" *6-Pgdh* allele (approximately 10% activity), significant differences in fitness between *G-6-pdh* genotypes emerged. The *G-6-pdh^B* allele has higher enzyme activity than the *G-6-pdh^A* allele (W. Eanes, pers. comm.). The high-activity G-6-PDH-B (males) G-6-PDH-BB (females) genotypes apparently build up a greater concentration of 6-phosphogluconate than the alternate low-activity genotypes, but only in the presence of low *6-PGDH* activity. This has the consequence of a greater inhibition of phosphoglucose isomerase in the main line of glycolysis (Figure 4) and thereby significantly decreases fitness. These results illustrate the fitness differences that may result from biochemical variations from enzyme polymorphism but emphasize the dependence of such differences on the genetic state of other genes with whose products there are functional interactions.

LACTATE DEHYDROGENASE IN *FUNDULUS HETEROCLITUS*

Powers and his collaborators have made a compelling case for the adaptive importance of catalytic efficiency differences between lactate dehydrogenase-B allozymes in the teleost fish *Fundulus heteroclitus*. The two common alleles of the heart-type, or lactate dehydrogenase-B, vary in frequency as a north–south cline along the Atlantic coast of the United States (Powers and Place, 1978). Because temperature has a profound effect on enzyme function, Powers' work has focused upon the temperature-dependent catalytic properties of the LDH allozymes and how these may significantly affect the physiology of energy metabolism.

At low pyruvate concentrations, there were significant differences in reaction velocities between LDH-B^bB^b and LDH-B^aB^a. At pH values below 8.00, the LDH-B^bB^b allozyme exhibited greater reaction rates at lower temperatures. This difference in relative reaction rate was reversed at temperatures greater than 25°C for pH values between 6.50 and 7.00. When reaction rates were compared at constant relative alkalinity, the same differences were observed. These findings correspond to the distribution of the two alleles among natural populations, where *Ldh-B^a* is common in colder, northern waters and *Ldh-B^b* is common to warmer latitudes (Place and Powers, 1979). This study is the only one other than for esterase-6 in *D. melanogaster* to have estimated k_{cat}/K_m.

What might be the adaptive importance of the differences in catalytic efficiency between *Ldh* alleles in *Fundulus*? Powers and co-workers have shown several important biological variables to be correlated with *Ldh* genotypes. For example, individuals of genotype *Ldh-B^aB^a* exhibit lower concentrations of intraerythrocytic ATP than alternate genotypes. ATP functions as an allosteric modifier of *Fundulus* hemoglobin, similar to the function of 2,3-diphosphoglycerate in human red blood cells; ATP decreases the affinity of hemoglobin for oxygen. These differences in ATP concentration lead to significant differences in oxygenation such that LDH-B^bB^b exhibits highest levels of intraerythrocytic ATP and lowest blood oxygen affinity (Powers et al., 1979).

Differences in blood oxygen affinity, correlated with *Ldh* genotype, appear to have several other important physiological consequences. Respiratory stress is known to trigger embryonic hatching in *Fundulus*. LDH-B^aB^a individuals have lowest intraerythrocytic concentrations of ATP, highest relative oxygen affinities, and, therefore, decreased ability to deliver oxygen to tissues. This is the apparent basis for the earlier mean hatching times for individuals of this genotype (DiMichele and Powers, 1982a). A similar mechanism is presumed to explain differences in swimming ability between individuals homozygous for different *Ldh* alleles. When fish are swum to exhaustion (at 10°C), the critical swimming speed of the LDH-B^aB^a genotype was 3.6 body lengths per second whereas the other homozygote had a critical swimming speed of 4.3 body lengths per second (DiMichele and Powers, 1982b). The increasing level of serum lactate during exercise leads to a significant decrease in blood pH. This further magnifies the oxygen unloading differences that are the consequence of *Ldh*-correlated ATP differences (DiMichele and Powers, 1982b).

Studies in *Fundulus heteroclitus* represent a compelling argument for the adaptive significance of the *Ldh* polymorphism in that species, but the strength of this argument depends upon establishing a functional connection between genotype-dependent LDH biochemistry and the correlations of these with other biological variables. These connections at present are still unknown. One would predict, for example, that if LDH biochemistry per se is important in producing differences in embryo hatching and/or swimming stamina, then the reported observations should be reversed at higher temperatures because the relative catalytic efficiencies of the allozymes are reversed at temperatures greater than 25°C (Place and Powers, 1979).

The model studies of alcohol dehydrogenase, α-GPDH, amylase, Esterase-6 in *Drosophila*, and LDH in *Fundulus* are beginning to provide detailed information on the potential adaptive significance of enzyme polymorphism. In each case, there is a tantalizing relationship between the biochemical, physiological, and population genetic variations; because each is yet incomplete, they serve only as the outlines for a detailed understanding of a molecular mechanism of adaptation.

Other studies, though even less complete, illustrate biochemical differences that occur between enzyme variants, including octopine dehydrogenase (Walsh, 1981) and phosphoglucose isomerase (Hoffman, 1981) in the sea anemone *Metridium senile*, phosphoglucose isomerase of *Colias* butterflies (Watt, 1977), and lactate dehydrogenases in salmonid fishes (Klar et al., 1979).

ENZYME POLYMORPHISM AND CELL VOLUME REGULATION

In osmoconforming marine crustaceans and mollusks, the first stage of response to hyperosmotic stress is the regulation of cell volume by a rapid accumulation of intracellular free amino acids. The production of this free amino acid pool is metabolically complex but is at least partially dependent upon the catabolism of intracellular proteins. Protein catabolism occurs in cellular lysosomes, which are known to be involved in catabolic turnover of intracellular proteins (Segal, 1975; Dean, 1977) and to have relatively high concentrations of free amino acids in comparison to the cytosol (Tappel, 1969; Ward and Mortimer, 1978; Koehn et al., 1980a). The amino acids that result from protein catabolism are transaminated to the specific family of residues that are important in cell volume regulation, including alanine, glycine, and (in crustaceans only) proline (Figure 5).

Burton and Feldman (1982) have investigated the effect of a glu-

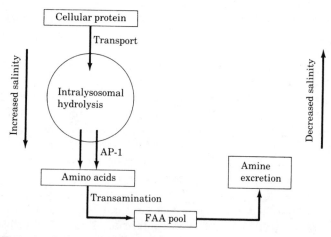

FIGURE 5. A schematic representation of how the cytosolic free amino acid pool originates (in part) from cellular protein during hyperosmotic stress in osmoconforming marine invertebrates such as *Tigriopus* and *Mytilus* (see text for details).

130

tamate-pyruvate transaminase polymorphism on variation in cell volume regulation in the intertidal copepod *Tigriopus californicus*. Glutamate-pyruvate transaminase (GPT, EC 2.6.1.2) catalyzes the final step of alanine synthesis, the most important amino acid in the early stages of response to hyperosmotic stress (Bishop, 1976). There are two common *Gpt* alleles in *Tigriopus* populations, Gpt^F and Gpt^S; Gpt^F has significantly higher specific activity (Burton and Feldman, 1983) in both the alanine-synthesizing and alanine-catabolizing directions of the catalyzed transamination. Under conditions of hyperosmotic stress, individual adult copepods of $Gpt^{F/F}$ and $Gpt^{F/S}$ accumulate alanine (though not glycine or proline) faster than $Gpt^{S/S}$ individuals. This slower rate of response to hyperosmotic stress in individuals results in significantly higher mortality of $Gpt^{S/S}$ larvae when they are exposed to hyperosmotic conditions. The significantly different biochemical activities of these two allozymes are reflected in specific physiological variations and differential genotypic survivorships under the experimental conditions employed. The genotypic differences in ability to cope with hyperosmotic conditions combine with the population structure of *Tigriopus* (many temporarily isolated tide pools) to produce a mosaic pattern of spatial genetic differentiation among populations (Burton and Feldman, 1981).

There are many similarities between the study by Burton and Feldman on *Tigriopus* and our own work on the aminopeptidase I polymorphism in the marine bivalve *Mytilus edulis*. Aminopeptidase I is also involved in cytosolic free amino acid production that occurs in response to hyperosmotic stress; but whereas GPT is involved in transamination, aminopeptidase I is a lysosomal enzyme that hydrolyzes a variety of neutral and aromatic NH_2-terminal amino acids of di-, tri-, and tetrapeptides (Young et al., 1979). The total activity of aminopeptidase I, as well as other lysosomal enzymes, is influenced by environmental salinity variations (Koehn, 1978; Moore et al., 1980), and the increase in total enzyme activity that occurs during hyperosmotic stress can be interpreted as an increased hydrolytic capacity of the lysosomes. This observation, together with the demonstration that salinity variations in the environment alter the rate of protein turnover in cell-free lysosome preparations, the permeability of the lysosome membrane to substrate, and the concentrations of lysosomal amino acids (Bayne et al., 1981), all constitute evidence that aminopeptidase I is important in the production of free amino acids during hyperosmotic stress. Following transamination (Greenwalt and Bishop, 1980), the products of this reaction constitute important cytosolic solutes for cell volume regulation.

131

In *Mytilus edulis,* there are three common aminopeptidase I alleles, designated Lap^{94}, Lap^{96}, and Lap^{98}. The catalytic properties of Lap^{98} and Lap^{96} cannot be shown to be different from one another, but the genotypes involving Lap^{94} exhibit significantly greater specific activities than those genotypes without this allele (Koehn and Immermann, 1981). Unlike the situation in *Tigriopus,* where their population structure leads to local differentiation of allele frequencies, allele Lap^{94} exhibits a high relative frequency that is homogeneous over great geographic areas of populations experiencing oceanic salinity. The higher activities of aminopeptidase I genotypes with the Lap^{94} allele are due to a higher apparent k_{cat} of this allele, compared to other alleles. These genotypes exhibit a 20% higher activity per unit enzyme protein than the other genotypes (Koehn and Siebenaller, 1981).

The relative differences in the rate of substrate conversion to product between genotypes with and without the Lap^{94} allele lead to different rates of accumulation of cytosolic free amino acids during hyperosmotic acclimation (Figure 6). Those genotypes with a higher catalytic efficiency accumulate free amino acids at a higher rate than low catalytic efficiency genotypes (Hilbish et al., 1982). The similarity between these results and those reported by Burton and Feldman (1983) for *Tigriopus* is obvious; biochemical variations in the synthetic pathway leading to the cytosolic free amino acid pool can produce differences in the rate of free amino acid synthesis.

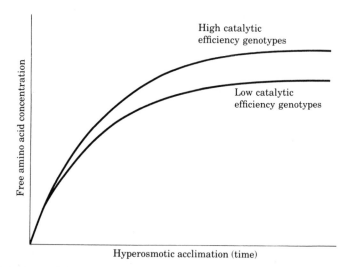

FIGURE 6. Genotype-dependent differences in accumulation of free amino acids for genotypes in *Mytilus edulis.* High catalytic efficiency genotypes refer to enzyme genotypes with the Lap^{94} allele (both homozygotes and heterozygotes), which have a higher catalytic efficiency than alternate genotypes. (Adapted from Hilbish et al., 1983.) See also Figure 5.

In *M. edulis,* fitness differences that might depend upon differences in hyperosmotic acclimation cannot be directly tested, as they have been in *Tigriopus.* Nevertheless, because of the population structure of *M. edulis,* differences in survival among aminopeptidase I genotypes can be demonstrated in natural populations. Larval dispersal of oceanic larvae into estuaries of reduced salinity occurs during the annual reproductive cycle of *M. edulis.* Following settlement of oceanically derived larvae in estuaries, there is differential mortality among aminopeptidase I genotypes that dramatically reduces the frequency of the Lap^{94} genotypes in low salinity environments (Figure 7). This differential mortality is accompanied by significant differences between the two genotypic groups in a measure of physiological condition—the relationship between dry tissue weight and shell length (Figure 8). Genotypes with Lap^{94} exhibit both higher relative mortality and poorer physiological condition (Koehn et al., 1980b).

Whether or not the foregoing genotype-dependent differences in catalytic efficiency and hyperosmotic acclimation ability constitute an adaptive genetic mechanism depends upon the environmental circumstances in which they occur. The high catalytic efficiency of allele Lap^{94} is at an apparent advantage in environments of high salinity combined with warm environmental temperatures. High temperatures elicit increased enzyme protein concentrations in catabolic enzymes (i.e., increased hydrolytic potential; Hazel and Prosser, 1974), and high

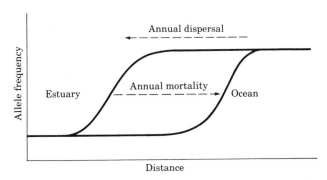

FIGURE 7. The formation and variation of the *Lap* allele frequency cline in Long Island Sound. Genotypes with Lap^{94} are subject to higher mortality than other genotypes. This results in a steep gene frequency cline that is initially displaced into the estuary (to the left in the illustration); however, mortality shifts the cline (to the right) to where it is correlated with spatial changes in salinity. (Adapted from Koehn et al., 1980a.)

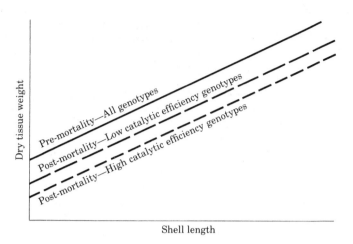

FIGURE 8. Relationship between dry tissue weight and shell size for pre-mortality and post-mortality individuals. Following settlement of oceanic *Mytilus edulis* larvae (see Figure 7) in an estuary such as Long Island Sound, but prior to differential mortality, the physiological condition (dry weight versus shell length) of all aminopeptidase I genotypes is the same. After mortality, which is greatest for genotypes with Lap^{94}, surviving individuals with the Lap^{94} allele are in poorer physiological condition than other genotypes. (Adapted from Koehn et al., 1980a.)

salinity induces an increase in aminopeptidase I activity (Koehn, 1978). It can be expected that conditions would favor the evolution of an enzyme (or allozyme) with higher relative catalytic efficiency where high temperature and high salinity conditions prevail together. Indeed, it is only where these two environmental conditions jointly occur that the Lap^{94} allele is found in high frequency (Koehn et al., 1976). The disadvantage of a higher reaction rate in regions of either low salinity and high temperature, or high salinity and low temperature, is apparently a consequence of the imbalance between the rate of mobilization of cellular protein reserves and the rate of utilization of these resources. In the face of a fluctuating salinity environment, which occurs at the boundary between oceanic and estuarine waters, the more efficient response of the high catalytic efficiency genotype apparently results in a loss of nitrogenous compounds during hypoosmotic acclimation (Hilbish et al., 1982). In a fluctuating environment, the higher amplitude in variations of cellular protein catabolism (during hyperosmotic acclimation) and excretion of amines (during hypoosmotic acclimation) results in the final depletion of nitrogen reserves. It is this more rapid depletion of energy reserves that is apparently reflected in the different rates of tissue loss that accompany differential mortality.

Despite the detailed biochemical and physiological characterizations of the genotypic components of cell volume regulation in both

Tigriopus and *Mytilus,* there are many factors that are still unknown. For example, differences in activity between *Gpt* genotypes in *Tigriopus* could be due to differences in kinetic parameters of the two allozymes or alternatively to closely linked regulatory genes. Further genetic and biochemical studies of this system would help clarify this point. Although $Gpt^{F/F}$ and $Gpt^{F/S}$ genotypes are more efficient in hyperosmotic acclimation, what might be the advantage of alternate genotypes that leads to a balanced polymorphism?

In *Mytilus,* the effect of salinity variations is understood, but the role of temperature has not been investigated. More importantly, differential survival among genotypes in natural populations has been attributed to the effect that different enzyme activities have on total nitrogen metabolism and no direct evidence is yet available that bears upon this interpretation.

POTENTIAL MECHANISMS OF ADAPTATION VERSUS ADAPTIVE MECHANISMS

We began this chapter with a query as to whether or not it is possible to describe a molecular mechanism for fitness differences among enzyme genotypes. The answer to this question, at present, must be an unequivocal "no." We are beginning to see a series of examples that provide a mechanistic description of how the biochemical diversity of enzyme polymorphism affects certain biological processes, though it is by no means clear if this biochemical diversity is adaptive. There is now considerable evidence that enzyme polymorphisms represent phenotypic diversity. In several cases, the biochemical variation has measurable physiological consequences. Why is it that, in the face of much information that bears upon our question of interest, we cannot demonstrate with greater certainty the adaptive significance of enzyme polymorphism?

Studies that we have reviewed represent detailed (but still incomplete) descriptions of *potential* mechanisms of adaptation; these are biochemical/physiological mechanisms by which adaptation *might occur.* They are potential mechanisms of adaptation, but not necessarily adaptive mechanisms. In each case, the biochemical properties of an allele appear to serve as an adaptation to a particular environmental circumstance (high temperature, low salinity, etc.). The potential adaptive significance of each enzyme polymorphism thus depends upon the range of environments to which each is exposed and the relative frequencies of alternate environments.

The potential adaptive significance of these polymorphisms is also very much dependent upon the genetic state of other genes with which they may interact, either epistatically or via linkage. The results of Eanes, where fitness differences of *G-6-pdh* genotypes depend upon genetic activity variations at the *6-Pgdh* locus, is an obvious example of epistatic interaction. A similar conclusion has been emphasized by Dykhuizen and Hartl (1980), who demonstrated in *E. coli* that 6-PGDH allozymes differing in apparent K_m are selectively neutral on a normal genetic background, but differentially selected in a genetic background lacking alternative metabolic routes for 6-phosphogluconate. In the latter case, selection occurred in the direction expected from the different apparent K_m values of the allozymes.

The results we have reviewed imply that the functional structure of the genome of these few species cannot be accurately represented by Figure 3. Although activity differences between different enzyme alleles amount to no more than 10–15%, this is often sufficient to produce large physiological effects. The results in *Tigriopus, Mytilus,* and *Fundulus* would each seem to correspond to an oversimplified view (Figure 1), where individual genotypes at a single locus have distinct, but different, physiological and fitness consequences. However, the studies in *D. melanogaster* correspond most closely to the situation illustrated in Figure 2. In each *Drosophila* study, additional genetic elements are known to influence the enzyme polymorphism. We suspect that the only reason why the studies in *Tigriopus, Mytilus,* and *Fundulus* appear simpler is because of the inability to experimentally dissect the genetic basis of phenotypic diversity in those species. The ability to understand the environmental circumstances of these three species more accurately is a distinct experimental advantage for interpreting the biochemical characterizations of the polymorphisms, but it is not possible to fully elucidate the genetic basis of phenotypic variations. We must conclude, at least for the present, that the functional structure of the genome can be most accurately represented by Figure 2; most allelic variants represent relatively large effects upon the biochemical phenotype, but the exact nature of this phenotype will be influenced by additional genetic elements. If there is a selective unit, it consists of a small number of interacting genes.

The study of the molecular basis of the microevolutionary process has become an exciting area of evolutionary biology. The study of enzyme polymorphism has begun to focus upon the mechanism of enzyme evolution and to thereby provide a better understanding for the functional and structural diversity of enzyme homologs among species. A number of different traditional fields of study have become important in this approach to problems of adaptation, including enzymology, metabolic biochemistry, physiological ecology, and comparative physiology. This will necessarily dictate a broader scientific perspective by evolutionary geneticists.

136

CHAPTER 7

DNA POLYMORPHISMS IN THE HUMAN β GLOBIN GENE CLUSTER

Haig H. Kazazian Jr., Aravinda Chakravarti, Stuart H. Orkin, and Stylianos E. Antonarakis

Natural populations contain a large amount of genetic variability that has been classically studied by using genetic and biochemical techniques. Most of the variability observed is that in coding regions of genes that lead to variation at the protein level. With the advent of DNA analysis, it has become possible to study genetic variability in any region of the genome as well as to identify causes of such variability. In the few years since Kan and Dozy (1978) first discovered a DNA polymorphism in a human population, population genetics and other fields have benefited greatly from information obtained through the study of polymorphism at the DNA level.

In this chapter we will discuss the nature of DNA polymorphism in general and the known polymorphisms in the β globin gene cluster in particular. These specific polymorphisms have been useful in (1) the demonstration of a small number of common sequence variations in the cluster; (2) the characterization of the molecular basis of inherited disorders of β gene expression, the β thalassemias; and (3) the discovery of multiple origins of identical β globin mutations.

Analyses of polymorphism of the β globin gene cluster in various populations have demonstrated that any two β gene clusters chosen at random differ by 1 nucleotide in every 500. Three common types of normal β genes, which we call frameworks, differ by 1, 4, and 5 nucleotides from each other in Mediterraneans. Our analysis indicates that the existence of these frameworks can be explained by neutral mutation followed by genetic drift.

MECHANISMS OF DNA POLYMORPHISM

Based on data from a number of systems, we can classify the mechanisms that generate DNA polymorphism into four categories.

Point mutations at a single nucleotide

These mutations can alter the amino acid sequence of a protein if they occur in a coding region. Examples are the mutations for β^S and β^E globin genes. Second, they may alter the third nucleotide of a codon and fail to produce an amino acid substitution, such as the C → T change at the second codon of the β globin gene (see below). Third, they may occur in an intervening or flanking sequence. However, a minority of these substitutions will alter a restriction endonuclease site, such as the mutation that eliminates an *Hpa*I restriction site 5 kilobases 3' to the β globin gene, and these can usually be detected by restriction endonuclease analysis.

Insertion/deletion of unique DNA sequences

Polymorphism due to insertion of a 200-base pair sequence has been demonstrated in a heat shock protein locus (*87A hsp-70*) in *Drosophila melanogaster* (Leigh Brown and Ish-Horowicz, 1981). Similarly, polymorphisms due to several insertion/deletion sequences have been observed in a 13-kilobase DNA region containing the alcohol dehydrogenase locus in *D. melanogaster* (Langley et al., 1982). Insertions and deletions are also presumed explanations for the variability in restriction fragment lengths seen in two regions of the α globin cluster in man (Higgs et al., 1981).

Insertion/deletion of tandem DNA sequences

This type of variability differs from that discussed in the preceding paragraph because it affects a specific short DNA sequence but not any random segment. Spritz (1981) has shown that there is polymorphism for the number of repeats (4, 5, or 6) of the sequence ATTTT in a region approximately 1.5 kilobases (kb) 5' to the β globin gene in man. Bell et al. (1982) and Ullrich et al. (1982) have also discovered a similar type of polymorphism based on variations in the number and sequence of a 14-base pair (bp) segment $ACAG_4TGTG_4$ that is found in a region 5' to the insulin gene in man. The lengths of this repeat region are polymorphic but fall into four size classes. Class I has a 109-base pair repeat region and accounts for 1% of insulin-bearing chromosomes in human populations. Class II has a 500- to 900-bp repeat region and accounts for 67% of chromosomes. Class III

(5%) has a repeat region of 1500 to 1900 base pairs and Class IV (27%) has a 2000- to 3000-bp repeat region (Ullrich et al., 1982). This type of polymorphism presumably occurs by unequal crossing-over between related sequences. These data indicate a mechanism whereby variability may be produced by recombination.

Gene conversion

In this type of polymorphism, a mutation at one locus may be moved into another homologous, or related, locus by a gene conversion type of recombination (Slightom et al., 1980). This mechanism is thought to begin with mispairing of homologous loci of two sister chromatids followed by replacement of the sequence in the homologous region of one sister chromatid by that of the other. Unequal crossing-over discussed in the preceding paragraph is an alternative result of the mispairing process. The polymorphism in a *Hind*III restriction site at the 3' end of the large intervening sequence of the $^{G}\gamma$ and $^{A}\gamma$ globin genes in man is presumably due to this mechanism (Jeffreys, 1979).

It is interesting that the globin clusters in man provide examples of each of these mechanisms. Perhaps all of these types of polymorphisms are ubiquitous in the genome.

POLYMORPHISM IN THE β GLOBIN GENE CLUSTER

The β globin gene cluster occupies approximately 50 kilobases of DNA in the short arm of chromosome 11 in humans (Efstratiadis et al., 1980). Embryonic (ε), fetal ($^{G}\gamma$ and $^{A}\gamma$), and adult (δ and β) globin genes are arranged in a 5' to 3' (left to right) order and the gene sequences account for approximately 15% of the DNA in this region (Figure 1). ψβ1 in Figure 1 is the β-like nonfunctional gene (pseudogene). The function of the flanking sequences, which contain certain repeated DNA regions, is not well understood.

Using the methods of restriction endonuclease analysis of DNA obtained from human peripheral leukocytes, a number of DNA poly-

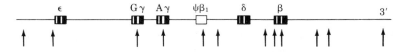

FIGURE 1. β Globin gene cluster and the location of 12 known polymorphic restriction sites (↑ , sites of cleavage by restriction endonucleases): ε, embryonic gene; $^{G}\gamma$, $^{A}\gamma$, fetal genes; ψβ1, pseudogene; δ, β, adult genes.

139

morphisms have been found in the β globin gene cluster (Kan and Dozy, 1978; Tuan et al., 1979; Jeffreys, 1979; Kan et al., 1980; Antonarakis et al., 1982a). In this discussion we will concentrate on sites that are polymorphic to the extent that the less common variant (whether presence or absence of the site) is present in one or more racial groups with a frequency greater than 5%. At present, by this definition, 12 polymorphic sites have been found in the β gene cluster (Figure 1). Nine of these sites are public (occur in all racial groups), and the frequency of the less common allele is greater than 5% in all racial groups studied to date. The remaining polymorphic sites, *Taq*I, *Hinf*I, and *Hpa*I, are private in that they are polymorphic in Blacks but not in other racial groups. It is of interest that certain private polymorphisms such as *Taq*I and *Hinf*I have attained quite high frequencies (25–35%) in Blacks even though they are very rare in other racial groups (H. H. Kazazian, unpublished). All polymorphisms on which data are available (6 of the 12) are single nucleotide substitutions. As mentioned earlier, significant polymorphic deletions and insertions have been found in the α globin gene cluster and in a region 5′ to the human insulin gene. Such deletion polymorphisms have not yet been observed in the β globin cluster. The polymorphisms in the β globin cluster are in the following locations: 7 of the 12 are in flanking DNA, 3 are in intervening sequences, 1 is in a pseudogene sequence, and 1 is in the coding region of the β gene.

Study of nucleotide sequences has also led to the elucidation of nucleotide polymorphisms that are not detectable by restriction site analysis. In the course of sequencing various mutant genes (especially β thalassemia), whose neighboring DNA contained different patterns of polymorphic restriction sites, nucleotide polymorphisms were found in the β gene (see below).

COMMON SEQUENCE VARIATION IN THE β GLOBIN GENE CLUSTER

We have studied the polymorphic restriction endonuclease sites of 200 normal and 200 mutant β globin gene clusters and have correlated the information with sequence analyses of a subset of 20 β globin genes (Orkin et al., 1982). These data have enabled us to reach some general conclusions about the sequence variation within this cluster and, by analogy, about that within other gene clusters.

The general picture that emerges is that there is a relatively small number of common haplotypes with respect to restriction site pattern (Figure 2). A haplotype is a particular sequence of restriction sites in the DNA region studied. The DNA region studied covers the entire β globin gene cluster from the embryonic ε gene to the 3′ region of the β globin gene. Each of these common haplotypes appears to represent

140

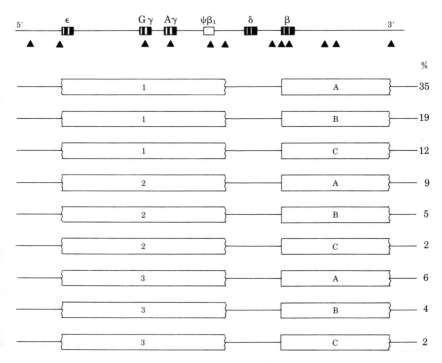

FIGURE 2. Nine common β globin gene region haplotypes observed in Mediterraneans. Two nonrandomly associated sequence clusters are shown; the 5′ cluster includes the ε, $^G\gamma$, $^A\gamma$, and ψβ1 genes, and the 3′ cluster includes the β gene and at least 18 kilobases 3′ to it. The three common sequence types in the 5′ cluster are designated 1, 2, and 3, and the three common sequence types in the 3′ cluster are called A, B, and C. The percentage of each haplotype among normal Mediterranean chromosomes is shown on the right. A recombination "hot spot" is tentatively located around the *Hinf* I site between the δ and β genes. ▲, sites of cleavage by restriction endonucleases.

a common DNA sequence. In the β globin gene and, by extrapolation, extending 18 kilobases 3′ to the gene, there are three common sequence types (frameworks 1, 2, and 3) in Mediterraneans and three common sequence types (frameworks 1, 2, and 3* Asian) in Asians (Figure 3). In the β gene, two of these common sequence types are present in both groups, whereas one varies by a single nucleotide between the groups. By restriction site analysis, Blacks are also shown to have the three β gene frameworks, but whether or not minor sequence differences exist between the Black frameworks and those of

141

Mediterraneans and Asians has not yet been established. How four different β gene sequence types have attained high frequencies in different racial groups and why these sequence types are so similar between the groups is a question for discussion later.

On the 5' side of the β gene spanning at least 32 kilobases from the ε gene to the 3' side of the ψβ1 gene there are also three common sequence types (Figure 2). These common sequence types are present in both Mediterraneans and Asians, but Blacks have common sequence types different (at least by restriction site analysis) from those sequence types common to the other racial groups. Between these two DNA regions in the β gene cluster (i.e., (1) the β gene and its 3' side and (2) the ε, $^G\gamma$, $^A\gamma$, ψβ1 genes) there is a segment of 11 kilobases within which randomization of the two DNA regions seems to occur (Figure 2). This segment is thought to be a relatively high recombination area, and a possible "recombination hot spot" is now thought to be located around a polymorphic *Hinf*I site 1 kilobase 5' to the β gene. Because the common sequence types of the two DNA regions are combined in a relatively random fashion, recombination within the β globin gene cluster may be restricted to this segment.

β THALASSEMIA MUTATIONS AND THE NORMAL SEQUENCE VARIATION IN THE β GLOBIN GENE CLUSTER

The number of different β^{thal} alleles characterized has been greatly expanded by a new strategy. This strategy depends upon determining the haplotype for each chromosome carrying a β thalassemia gene by using restriction endonuclease analysis (Figure 4). One can then sequence only those mutant genes present in different haplotypes. This strategy has produced a high yield of previously undiscovered mutations (Orkin et al., 1982). The number of different β^{thal} alleles now known is approaching 20. Up to now only Mediterranean peoples and Asian Indians have been studied in detail, and the mutations in each group are unique to that group. After Chinese, Black Africans, and other groups with a high incidence of this disease are studied, we expect that the number of different β^{thal} alleles may be 60 or more. Of course, many different kinds of mutations could reduce gene expression and produce β thalassemia. However, any β^{thal} mutation that occurs in a population could attain high frequency due to the selective advantage of β thalassemia heterozygotes in a malarial environment.

In general, β gene mutations appear to be more recent events than the sequence variation associated with the observed DNA polymorphisms. When the mutations are found associated with a particular common gene framework, they tend to be retained on that framework. However, it is possible for them to lose their association with the

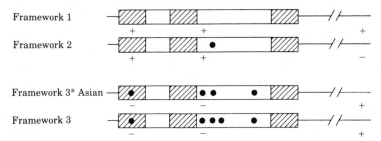

FIGURE 3. Normal β globin gene frameworks. Filled areas are coding regions in the β gene and open areas are intervening sequences. Frameworks 1, 2, and 3 are found in Mediterraneans, and their patterns of polymorphic restriction sites shown as +++, ++−, −−+ correspond to 3' sequence types A, B, and C of Figure 2, respectively. Black dots in frameworks 2 and 3 indicate the locations of single nucleotide differences from the sequence of framework 1. Framework 3* Asian is found only in Asians and lacks one of the nucleotide substitutions found in framework 3. The space between framework 2 and framework 3* Asian is meant to suggest the possible existence of evolutionary intermediates.

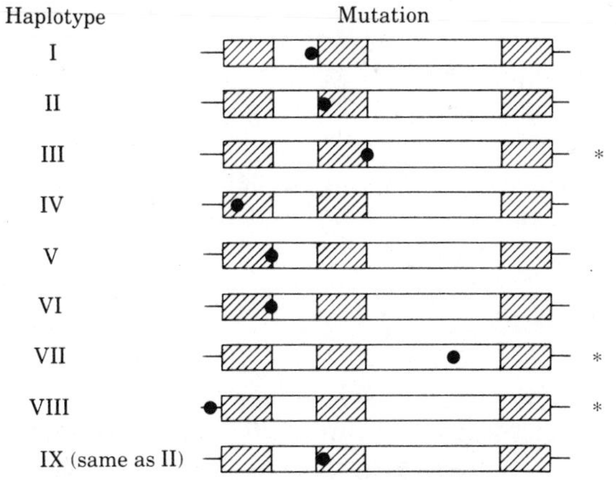

Haplotype | Mutation
I
II
III   *
IV
V
VI
VII   *
VIII   *
IX (same as II)

FIGURE 4. β Thalassemia mutations associated with different haplotypes in Mediterraneans. Nine different haplotypes for seven polymorphic restriction sites were found associated with β^{thal} chromosomes. At least one β gene associated with each haplotype was analyzed, and a different mutation was associated with eight of the haplotypes. Haplotype IX had the same mutation as haplotype II (Orkin et al., 1982). Mutation sites are shown as black dots. Asterisks at the right of haplotypes III, VII, and VIII indicate that these mutations can be detected directly by restriction site analysis.

sequence 5' to the β gene by recombination. Thus, discovery of a gene mutation—say, to β^E—on two or three different β gene frameworks is good evidence that the mutation has had multiple independent origins. Indeed, the β^E mutation has been found on two different β gene frameworks in Southeast Asia (Antonarakis et al., 1982b). On the other hand, if the same mutation is associated with one β gene framework and two different 5' sequence types, it suggests that it originated from a single mutation and subsequently it was recombined with a different 5' sequence type.

This description of the variation in common sequence types and in recombination rates within a gene cluster has implications for the study of genetic diseases. In order to provide a "complete" description of the molecular basis of a genetic disease, the following plan would be useful, given our experience with β thalassemia and hemoglobinopathies. (1) Obtain genomic clones containing the gene in question and neighboring sequences. (2) Discover DNA polymorphisms within and adjacent to the gene, and by family studies find nonrandom associations such as those with only three combinations for two or three restriction site polymorphisms. (3) Study genomic DNA of ten or more patients "homozygous" for the disease in question representing each affected ethnic group to discover the mutant gene on various chromosome backgrounds. (4) Sequence one mutant gene associated with each of these backgrounds to discover different mutant alleles. One might expect that each chromosome background of each ethnic group would be associated with one or more different mutant alleles. If there exist many possible mutations that eliminate protein function or synthesis and if positive selection for heterozygotes is weak (as is expected for most mutations), then many different mutant alleles should be found. Molecular characterization of genetic disease may someday be important in the construction of rational therapy.

NUCLEOTIDE VARIABILITY IN THE β GLOBIN GENE CLUSTER

As mentioned earlier, we call each sequence of restriction sites in the entire β globin gene cluster a haplotype. Nine common haplotypes have been found associated with normal β genes in Mediterraneans, Indians, and Cambodians. A small number of rarer haplotypes have also been found in these groups. The data can be used to determine the proportion of individuals with two different haplotypes in the β globin cluster. In addition, variability in DNA sequences in the region can be estimated.

Haplotype heterozygosity or the probability that two haplotypes chosen at random from a population are different [Nei and Tajima's (1981) nucleon diversity] was determined from seven polymorphic sites

144

for Greeks, Italians, and Indians and for five sites for American Blacks. Ninety-two percent of Greeks, 65% of Italians, 71% of Indians, and 83% of Blacks were found to be heterozygous for the haplotypes.

Nucleotide diversity (π) or the proportion of different nucleotides between two randomly chosen haplotypes (Nei and Tajima, 1981) can be computed from the above polymorphic restriction site data. Under the neutral mutation–genetic drift theory, $\pi = 4N_e\mu$ where $N_e =$ effective population size and $\mu =$ mutation rate per nucleotide. Estimates of π by three different methods are 0.0029 for Greeks, 0.0015 for Italians, 0.0018 for Indians, and 0.0015 for Blacks. The average π for these population groups is approximately 0.002, which suggests that in the intervening and flanking sequence DNA of the two β globin gene clusters of any individual, 1 in every 500 nucleotides is different. Our value is twice that previously computed by Ewens et al. (1981), who used three restriction sites in the same gene cluster. Whether or not this degree of nucleotide variability is representative of other regions of the genome is unknown.

It is also interesting to use the β gene nucleotide sequence data to compute π for the second intervening sequence (IVS2) of the β gene. The frequencies of frameworks 1, 2, and 3 in Mediterraneans are 0.53, 0.28, and 0.19, respectively. π for the 850-nucleotide IVS2 is approximately 0.0017. Similar values are obtained for the IVS2 segments in Indians and Cambodians. Thus, both of our estimates of π—one derived from restriction site haplotypes and the other from nucleotide sequence data—are approximately 0.002.

ORIGIN OF β GENE FRAMEWORKS: SELECTION VERSUS DRIFT

We have described three common β gene frameworks in Mediterraneans. Two of these frameworks are also present in Asians, but framework 3* Asian is intermediate in sequence between frameworks 2 and 3 in Mediterraneans. It is probable that other intermediate frameworks were once common, but they became rare or were lost. The fundamental question is whether these frameworks are maintained by selection or whether they can be explained through genetic drift. Selection is an attractive hypothesis because framework 3 is rather different from frameworks 1 and 2. However, selection is extremely difficult to infer without knowledge of the functional differences of the frameworks. On the other hand, we can show that the observed frameworks and their frequencies are compatible with predictions of neutral theory.

145

Kimura and Crow (1964) have defined the effective number of alleles (n_e) in a population as $n_e = 1/F$, where F is the probability that an individual will be homozygous at the locus in question. For the β gene frameworks in Mediterraneans, $F = 0.395$, so $n_e = 2.53$. The effective number of alleles is equal to the actual number of alleles if all alleles are equally frequent, but otherwise it is smaller. In an equilibrium population, n_e is given by $1 + 4N_ev$, where N_e = effective population size and v = mutation rate. We have previously calculated $\pi = 0.002$ for the frameworks from IVS2 data only. Because $\pi = 4N_e\mu$ (μ = mutation rate per nucleotide) and because IVS2 has 850 nucleotides $v = \mu \times 850$, $4N_ev = 0.002 \times 850 = 1.70$. Therefore, $n_e = 2.70$. Thus, the calculated value (2.70) of n_e is very close to the observed value (2.53) of n_e, suggesting that the β gene frameworks could have arisen by neutral mutation and genetic drift. Similar computations for IVS2 of Indian and Cambodian β genes also show agreement between the two values of n_e. We are only saying that the results are compatible with neutral theory. We have previously stated that selection is an attractive model.

CHAPTER 8

POLYMORPHISM OF MITOCHONDRIAL DNA IN POPULATIONS OF HIGHER ANIMALS

John C. Avise and Robert A. Lansman

INTRODUCTION

The complete nucleotide sequence of the mitochondrial genome has been determined for one individual in each of three species—mouse (*Mus musculus*) (Bibb et al., 1981), cow, and human (Anderson et al., 1981, 1982). Mitochondrial DNA (mtDNA) has also been partially characterized for coding functions, and major features of gene organization are exemplified in Figure 1. Mitochondrial DNA is a closed circular molecule, of length 16,295 base pairs in the mouse. Fully 94% of the molecule encodes functional RNA. Genes for 22 transfer RNAs are interspersed between genes for the 12S and 16S ribosomal RNAs; cytochrome *c* oxidase subunits I, II, and III; ATPase subunit 6; cytochrome *b*; and eight as yet unidentified proteins. Aside from 32 nucleotides scattered about the mouse mtDNA genome, only two significant regions of the molecule lack defined coding function—the 879- and 32-base pair regions around the origins of heavy (D-loop) and light strand replication, respectively. Some adjacent genes exhibit terminal overlap in coding sequence.

Human mtDNA is slightly longer (16,569 base pairs) almost entirely because of an additional 243 nucleotides in the D-loop region. Gene order appears identical to that of the mouse. These results sup-

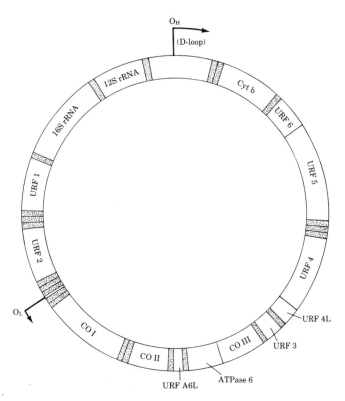

FIGURE 1. Major features of gene organization in mouse mitochondrial DNA. Transfer RNA genes are indicated by stippled regions. CO I, CO II, and CO III are genes for cytochrome oxidase subunits, Cyt b is cytochrome *b*, and URFs are unidentified reading frames. O_H and O_L are origins of heavy and light strand DNA replication. (Modified from Bibb et al., 1981.)

port the general consensus reached for other higher animal mtDNAs not so thoroughly characterized, namely, that mtDNA is remarkably conservative in size, function, and organization (Brown, 1981). The overall sequence divergence between man and mouse mtDNA is approximately 30%, but this is variable across genes. Greatest sequence homology (approximately 70–90%) occurs in the tRNA and rRNA genes, and lowest homologies (approximately 60–70%) are observed at the unidentified protein-coding loci. Portions of the D-loop region also show low homology.

Obtaining these complete mtDNA sequences were monumental efforts. From the perspective of a population geneticist interested in mtDNA sequence polymorphisms among large arrays of individuals, quicker and easier means of sampling nucleotide sites within the mtDNA genome are required. Fortunately, the discovery of Type II

148

restriction endonucleases has permitted the development of such assays. The purpose of this chapter is to summarize conclusions from research on mtDNA polymorphisms within species of higher animals. For a complementary review of between-species differences and longer-term mtDNA evolution, see Chapter 4 by Brown.

Considering the conserved function of mtDNA in higher animals and the exceptional economy of organization of the molecule, one might expect the primary sequence also to be well conserved. This has not proved to be the case: mtDNA may evolve 5 to 10 times faster than single copy nuclear DNA (Brown et al., 1979; Chapter 4 by Brown). Many of the findings about levels and patterns of mtDNA polymorphism within species are equally surprising and would not likely have been predicted from general knowledge of mitochondrial biology. We will organize this review by presenting, one at a time, the major tentative conclusions reached from early restriction enzyme studies of mtDNA. We will critically reexamine each of these conclusions in the light of more recent data and theory, some of which are still unpublished.

RESTRICTION ENZYME ASSAYS

Approximately 350 restriction endonucleases, involving more than 85 distinct recognition specificities, are now known (Roberts, 1982). A given enzyme recognizes a specific oligonucleotide sequence that is four, five, or six base pairs in length and cleaves two phosphodiester bonds within or near the sequence, one in each strand of the duplex. DNA fragments produced by restriction enzyme cleavage can be separated by molecular weight by gel electrophoresis and can be visualized by any of several staining or autoradiographic techniques (Figure 2). In a typical population survey, mtDNA is purified from tissues of individual organisms and then digested separately by each of a battery of restriction enzymes to obtain "digestion profiles" for the mtDNA molecules. Detailed procedures for isolation and assay of mtDNA have been reviewed (Lansman et al., 1981).

Because mtDNA is a closed circular molecule, the number of linear digestion fragments produced by a restriction enzyme is equal to the number of recognition sites in the mtDNA. For a given enzyme, the approximate anticipated numbers of such fragments can be readily calculated if the size and G + C content of the mtDNA is known and if the frequencies of cleavage sequences in mtDNA are similar to those expected in random sequences of the same base composition (Nei and Li, 1979). As generally predicted, "6-base" enzymes produce 1–8 frag-

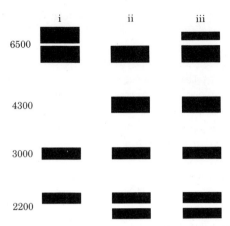

FIGURE 2. Diagrammatic representations of mtDNA "digestion profiles." The pattern in lane iii could result from restriction site heterogeneity or from incomplete digestion. The mtDNAs in lanes i and ii most likely differ by a restriction site that cleaves the 6500-base pair fragment in lane i to the 4300- and 2200-base pair fragments in lane ii.

ments in a typical mtDNA digestion, whereas "4-base" enzymes usually produce as many as 20 or more detectable fragments. Because the potential benefits of increased data from "4-base" enzymes is, in our opinion, offset by increased difficulties of scoring gels and determining fragment homology, for most purposes we prefer to employ 5- and 6-base enzymes (but see Brown, 1980, for a counterargument).

An index to the genetic similarity between mtDNAs from different organisms is the proportion of fragments shared in their digestion profiles. In some cases, further refinement can be obtained by mapping positions of restriction sites relative to one another or to known landmarks of the molecule (Brown and Vinograd, 1974; Nathans and Smith, 1975). Several statistical procedures have been developed to convert raw "fragment" or "site" data to quantitative estimates of nucleotide sequence divergence (p) between mtDNAs (Engels, 1981; Gotoh et al., 1979; Kaplan and Langley, 1979; Kaplan and Risko, 1981; Li, 1981; Nei and Li, 1979; Nei and Tajima, 1981; Upholt, 1977). Qualitative methods of data analysis have also been advocated (Avise et al., 1979a, b; Lansman et al., 1981; Templeton, 1983), as shown later.

BIOLOGY OF MITOCHONDRIAL DNA

Mitochondrial DNA replication and transcription occur within the organelle, autonomously from nuclear DNA. In vertebrate somatic

150

cells, rate of mtDNA turnover is probably higher than that of nuclear DNA (Rabinowitz and Swift, 1970). A vertebrate cell contains many, often thousands, of mtDNA molecules (Birky, 1978; Gillham, 1978; Potter et al., 1975). Mature oocytes are particularly rich in mtDNA. For example, the exceptionally large *Xenopus* egg carries an estimated 10^8 mtDNA molecules (Dawid and Blackler, 1972). The midpiece of mature sperm also carries small numbers (approximately 100) of mtDNA molecules. In some animals, these mtDNAs disperse into zygote cytoplasm during fertilization (Gresson, 1940; Friedlander, 1980), but in other species sperm mitochondria do not penetrate the egg (Ursprung and Schabtach, 1965). The sheer preponderance of egg mtDNA in zygotes can certainly account for the common observation that most mtDNA is maternally inherited.

Because an individual animal contains many billions of copies of mtDNA, the evolutionary dynamics of mtDNA in an animal population are interrelated with and partially dependent upon the underlying dynamics of the population of mtDNA molecules in each animal. Empirical surveys of mtDNA composition usually employ somatic tissue, so understanding of somatic cell mtDNA segregation is important. However, mtDNA dynamics in germ-cell lineages are of particular interest because only here are results directly relevant to mtDNA evolution. Unfortunately, virtually nothing is known about fundamental characteristics of mtDNA dynamics in germ cells, for example, the number of rounds of mtDNA replication during gametogenesis or the effective population sizes of mtDNA in intermediate germ cells. Recent theoretical models point out the need for such information (Chapman et al., 1982; Birky et al., 1982, 1983; Takahata and Maruyama, 1981).

INTRAINDIVIDUAL MITOCHONDRIAL DNA SEQUENCE HOMOGENEITY

Contrary to some earlier expectations, virtually all studies to date have concluded that the huge populations of somatic cell mtDNAs within an individual organism appear homogeneous in nucleotide sequence. The evidence is as follows. In mtDNAs that are digested to completion with a restriction enzyme, the total molecular weight of the fragments produced typically equals the known genome size of higher animal mtDNA, approximately 16,000 base pairs. Significant sequence heterogeneity in a sample would be evidenced by additional fragments exceeding the total anticipated weight. Furthermore, in gels stained with ethidium bromide, the fluorescence of each band in the digestion pattern is proportional to the molecular weight of the fragment. Nonstoichiometric bands should appear in digests of tissue heteroplasmic for mtDNA. Studies listed in Table 1 include a total of more than 3000 different single-restriction-enzyme digests of a dozen mammalian species, without report of a case of intra-individual mtDNA sequence heterogeneity. In addition, a few studies have explicitly compared mtDNAs from two or more different tissues (i.e., kidney, heart, liver, platelets, skin fibroblasts) of the same animal (Denaro et al., 1981; Francisco et al., 1979; Potter et al., 1975). No differences among tissues have been documented.

Critique

The above case for homoplasmic intraindividual mtDNA may appear stronger than it really is. A minority mtDNA sequence would have to constitute at least 1–5% of the total mtDNA in a sample to reach limits of normal empirical detectability, even when fragments are assayed by the more sensitive autoradiographic methods. It is thus conceivable that many rare mtDNA sequences commonly coexist with the predominant sequence in a given tissue but remain undetected. Heterogeneity of mtDNA in cultured mammalian cells has been demonstrated (R. Slott, R. O. Shade, and R. A. Lansman, unpublished).

There is another seldom-acknowledged bias operating against reports of possible heteroplasmicity of mtDNA sequence. Not uncommonly, additional faint mtDNA bands are observed on restriction digest gels. These are usually attributed to "incomplete digestions," presumably resulting from partial loss of restriction enzyme activity, low concentration of enzyme relative to mtDNA, too short an incubation period, or to the known several-fold difference in the kinetics of site cleavage (Thomas and Davis, 1975). Such supernumerary bands usually disappear upon more complete redigestions (but see Potter et al., 1975). However, in routine animal surveys not specifically concerned with the issue of individual mtDNA heterogeneity, gel scoring is often done without redigestion, under the assumption that the faint bands represent mtDNA segments not yet fully cleaved. True mtDNA sequence heterogeneity would produce digestion profiles that would not be readily distinguished from those of incomplete digests (Figure 2, lane iii).

Coote et al. (1979) do report heterogeneity of mtDNA in different tissues from a single ox, but their interpretation is also complicated by the possibility of incomplete digestion. The general problem of intraindividual mtDNA sequence variation needs much more empirical study. Because between-individual mtDNA divergence is extensive (see later), the eventual documentation of at least some individual heteroplasmicity seems inevitable. Such a discovery will be especially

152

TABLE 1. Estimated genetic distances (p, base substitutions per nucleotide) in mtDNAs between conspecific mammals.

Species	Animals surveyed	Geographic source of samples	Restriction enzymes	Approximate $\bar{p}$*	Maximum p	Source‡
Ovis aries	2	Domestics	3	0.02	0.020	1
Capra hircus	3	Domestics	3	0.01	0.010	1
Geomys pinetis	87	Several states, Southeast United States	6	0.02	0.047	2
Peromyscus polionotus	36	Several states, Southeast United States	6	0.01	0.030	3
Peromyscus maniculatus	135	Entire N. Amer. continent	8	0.02	0.070	4
Rattus norvegicus	21	N. Amer.; Puerto Rico; Japan	6	0.01	0.018	5
Rattus rattus	26	N. Amer.; Puerto Rico; East Asia	6	0.04	0.096	5
Mus musculus	≅30	Diverse lab strains; native subspecies worldwide†	10–15	0.03	0.096	6, 7
Pongo pygmaeus	5	Sumatra and Borneo	25	0.02	0.050	8
Pan troglodytes	10	Captives, origin uncertain	25	0.01	0.045	8
Pan paniscus	3	Captives, origin uncertain	25	0.01	0.015	8
Gorilla gorilla	4	Captives, only one subspecies represented	25	0.006	0.009	8
Homo sapiens	21	Three major races, worldwide	18	0.004	<0.01	9, 10

* Usually calculated by the approach of Nei and Li (1979). Mean values are approximate only; actual genetic distances in any given comparison are heavily dependent upon geographic closeness of collection, and individual papers should be consulted for details. Ferris et al. (1982, 1983) also studied mtDNA variation in *M. musculus* and *M. domesticus*, two forms considered subspecies in the work of Yonekawa et al. (1980, 1981).

† Taxonomy of *Mus musculus* is controversial, with some workers recognizing two distinct species.

‡ Source: 1. Upholt and Dawid (1977), 2. Avise et al. (1979b), 3. Avise et al. (1979a), 4. Lansman et al. (1983a), 5. Brown and Simpson (1981), 6. Yonekawa et al. (1981), 7. King et al. (1981), 8. Ferris et al. (1981b), 9. Brown and Goodman (1979), 10. Brown (1980).

153

significant if it occurs in an organism in which the pattern of mtDNA segregation in somatic tissues, and across animal generations, can be experimentally followed. A promising case of a segregating mtDNA lineage has been reported in a herd of cows descended from a common female parent (Laipis and Hauswirth, 1980). As yet, however, no individual cows have been found that simultaneously exhibit both of the mtDNA sequences. S. Hechtel and W. M. Brown (pers. comm.) have also observed mtDNA sequence heterogeneity in batches of un-fertilized eggs from individual sea urchins.

MITOCHONDRIAL DNA POLYMORPHISM WITHIN SPECIES

Table 1 lists the major published studies of mtDNA polymorphism within species. The amount of sequence heterogeneity is striking. Mean values of p (estimated nucleotide sequence divergence) between conspecific individuals commonly range from 0.3 to 4%, and maximum observed p's sometimes approach 10%. Most major studies to date have dealt with mammals, and in the future it will be especially important to survey other groups of animals. Preliminary results from some reptiles, birds, and other groups have been reported (Brown and Wright, 1979; Glaus et al., 1980; Fauron and Wolstenholme, 1980b), and more extensive surveys are underway.

We can briefly summarize results of one typical mammalian study. Nei and Tajima (1981) have suggested the term *nucleomorph* for different restriction-site patterns for a DNA segment such as mtDNA. Among 87 pocket gophers (*Geomys pinetis*) collected from across the species range and assayed by six restriction enzymes, a total of 23 mtDNA nucleomorphs were observed (Avise et al., 1979b). Mean and maximum estimates of sequence divergence between pairs of nucleo-morphs were $\bar{p} \simeq 0.020$ and $p = 0.047$, respectively. Within a given local collection, 1–3 nucleomorphs were observed, and estimates of p were less than 1%.

The only reported exception to extensive within-species mtDNA sequence divergence involves humans (gorillas also exhibit limited mtDNA sequence variation, but only a few individuals of one sub-species were assayed; Ferris et al., 1981b). Among 21 humans of diverse ethnic and geographic origin, $\bar{p} \simeq 0.004$. Brown (1980) raises the possibility that all living humans may have evolved from a small, mitochondrially monomorphic population that existed as recently as 200,000 years ago.

Critique

Some of the larger within-species estimates of p in Table 1 are one-fourth to one-third as great as the evaluated sequence divergence of

154

30% between man and mouse mtDNAs, which have been completely sequenced (Anderson et al., 1981)! Two classes of explanation could account for this surprising result. First, it is conceivable that within-species estimates of p are seriously in error, for some reason biased toward large values. In estimating p from raw fragment or site data, several assumptions are made—that mtDNA polymorphism reflects base substitution differences only; that the frequencies and distributions of cleavage sequences in the mtDNA are similar to those expected in random sequences; and that fragment and site homologies are correctly determined. Justifications for these assumptions have been advanced (Upholt, 1977; Nei and Li, 1979; but see later and also Adams and Rothman, 1982).

Alternatively, if within-species estimates of p are valid, then observed intraspecific polymorphism in mtDNA must not reflect long-term evolutionary rate of accumulation of nucleotide substitution (Upholt and Dawid, 1977). Perhaps some slightly deleterious substitutions contribute to intraspecific polymorphism but do not generally persist over evolutionary time. More likely, certain nucleotides (i.e., in silent positions) may be especially free from evolutionary constraints. Brown and Simpson (1982) report that 94% of the nucleotide substitutions in sequenced portions of the mtDNA genomes from *Rattus norvegicus* and *R. rattus* are silent. Mitochondrial DNA molecules might rapidly become saturated with changes at these positions, whereas the remainder of the molecule evolves much more slowly. Either of these possibilities should result in a curvilinear relationship between p and time. Such empirical curvilinear relationships have been presented by Brown et al. (1979, 1982) and Ferris et al. (1981b).

The issue of the validity of current estimates of p can be finally resolved by use of actual nucleotide sequence data, which for large portions of the mtDNA molecule are now being obtained for conspecific organisms. Complete nucleotide sequences are available for a 900-base pair region surrounding the mtDNA D-loop region in seven humans (Greenberg et al., 1982). Among pairwise comparisons of these seven mtDNAs, evaluated levels of p range from 0.004 to 0.032 ($\bar{p} = 0.017$). If these same regions of mtDNA had, in fact, been accurately mapped for restriction sites by approximately 25 commercially available restriction enzymes, summary estimates of p for the region would have ranged from 0.006 to 0.034 ($\bar{p} = 0.020$) (Aquadro and Greenberg, 1983). This strong agreement between restriction site estimates of $\bar{p}$ and sequence evaluation of $\bar{p}$ suggests that current estimates of mtDNA polymorphism within species are not grossly biased toward larger values by the restriction enzyme approach. [Levels of $\bar{p}$ for this 900-base pair region in humans are several-fold higher than estimated nucleotide differences for the whole human mtDNA genome (Brown, 1980); presumably this is due to the hypervariability of portions of the D-loop region of the molecule (see Aquadro and Greenberg, 1983).]

In any event, mtDNA polymorphism within species is clearly extensive. The joint observations of common individual homoplasmicity and between-individual sequence divergence can be accommodated under a neutral model involving chance sampling drift of mtDNA nucleomorphs in germ-cell lineages. Chapman et al. (1982) provide computer simulations of how the process may occur. In their models, each germ-cell lineage has a fixed number (n_M) of mtDNA molecules, each of which can mutate at a prescribed rate μ each cell generation. Multinomial sampling of mtDNA molecules occurs in heteroplasmic germ cells; and in every animal generation (assumed to be 50 germ-cell generations), each adult female produces female progeny according to a Poisson distribution with mean equal to 1. In the simulations, H_1 (the probability that two randomly drawn mtDNA sequences from the same individual are different) and H_2 (the corresponding probability for the population) are monitored through time: some populations are initiated with all individuals homoplasmic and identical (Pop. A); and some populations are initiated with each individual homoplasmic for a unique mtDNA sequence (Pop. B). Thus, H_1 and H_2 register within- and between-individual mtDNA sequence heterogeneity, respectively, and their ratio is of particular interest.

Figure 3 shows two typical runs of the simulation. As expected, H_2 in populations A and B converges due to the decay of heterogeneity by drift, balanced by the accumulation of variation by mutation. H_1 is appreciably lower with smaller n_M (Figure 3). Both H_1 and H_2 are lower with lower μ, all else being equal (Chapman et al., 1982). Overall, H_1 remains extremely low and the ratio H_2:H_1 high only when $n_M\mu$ is small. Thus, the models predict that effective population sizes of mtDNA molecules in germ-cell lineages should be small (i.e., 10–100), and they point out the need for empirical data in this area. Census sizes of mtDNAs in mature oocytes are huge, but this may be misleading.

Alternatively, if mtDNA population sizes remain consistently large in germ-cell lineages, some model involving selection might be required. The selection pressure would, however, have to be of a rather peculiar nature because it would require active conservation of within-individual sequence homogeneity while still allowing between-individual differences to accumulate rapidly. Some form of intense intracellular competition among mtDNA nucleomorphs is perhaps conceivable. Another related possibility is that, as in yeast, random choice of a small number of mtDNA molecules for replication may cause cells to move rapidly toward homoplasmicity (Birky et al., 1983).

156

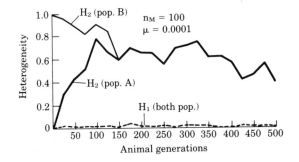

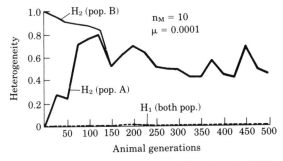

FIGURE 3. Examples of computer-simulated levels of mtDNA sequence heterogeneity for populations A (initiated with all individuals homoplasmic for same sequence) and B (initiated with each individual homoplasmic for unique sequence). H_1 and H_2 register within- and between-individual mtDNA sequence heterogeneity, respectively. n_M is the number of mtDNA molecules per germ cell, and μ is the probability of mutation per mtDNA molecule per germ-cell generation. (From Chapman et al., 1982.)

GENETIC BASIS OF MITOCHONDRIAL DNA POLYMORPHISM

Mitochondrial DNA ranges in size from 15,700 to 17,700 base pairs among assayed species as diverse as sea urchins, amphibians, reptiles, birds, and mammals (Brown, 1981). Among species of *Drosophila*, the size range is even greater (15,600–19,400 base pairs), and as in other animals most of the length differences occur in the region surrounding the origin of heavy strand replication (Fauron and Wolstenholme, 1976; Shah and Langley, 1979b). Within species, some mtDNA size variation has also been reported (Brown, 1980; Brown and Simpson, 1981; Fauron and Wolstenholme, 1980b). For example, one geographic population of lizards (*Cnemidophorus sexlineatus*) differs from others by an additional 1200 base pairs (Brown, 1981). Genome size variation provides straightforward evidence that additions and/or deletions have played a role in mtDNA evolution.

Nonetheless, the consensus is that the majority of mtDNA polymorphism within species is attributable to gain or loss of particular restriction sites, without detectable alteration of genome size. For example, all differences among a total of 61 mtDNA nucleomorphs observed in geographic samples of the mouse *Peromyscus maniculatus* resulted from changes of individual restriction sites that were mapped on a molecule that appeared constant in size (Lansman et al., 1983a). Particularly within coding regions of genes, restriction site changes are far more likely to arise from base substitutions than from small nucleotide additions or deletions, most of which would alter the reading frame.

Critique

An alternative explanation for apparent gain or loss of restriction sites is differential methylation of cytosine in mtDNA from different animals. Methyl-modified cytosine (5-MeCyt) is the only modified base known in vertebrate DNA, and it protects DNA from cutting by some restriction enzymes (Singer et al., 1979; Bonen et al., 1980). However, several lines of evidence argue that differential methylation does not account for mtDNA polymorphism. First, the maximum amount of base methylation in mouse or hamster mtDNA is very low (<1%), far less than the 21% C content of mouse mtDNA (Nass, 1973; Brown and Goodman, 1979; Singer et al., 1979). Second, restriction enzymes lacking the dinucleotide C-G in their recognition sites uncover at least as much mtDNA restriction site polymorphism as do enzymes that contain it (Brown, 1980). Third, mtDNAs cloned in *Escherichia coli* yield the same fragment patterns as do native mtDNAs when digested with several restriction enzymes (Castora et al., 1980; Chang et al., 1975), despite the fact that cloning has been observed to increase probabilities of altered methylation (Gautier et al., 1977).

Direct confirmation of the common occurrence of base substitution in mtDNA polymorphism has come from recent sequencing studies. From the mtDNAs of each of five humans, Greenberg et al. (1982) cloned and sequenced an approximately 900-base pair noncoding region surrounding the origin of H-strand replication; sequences were further compared to those of three other human mtDNAs previously sequenced (Anderson et al., 1981; Crews et al., 1979; Walberg and Clayton, 1981). Four instances of mtDNA length variation were revealed among the eight samples, each involving additions or deletions of one or two nucleotides. All of the remaining 50+ sequence altera-

158

tions (>90%) were due to simple base substitutions (Greenberg et al., 1982; Aquadro and Greenberg, 1983). The percentage of mtDNA polymorphism attributable to base substitution may be even higher for the remainder of the mtDNA genome, most of which has active coding function. Limited sequence analysis of four cloned *Peromyscus* mtDNA molecules reveals no additions or deletions outside the D-loop region (J. F. Shapira et al., unpublished).

All available data from closely related populations and species have revealed a great preponderance of transition over transversion substitutions (Aquadro and Greenberg, 1983; Brown et al., 1982; Brown and Simpson, 1982). For example, in the study of Greenberg et al. (1982), 53 of 55 substitutions (96%) were transitions. Furthermore, the distribution of variable sites was highly nonrandom, with the majority of site changes occurring in two regions roughly flanking the D-loop.

MATERNAL INHERITANCE OF MITOCHONDRIAL DNA

In higher animals, mtDNA appears to be inherited maternally—transmitted to progeny through egg cytoplasm. This conclusion stems from examination of mtDNA in progeny of crosses between parents differing in mtDNA restriction fragment pattern. For example, horse and donkey mtDNAs differ in position of *Hae*III restriction sites. Hinnies (progeny of crosses between male horse and female donkey) exhibit the donkey mtDNA pattern, and mules (progeny of the reciprocal cross) exhibit the horse pattern (Hutchison et al., 1974). Similar evidence for maternal inheritance has been reported in experimental crosses of *Peromyscus* (Avise et al., 1979a), *Rattus* (Francisco et al., 1979; Hayashi et al., 1978), *Xenopus* (Dawid and Blackler, 1972), humans (Giles et al., 1980), and others.

Critique

As already mentioned, an mtDNA sequence would have to constitute at least 1–5% of total mtDNA to be noticed in conventional restriction assays. Thus, any "paternal leakage" (*P*, the proportion of progeny mtDNA derived from the immediate father) less than this amount would normally escape detection. The question of paternal leakage is important, because even very low *P* can have important evolutionary consequences as a "gene-flow bridge" between female lineages otherwise completely isolated from one another with respect to mtDNA sequence (Chapman et al., 1982).

In an attempt to increase power to detect low-level P, Lansman et al. (1983b) utilized a uniquely favorable backcross strain of tobacco budworms (*Heliothis*). Two budworm species that differ in *Xba*I and *Eco*RI digestion profiles had been hybridized, and the fertile female progeny successively backcrossed to males of one of the species for 91 consecutive generations. In insects, sperm mitochondria have been observed within the zygote, but their fate remains unknown (Friedlander, 1980; Chapman et al., 1982). If effective paternal leakage occurs in *Heliothis* at constant rate P per generation, the total accumulated proportion Q of paternal mtDNA expected in backcross generation i is $Q_i = 1-(1-P)^{i+1}$. Thus, for P as low as 1 molecule in 10,000 per generation, $Q_{91} \cong 1\%$, a level that would be detectable by the autoradiographic assay employed. Nonetheless, absolutely no paternally derived mtDNA was detected in these 91-generation backcross progeny.

Thus, all available information is consistent with strict maternal inheritance of mtDNA. Given the small ratio (S/E) of sperm to egg mtDNAs in zygotes, this result is not too surprising. However, if sperm mtDNAs in the zygote do survive and replicate at rates equal to those of egg mtDNAs, and if random germ-cell lineage drift of mtDNAs does occur as envisioned in the models of Chapman et al. (1982; see also preceding discussion on mtDNA polymorphism within species), then a tiny fraction (S/E) of gametes of the following generation should carry only paternally derived mtDNA. It has also been suggested, however, that sperm mtDNAs are actively altered or degraded and not utilized or transmitted through the zygote. Some evidence for physical alteration of chloroplasts and mitochondria during microsporogenesis in higher plants lends support to this conjecture (Vaughn et al., 1980).

ESTIMATION OF MATRIARCHAL PHYLOGENY

The observed high levels of within-species mtDNA polymorphism provide opportunities for reconstructing evolutionary relationships among conspecifics. If mtDNA is indeed strictly maternally inherited, the phylogenies derived from either mtDNA restriction fragment or site data will represent estimates of matriarchal relationship. Furthermore, because mtDNA genotypes are clonally transmitted (barring mutation) from female to progeny and are not recombined during sexual reproduction, the mtDNA genotype of an individual organism can in theory provide definitive information about the female lineage to which it belongs (Lansman et al., 1981). Attempts to capitalize upon these unique advantages of mtDNA genotypes have only begun (Avise et al., 1979a,b; Brown, 1980; Brown and Simpson, 1981; Lansman et al., 1981, 1983a; Templeton, 1983; Yonekawa et al., 1981).

One general method of data analysis involves manipulation of a quantitative matrix of estimates of nucleotide sequence divergence (p) between mtDNA nucleomorphs (clones). Many different analyses can be performed upon any such matrix, and the method of choice may depend on goals of the analysis. Thus, one could assume that evolutionary rates are roughly constant in all lineages, in which case phenetic clustering procedures might be employed (Sneath and Sokal, 1973). Or one could search for a network whose branch-lengths connecting clones most closely match distances in the matrix from which the network was formed (Farris, 1972; Prager and Wilson, 1978). Some of these approaches have been applied to mtDNA data (Avise et al., 1979a,b; Lansman et al., 1983a; Yonekawa et al., 1981). However, it should be realized that estimates of p involve a large sampling variance when the number of restriction sites assayed is fairly small (Nei and Tajima, 1981; see also Engels, 1981). For example, a typical survey involving five to ten 6-base restriction enzymes and a total of approximately 50 restriction sites might yield an estimate of $p = 0.02$ between two particular mtDNA clones. The standard deviation of this estimate is approximately 0.014 (from formula 19, Nei and Tajima, 1981; Li, 1981). Thus, in population surveys of the size typically conducted to date, many estimates of p between mtDNA clones within a species are not significantly different from one another.

A much stronger method of analysis takes advantage of the qualitative relationships among particular fragment patterns or restriction maps (Lansman et al., 1981). For example, mtDNA restriction maps can often be arranged into a transformation series that reflects the probable steps along which restriction patterns were interconverted during evolution. Figure 4 shows the probable transformations among the assemblage of 11 *EcoRI* restriction maps observed in a study of 135 *Peromyscus maniculatus* collected across North America (Lansman et al., 1983a). In this study, similar networks were likewise developed from the restriction maps for each of seven other endonucleases, and a final composite network was constructed which simultaneously incorporated all data.

For conspecific populations, the composite phylogenetic network (derived solely from the genetic data) can subsequently be superimposed over the geographic sources of collections. In the two cases where this approach has been attempted, the results yielded highly plausible geographic patterns of mtDNA clonal relationships. Thus, both within *Peromyscus maniculatus* (Lansman et al., 1983a) and *Geomys pinetis* (Avise et al., 1979b), mtDNA clones that appear closely related genetically are also usually geographically contiguous. Rare or local

161

clones are usually related to the widespread clone of that geographic region by one or two assayed site changes. In each species, major genetic "breaks" in mtDNA composition separate clonal assemblages that occupy different portions of the total species range.

Critique

Attempts to infer phylogeny from mtDNA genotypes could be compromised if homoplasy (the evolutionary convergence or reversal of restriction sites) were common. In qualitative networks, homoplasy would have the effect of introducing ambiguity into placement of some clones and would result in total network lengths that exceed the observed minimum mutation distances between clones. By both of these criteria, homoplasy was observed to be fairly common in the extensive mtDNA data for *Peromyscus maniculatus* (Lansman et al., 1983a).

One straightforward approach to document site convergence involves joint examination of restriction maps for pairs of enzymes. Suppose, for example, that *Eco*RI yields restriction map patterns arbitrarily labeled A and B, and *Hinc*II produces patterns C and D.

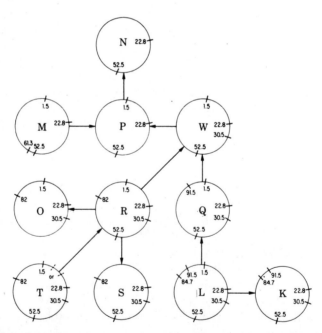

FIGURE 4. Probable evolutionary transformations among the 11 *Eco*RI restriction maps observed in *Peromyscus maniculatus*. Arrows indicate direction of site loss and not necessarily direction of evolution. (From Lansman et al., 1983a.)

162

Further assume that in a collection of mtDNA clones, all four two-enzyme genotype combinations (AC, AD, BC, BD) were observed. Barring recombination, at least one instance of evolutionary convergence must be invoked to account for this "dilemma," no matter which genotype is considered ancestral or where the four genotypes are placed in a larger phylogenetic network. In the *P. maniculatus* data, several two-enzyme dilemmas were observed. However, each such dilemma involved pairs of restriction maps that were closely related, usually by a single site change. Thus, homoplasy was certainly bounded, and no single-enzyme restriction map appeared to have arisen independently from other restriction maps that were not genetically close to it.

Analysis of the *P. maniculatus* data also revealed a few instances of sites that appeared to "blink" on and off repeatedly during evolution. For example, the presence or absence of one particularly hypervariable *Hinc*II site was all that distinguished members of each of four pairs of *Hinc*II restriction maps that were otherwise genetically very distinct from each other (Lansman et al., 1983a).

Significant homoplasy was further confirmed in recent analyses of nucleotide sequences in human mtDNAs. A minimum-length phylogenetic tree, constructed from seven sequences each approximately 900 nucleotides long, required convergent base substitutions at five of the 45 nucleotide replacement sites (11%) (Aquadro and Greenberg, 1983). This relatively high level of homoplasy is apparently attributable to the hypervariability of particular areas of the mtDNA molecule flanking the D-loop region and to the empirical preponderance of transitional substitutions. The observations that rates of base substitution vary considerably along the mtDNA molecule and that most substitutions are transitions will have to be taken into account in future mathematical or statistical models of mtDNA sequence evolution.

CONCLUSION

Several findings on mtDNA polymorphism within species of higher animals were unanticipated: the high level of polymorphism itself; the apparent rarity of individual heteroplasmicity; the attribution of most polymorphism to base substitution; the preponderance of transitions; and the frequency of convergent site evolution. As stressed in this chapter, by stringent criteria some of these conclusions are inadequately documented. Nonetheless, a general outline for evolution of higher-animal mtDNA is now apparent.

163

A further recent surprise is the different scenario emerging for plant mtDNA (Sederoff et al., 1981). For example, maize mtDNA is approximately 30 times larger than most animal mtDNA and consists of at least seven classes of molecules that vary in size and abundance. Sederoff et al. (1981) conclude that in maize and teosinte mtDNA sequence homology is generally conserved and most polymorphism is attributable to major reorganizations of sequence. The evolutionary significance of the different pattern of plant mtDNA polymorphism is not known.

Recent studies suggest that the level of mtDNA polymorphism in higher animals is several-fold higher than that of single-copy nuclear DNA (Engels, 1981; Nei and Li, 1979; but see Shah and Langley, 1979b). Some of these conclusions resulted from indirect comparisons of protein-electrophoretic data with data from mtDNA restriction digests. Because restriction enzyme approaches can also be applied to measure polymorphism in the nuclear genome (provided homologous sequences of appropriate size can be isolated and cloned from different animals), direct answers to the question of relative levels of mtDNA versus nuclear DNA polymorphism should be forthcoming (Engels, 1981). Whatever the final outcome, mtDNA polymorphism will continue to be of special interest in its own right. The maternal inheritance and high polymorphism of mtDNA provide unique opportunities for population analysis.

CHAPTER 9

GENETIC POLYMORPHISM AND THE ROLE OF MUTATION IN EVOLUTION

Masatoshi Nei

INTRODUCTION

In the last several decades the balance theory of evolution (Dobzhansky, 1955) has remained a dominant force in evolutionary biology. According to this theory, genetic variability in natural populations is maintained mainly by balancing selection, and this variability is used as the material for future evolution when the environment changes. Although mutation is regarded as the ultimate source of genetic variation, its role in evolution is considered to be minor. This is because mutation occurs repeatedly at the phenotypic level and most natural populations seem to carry sufficient genetic variability, so that almost any genetic change can occur by natural selection whenever the change is needed. Namely, natural selection plays a creative role (e.g., Mayr, 1963; Dobzhansky, 1970). This is in sharp contrast to the mutationism held by Morgan (1925, 1932) (not de Vries') or the classical theory maintained by Muller (1929, 1950). In these theories, natural selection plays a less important role than mutation, and its chief role is to preserve useful mutations and eliminate unfit genotypes (purifying selection); the creative role is given to mutation.

In the 1960s and 1970s, spectacular progress occurred in molecular biology, and this progress made it possible to study evolution at the most fundamental level, that is, at the codon (amino acid) or nucleotide level. This development led to many new discoveries in evolutionary

165

biology. Two of the most important discoveries are (1) that there is approximate constancy of the rate of amino acid substitution in each protein (Zuckerkandl and Pauling, 1965) and (2) that there is a large amount of genetic polymorphism at the protein level in many natural populations (Harris, 1966; Lewontin and Hubby, 1966). These discoveries have led a number of authors to emphasize the importance of mutation in evolution. Particularly, Kimura (1968a) and King and Jukes (1969) proposed the neutral theory, in which evolution occurs mainly by random fixation of neutral or nearly neutral mutations. Ohta (1973, 1974) extended this hypothesis to the case of slightly deleterious mutations. Nei (1975, 1980) suggested that mutation is the primary force of evolution even for morphological and physiological characters.

This new form of mutationism or neoclassical theory (Lewontin, 1974), particularly the neutral theory, has generated many controversies among evolutionary biologists in the last decade, and many biologists are still skeptical of the validity of this view. At the same time, a large amount of evidence has accumulated on both polymorphism and long-term evolution at the molecular level, so that we can examine the mechanism of evolution more objectively. In this chapter, I shall review the recent progress in the statistical study of protein and DNA polymorphisms and discuss its implications for the general theory of evolution. Specifically, I shall argue that mutation plays a much more important role in evolution than many evolutionists believe. In this review special attention will be given to the consistent explanation of polymorphism and long-term evolution. As emphasized by Kimura and Ohta (1971b), the polymorphism we observe in current populations is a snapshot picture of long-term evolution. Therefore, any viable theory for explaining polymorphism must be able to explain long-term evolution as well.

PROTEIN POLYMORPHISM

When Harris (1966) and Lewontin and Hubby (1966) reported their discovery that natural populations are highly polymorphic at the protein level, many population geneticists attempted to explain the polymorphism in terms of overdominant selection or truncation selection with overdominant gene action (e.g., Sved et al., 1967; King, 1967; Milkman, 1967). They thought that such a high degree of polymorphism could not be maintained without some kind of balancing selection. The following year, Kimura (1968a) attempted to explain the level of protein polymorphism in terms of the neutral mutation theory, but he was not very successful because the mathematical theory of neutral mutations was not well developed at that time (see also Crow, 1968). In the last 14 years, however, the mathematical theory has been refined extensively (Kimura and Ohta, 1971b; Nei, 1975; Ewens, 1979), and we can now examine the agreement between observed data and neutral predictions. Furthermore, substantial progress has occurred in the mathematical theory of multiallelic mutations with selection, so that we can study several alternative hypotheses.

Extent of polymorphism

Probably the most appropriate measure of protein polymorphism is the average heterozygosity over all loci examined. Heterozygosity (gene diversity) at a locus is defined as $h = 1 - \Sigma x_i^2$, where x_i is the frequency of the ith allele. Average heterozygosity ($\hat{H}$) is simply the average of h over all loci. In the theory of neutral mutations, the expectation of $\hat{H}$ in an equilibrium population is given by

$$H = 4N_e v/(1 + 4N_e v) \tag{1}$$

where N_e and v are the (species, not local) effective population size and the rate of neutral mutations per generation, respectively (Kimura and Ohta, 1971b). For neutral alleles, the rate of gene substitution is equal to the mutation rate (Kimura, 1968a). Therefore, under the "null" hypothesis of neutral mutations we can estimate v from the rate of amino acid substitution in proteins. Using this method and considering the molecular size of the proteins (which are studied by electrophoresis), Kimura and Ohta (1971b) and Nei (1975) have estimated that the rate of neutral mutations for electrophoretically detectable alleles is approximately 10^{-7} per locus per year. Therefore, we can compute v if we know the generation time. Estimation of N_e is not easy, but it is possible to make crude estimates of actual population size in certain species. The actual size, N, is usually larger than N_e, so that H obtained by using N instead of N_e in Equation (1) would give the upper limit of the expected heterozygosity. Therefore, if the neutral theory is valid, the observed heterozygosity ($\hat{H}$) is expected to be equal to or lower than H.

With this strategy, M. Nei and D. Graur (unpublished) examined the relationship between $\hat{H}$ and H for various species, including *Escherichia coli, Drosophila*, fishes, reptiles, and mammals. In this study we included only those species for which an estimate of N was obtainable and for which there were gene frequency data for at least 20 loci studied by electrophoresis. Estimates of N were obtained by census (e.g., human, Japanese macaques) or by the multiplication of local population density by the geographic distribution. The results obtained are presented in Figure 1. It is clear that in all species except

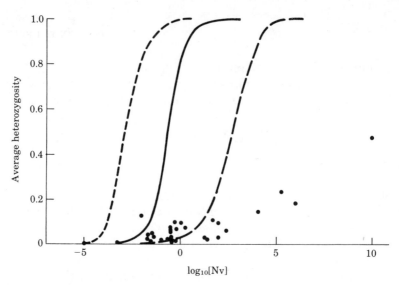

FIGURE 1. Relationship between average heterozygosity (gene diversity) and Nv for 30 species (•). The mutation rate (v) is assumed to be 10^{-7} per locus per year. Solid line: Expected relationship for neutral alleles. Broken line: Expected relationship for slightly deleterious alleles with a mean selection coefficient of $\bar{s} = 0.002$. This curve was obtained by using Kimura's (1979) formulae (23) and (24) with $\beta = 0.5$. Dashed line: Expected relationship for overdominant alleles with $s = 0.001$. The curve for overdominant alleles does not change appreciably even if s varies from homozygote to homozygote as long as the mean of s remains as 0.001.

one the observed heterozygosity is lower than the expected value (solid line), and thus the data can be accommodated with the neutral theory in most species. The exceptional species in Figure 1 is *Drosophila engyochracea*, which is believed to inhabit a very small area of Hawaii Island (Fontdevila and Carson, 1978). However, H. L. Carson (pers. comm.) recently collected specimens of this species in far removed locations of Hawaii Island, and thus the higher value of $\hat{H}$ than H in this species may be due to our underestimation of N.

An important message from Figure 1 is that although many selectionists tend to believe that the genetic variability of natural populations is too high to be neutral, actually it is too low compared with the neutral expectation under the assumption of equilibrium population. It is therefore clear that to explain the level of protein polymorphism we must consider the factors that reduce genetic variability. Overdominant selection or other types of balancing selection are clearly inadequate, because they tend to increase heterozygosity (see Nei, 1980, for other problems). The dotted line in Figure 1 represents the expected heterozygosity when overdominant selection with a selection coefficient of $s = 0.001$ is operating (Li, 1978; Maruyama and Nei, 1981). Although the selection coefficient is very small, the expected heterozygosity rises very rapidly as Nv increases and is generally much higher than that for neutral mutations. It should also be noted that the highest gene diversity so far observed is from *E. coli*, where overdominant selection cannot occur because of haploidy (Selander and Levin, 1980). [In a similar study, Soulé (1976) claimed that in some groups of organisms the observed heterozygosity is higher than the neutral expectation. However, he apparently used local effective sizes rather than species effective sizes for this group of organisms (M. E. Soulé, pers. comm.) and did not pay attention to long-term evolution. Therefore, his conclusion is unjustified.]

Although the observed heterozygosity is certainly lower than the expected, the difference between them is very large when Nv is large. Under the framework of neutral theory, this can be explained by the following two factors. First, a high value of Nv occurs mainly for small organisms such as *Drosophila* and *E. coli*, and in these organisms the effective population size is expected to be much smaller than the actual size because of frequent extinction and replacement of colonies (Maruyama and Kimura, 1980). Second, average heterozygosity is affected drastically by the bottleneck effect, and this effect is expected to last for a long time in large populations—often millions of generations (Nei et al., 1975). Actually, in the Pleistocene (10^4–2×10^6 years ago), there were several glaciations, and in these periods many organisms apparently went through bottlenecks. It is known that more than 60% of mammalian species became extinct and many new species appeared in the Pleistocene (Kurtén and Anderson, 1980). Therefore, the difference between $\hat{H}$ and H in Figure 1 can be explained by the assumption that the long-term effective size is much smaller than the actual size.

There are at least two other explanations for the finding that the observed heterozygosity is lower than the neutral expectation. One is random fluctuation of selection intensity. There are many mathematical models for this type of selection (see Karlin and Levikson, 1974; Gillespie, 1978), but in finite populations random fluctuation of selection intensity generally reduces genetic variability (Nei and Yokoyama, 1976; Takahata, 1981). The other is Ohta's (1974) hypothesis of *slightly* deleterious mutations, which postulates that the majority of polymorphic alleles are not really neutral but slightly deleterious and because of these mutations there is an upper bound for average heterozygosity. [Note that in most organisms many definitely deleterious mutations occur every generation, but they are quickly eliminated from the population without contributing to polymorphism or gene substitution. Therefore, these mutations are not considered in the neutral theory. Ohta's slightly deleterious mutations are supposed to have very small selection coefficients and contribute to polymorphism and gene substitution.] Extending Ohta's model, Li (1978) and Kimura (1979) developed a mathematical formula for the relationship between H and $N_e v$ for slightly deleterious genes. This relationship for the case of $s = 0.002$ (selection coefficient against deleterious alleles) is presented in Figure 1. It shows that average heterozygosity is certainly reduced in the presence of slightly deleterious genes; however, when $N_e v$ is large, the expected value is again much higher than the observed value. Therefore, without invoking the bottleneck effect, this model cannot explain the observed data. Furthermore, when $N_e v$ is small, the expected heterozygosity is significantly lower than the observed value in many species. Thus, this hypothesis seems to be less satisfactory than the neutral theory (see Nei, 1980, for other problems in this hypothesis).

Internal consistency of population parameters

It should be remembered that in the above test we have not really estimated the *effective* population size, and thus our conclusion is only qualitative. However, there are several other tests of the neutral theory and its alternatives where estimates of N_e are not required. In these tests the internal consistency of various population parameters for a particular theory to be tested is examined. For example, the distribution of single-locus heterozygosity (h) for neutral genes is a function of $4N_e v$ only. Therefore, if we know $M \equiv 4N_e v$ rather than N_e and v separately, we can compute the theoretical distribution of h and compare this with the observed distribution. An estimate ($\hat{M}$) of M can be obtained from average heterozygosity by using Equation (1), that is, $\hat{M} = \hat{H}/(1 - \hat{H})$. Using this method, Nei et al. (1976) and Fuerst et al. (1977) examined the agreement between the expected and observed distributions of h for 68 species. These studies showed that the observed distribution does not deviate significantly from the expected distribution under the neutral theory. Some examples are shown in Figure 2. It is noted that if the majority of alleles show overdominance, single-locus heterozygosity is expected to be either 0 or very close to 0.5 when H is approximately 0.1 (Maruyama and Nei, 1981). None of the species examined showed this pattern. We also tested whether the relationship between the mean and variance of single locus heterozygosity agrees with the neutral expectation or not. This test also did not reject the "null hypothesis" of neutral mutations, except in a few species.

In another study we examined the distribution of allele frequencies

170

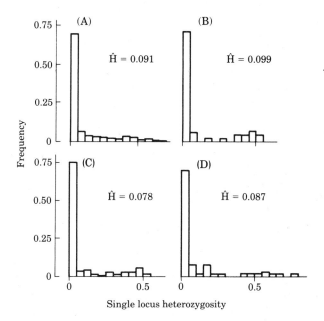

FIGURE 2. Distributions of single locus heterozygosities for man (B), the house mouse (C), and the *Drosophila mulleri* species group (D). The theoretical distribution (A) was obtained by computer simulation with the assumption of $H = 0.091$. (After Nei et al., 1976.)

or frequency spectrum (Chakraborty et al., 1980). This study has shown that in all species examined the observed distribution is U-shaped and is in approximate agreement with the neutral expectation. The theoretical distribution for neutral alleles is given by

$$\Phi(x) = M(1 - x)^{M-1}x^{-1} \qquad (2)$$

(Kimura and Crow, 1964), where x stands for the allele frequency. Figure 3 shows one example obtained from *Drosophila engyochracea*. Clearly, the observed distribution (A) is very close to the neutral expectation (B) but quite different from the expected distribution for overdominant alleles (C), which was obtained under the condition that the expected heterozygosity is equal to the observed value, that is, $\hat{H} = 0.127$. Although the observed distribution was always U-shaped, there was an excess of rare alleles (with frequencies of less than 0.01) in a substantial number of species. Ohta (1977b) took this as evidence for her hypothesis of slightly deleterious mutations. However, an ex-

171

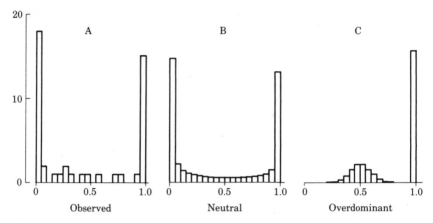

FIGURE 3. Observed and expected distributions of allele frequencies. A. Observed distribution for *Drosophila engyochracea* ($\hat{H}$ = 0.127). B. Expected distribution for neutral alleles with $\hat{H}$ = 0.127. C. Expected distribution for overdominant alleles with $\hat{H}$ = 0.127.

cess of rare alleles is also expected to occur as a result of the bottleneck effect (Chakraborty et al., 1980).

In addition to the intrapopulational parameters, we have also examined interpopulational parameters such as the distribution of genetic distance, the mean and variance of genetic distance, and the correlation of single locus heterozygosities between related species (Chakraborty et al., 1978). The results of these studies again showed that the observed data do not deviate far from the neutral expectations. [Skibinski and Ward (1982) recently claimed that their data on the relationship between genetic distance and heterozygosity cannot be explained by mutation and genetic drift alone. In our view (M. Nei and R. Chakraborty, unpublished), however, one of their assumptions is unwarranted; and if we remove this assumption, their data are again consistent with the neutral theory.] We can therefore conclude that the pattern of protein polymorphism within and between populations is in approximate conformity with the neutral expectation but substantially deviates from that for overdominant selection. Of course, the statistical methods so far used are generally very crude and would not detect small differences in prediction between alternative hypotheses. This is particularly so when selection coefficients are extremely small. However, it now seems clear that the population dynamics of protein polymorphism is largely controlled by the stochastic factors whether some weak selection is involved or not (see also Chapter 5 by Selander and Whittam).

In the past two decades many authors have argued for various

172

types of selection involving ecological factors. Examples are Levins' (1968) hypothesis of adaptive strategy and Nevo's (1978) specialist–generalist theory. However, these types of selection are generally ill-defined, and it is very difficult to study a corresponding mathematical model and derive any testable predictions analogous to those for neutral mutations. One common feature for these hypotheses is that genetic variability is actively maintained by selection and thus leads to a heterozygosity higher than the neutral expectation. As mentioned earlier, however, actual data indicate that the observed heterozygosity is lower than the neutral expectation in an equilibrium population. Therefore, these types of selection do not appear to be very important.

A number of workers (e.g., Tracey et al., 1975; Singh and Zouros, 1978) have reported that in marine mollusks the frequencies of heterozygotes at some enzyme loci are lower than the Hardy-Weinberg proportions in juvenile stage but that as the organisms grow the deviations from Hardy-Weinberg proportions gradually diminish. These observations have led some authors to believe that overdominant selection is operating in the developmental process. However, the average heterozygosities for these organisms are not particularly high, though their population sizes are very large; and thus, it is unlikely that overdominant selection is operating at the entire fitness level (viability plus fertility).

DNA POLYMORPHISM

In the study of evolution, DNA sequences are much more informative than protein sequences because a large part of DNA sequences are not encoded into protein sequences and there is degeneracy of the genetic code. Thus, genetic variation in the noncoding regions of DNA (introns, flanking regions, etc.) or silent nucleotide substitutions can be studied only by examining DNA sequences. Examination of DNA sequences also reveals detailed information on the mechanisms of deletion, insertion, unequal crossing over, transposition of genes, gene conversion or even horizontal gene transfer (Busslinger et al., 1982). Unfortunately, however, techniques for studying DNA sequences were developed only recently, so that data on DNA polymorphism are still scanty compared with those on protein polymorphism. Nevertheless, many interesting results have already been obtained. In the following sections, I shall consider only those studies that are directly related to our problem.

Extent of polymorphism

For measuring the extent of DNA polymorphism, the average heterozygosity per locus is not appropriate for the following two reasons. (1) The size of a gene or locus varies extensively from gene to gene because of the varying nature of the number and size of introns and flanking regions. In immunoglobulin genes, the definition of a locus itself is unclear because the final polypeptides of immunoglobulins are produced by an extensive DNA rearrangement that occurs during the differentiation of antibody-producing cells (Leder, 1982). (2) At the DNA level, the average heterozygosity per locus (or per nucleon; Nei and Tajima, 1981) seems to be close to one in many populations; therefore, this quantity is not appropriate for measuring the extent of DNA polymorphism.

In view of this situation, Nei and Li (1979) proposed a measure called nucleotide diversity (π), which is the average number of nucleotide differences per nucleotide site between two randomly chosen DNA sequences (heterozygosity per nucleotide site) and can be estimated by

$$\pi = \Sigma \pi_{ij}/n_c \tag{3}$$

where π_{ij} is the proportion of different nucleotides between the ith and jth DNA sequences in the populations and n_c is the total number of comparisons available. This quantity usually cannot be larger than $3/4$, because there are only four different types of nucleotides. In an equilibrium population, the expected value of π is given by $4N_e\mu/(1 + 16N_e\mu/3)$, where μ is the mutation rate per nucleotide site per generation (Kimura and Ohta, 1971).

Table 1 shows examples of the estimates of this quantity (π). Most of these estimates were obtained by an indirect method, that is, the restriction enzyme technique (Nei and Li, 1979; Nei and Tajima, 1981). This table also includes examples of the proportion of nucleotide differences (π_{ij}) between a selected pair of alleles (DNA sequences). Table 1 indicates that nucleotide diversity is 0.004–0.013 in eukaryotic organisms and nearly the same for both mitochondrial DNA and nuclear genes, though the number of nuclear genes studied is very small. If we note that the coding region of structural genes usually consists of approximately 1000 nucleotide pairs, this result suggests that the nucleotide sequences of two structural genes randomly chosen from a population are rarely identical. Mitochondrial DNAs (mtDNAs) are maternally inherited and exist in the haploid form. Therefore, the effective population size for mtDNAs is expected to be approximately $1/4$ of that of nuclear genes. However, the mutation rate for mtDNA is apparently considerably higher than that for nuclear genes (Chapter 4 by Brown). These two compensating factors probably make π for mtDNA nearly equal to that of nuclear genes. Unlike eukaryotic

174

TABLE 1. Estimates of nucleotide diversity (π) or the proportion of nucleotide differences between a selected pair of DNA sequences (π_{ij}).

DNA or gene region	Organism	Method*	Sample size	No. of base pairs in region	π or π_{ij}	Source†
Nucleotide diversity (π)						
mtDNA	Human	R	100	16,500	0.004	1
mtDNA	Chimpanzee	R	10	16,500	0.013	2
mtDNA	Gorilla	R	4	16,500	0.006	2
mtDNA	Peromyscus	R	19	16,500	0.004	3
mtDNA	Fruitfly	R	10	11,000	0.007	4
β Globin family	Human	R	50	35,000	0.002	5
Adh gene region	Fruitfly	R	18	12,000	0.006	6
Hemagglutinin	Influenza virus	S	12	320	0.510	7
Selected pair of DNA sequences (π_{ij})						
Insulin	Human	S	2	1,431	0.003	8
Adh coding region	Fruitfly	S	2	765	0.009	9
Immunoglobulin C_κ	Rat	S	2	1,172	0.018	10
IgG2a	Mouse	S	2	1,114	0.100	11

* R, Restriction enzyme technique; S, Sequencing.
† Source: 1. Cann et al. (1982), 2. Ferris et al. (1981b), 3. Avise et al. (1979a) (intrapopulational nucleotide diversity), 4. Shah and Langley (1979a), 5. Kazazian et al. (Chapter 7), 6. Langley et al. (1982), 7. Air (1981), 8. Ullrich et al. (1980), 9. Benyajati et al. (1981), 10. Sheppard and Gutman (1981), 11. Schreier et al. (1981).

genes, the hemagglutinin gene of influenza A virus shows an extremely high nucleotide diversity. As will be discussed later, this high nucleotide diversity is apparently due to an unusually high mutation rate in this organism.

Although the π_{ij} for a selected pair of alleles would not be a good estimate of nucleotide diversity, data in Table 1 indicate that the nucleotide diversity estimated by the restriction enzyme technique is in rough agreement with that obtained by the sequencing method. There is, however, one exception. Namely, the π_{ij} for the $\gamma2a$ heavy-chain constant region gene of immunoglobulin in the mouse (see Chapter 2 by Li) is one order of magnitude higher than the usual π value for eukaryotic genes. This high value of π_{ij} has been suspected to be due to gene conversion between this gene and its neighboring gene (Schreier et al., 1981).

Gene conversion seems to occur quite frequently in multigene families such as immunoglobulin genes and major histocompatibility complex (MHC) genes (Clarke et al., 1982; Robertson, 1982). Because gene conversion may convert parts of the nucleotide sequence of an allele at a locus into those of an allele at another locus, it has an effect to increase allelic diversity within loci but to decrease gene differences between neighboring loci. A similar effect may be generated by unequal crossing over in multigene families.

If the population is in equilibrium and no gene conversion or unequal crossing over occurs, the expected value of π is a function of $N_e\mu$. Therefore, it is possible to study the relationship between the observed value of π and $N_e\mu$ as in the case of average heterozygosity. However, there are not enough data to conduct such a statistical study at the present time.

Silent polymorphism

From the standpoint of the neutral theory, it is interesting to examine the extent of DNA polymorphism that is not expressed at the amino acid level. Under the neutral theory, this silent polymorphism is expected to be high compared with the polymorphism expressed at the amino acid level, because silent mutations are subject to purifying selection less often than nonsilent mutations (Chapter 11 by Kimura). On the other hand, if polymorphism is actively maintained by natural selection and the effect of genetic drift is unimportant, one would expect that silent polymorphism is lower than nonsilent polymorphism. One way of testing this hypothesis is to compare the extents of polymorphism for the coding and noncoding regions of genes. Nucleotide changes in the noncoding regions do not affect amino acid sequences, so that we would expect a higher degree of polymorphism in the latter than in the former. This expectation is generally fulfilled in available data. For example, between the two alleles of the human insulin gene studied by Ullrich et al. (1980), there are four nucleotide differences in the noncoding regions but none in the coding regions. A higher degree of polymorphism in the noncoding regions than in the coding regions has also been observed in the globin genes in man (Slightom et al., 1980; Orkin et al., 1982) and the alcohol dehydrogenase (*Adh*) gene in *Drosophila* (Kreitman, 1982).

Another test of this hypothesis is to examine the polymorphism at the first, second, and third positions of codons in the coding regions. In nuclear genes, all nucleotide changes at the second positions lead to amino acid replacement, whereas at the third positions only approximately 28% of changes are expected to affect amino acids because of degeneracy of the genetic code. In the first position, approximately 95% of nucleotide changes lead to amino acid changes. Therefore, if the neutral theory is valid, the extent of DNA polymorphism is expected to be the highest at the third position and the lowest at the second position. Available data indicate that this is indeed the case except in some immunoglobulin genes. For example, the Adh^F and Adh^S alleles at the alcohol dehydrogenase locus in *Drosophila melanogaster* are electrophoretically distinguishable because of the amino acid difference (threonine versus lysine) at residue 192 of this enzyme (codon change from ATG to AAG). The nucleotide sequences of approximately a dozen alleles at this locus indicate that there is no other amino acid substitution in the enzymes but that there are many third position substitutions that are silent (Benyajati et al., 1981; Kreitman, 1982). Similar examples are found in the albumin gene (Lawn et al., 1981; Dugaiczyk et al., 1982) and the β globin gene (Orkin et al., 1982) in humans.

However, this is not necessarily true in immunoglobulin genes. Sheppard and Gutman (1981) showed that two alleles (*LEW* and *DA*) at the rat κ light-chain constant region gene show 12 nucleotide differences in the coding region (see Chapter 2 by Li for the structure and synthesis of immunoglobulins). Eleven of these 12 nucleotide differences (92%) have caused amino acid differences. This is considerably higher than the expected value of 75% under random nucleotide substitution, though the difference is not statistically significant. In the case of the constant region of the mouse immunoglobulin γ heavy chain gene (γ2a), two alleles $IgG2a^a$ and $IgG2a^b$ show a total of 111 nucleotide differences plus 15 additional differences due to insertion and deletion when 1114 nucleotides are examined (Schreier et al., 1981). Of the 111 nucleotide differences, 18 (16%) were silent and the rest were amino acid-altering substitutions. These two observations

177

suggest that in immunoglobulin genes amino acid-altering substitutions might be favored by selection because in immunoglobulins variability is needed.

To examine this possibility in more detail, I compared the constant and variable region genes of the κ light chains from humans and mice and computed the rates of nucleotide substitutions for the first, second, and third positions of codons separately, assuming that the time since divergence between human and mouse is 8×10^7 years. (The corresponding data for the γ heavy chain genes are not available.) The results obtained are presented in Table 2. Unlike our expectation, this table shows that in the constant region gene of the κ light chain the rate of nucleotide substitution is the highest at the third position and the lowest at the second position, as is usually the case with other genes. This suggests that the relatively high frequency of nonsilent differences between the two alleles of the κ chain gene mentioned above occurred by chance, not by selection.

However, the ratio of the second position rate to the third position rate is higher than that in other genes such as globin genes (Chapter 2 by Li). Furthermore, at the hypervariable sites of the variable region gene, all the three nucleotide positions show essentially the same substitution rate; and this rate is as high as the pseudogene rate, which is supposed to be equal to the intrinsic mutation rate at the nucleotide level (Chapter 11 by Kimura). [This parallels Ohta's (1978) finding that the rate of amino acid substitutions at the hypervariable sites is equal to that of fibrinopeptides, the highest known rate for proteins.] Unexpectedly, however, the first and second positions of the framework variable region show a substitution rate that is similar to

TABLE 2. Rates of nucleotide substitution per site per year for the constant and variable region genes of the immunoglobulin κ light chain.*

		Variable region	
Position in codon	Constant region (106)	Hypervariable (25)	Framework (70)
First	2.13 ± 0.43	3.06 ± 1.16	0.67 ± 0.26
Second	1.94 ± 0.40	5.54 ± 2.04	0.99 ± 0.32
Third	3.65 ± 0.65	4.79 ± 1.74	2.39 ± 0.57

* These rates were obtained from comparison of human and mouse sequences on the assumption that the time since divergence between man and mouse is 8×10^7 years. The figures in parentheses refer to the numbers of codons used. All values should be multiplied by 10^{-9}. Data from Hieter et al. (1980) and Bentley and Rabbitts (1980) were used.

that of globin genes. These results suggest that the high variability of immunoglobulins is generated by exploiting germline mutations in addition to the now well-known mechanism of DNA rearrangement in the developmental process (Leder, 1982) and the somatic mutations mentioned below.

Somatic mutations for immunoglobulin genes

Population geneticists usually pay little attention to somatic mutations because they are not inherited. To understand the nature of the genetic variation of immunoglobulins, however, we cannot ignore somatic mutations because these are apparently an important source of antibody variation. Recently a number of authors compared the nucleotide sequences of variable region genes in myeloma and hybridoma (mature cells) with those of germline genes in the mouse. Table 3 shows the somatic mutation rates estimated from these data. It is clear that in both the heavy-chain and light-chain genes the rate of somatic mutations is extremely high. It should be noted that these somatic mutations are scattered all over the variable region, not just the hypervariable region. At the present time there are no data for computing the nucleotide diversity for the immunoglobulin variable region, but if we note that the rate of nucleotide substitution for the hypervariable region is only approximately two times higher than that for the constant region (Table 2) and π_{ij} is approximately 0.02 (Table 1), it seems that the antibody variation generated by somatic mutation is nearly as great as that due to germline mutations (see

TABLE 3. Somatic mutation rates for immunoglobulin variable-region genes in the mouse.

Gene	Nucleotides examined	Mutation rate (%)*	Source
V_H (V_{B1-8})	294	3.4	Bothwell et al. (1981)
V_H (V_{T15})	303	4.0	Kim et al. (1981)
V_κ (V_{L6})	285	2.1	Pech et al. (1981)
V_κ (V_{L7})	285	2.1	Pech et al. (1981)

* Mutation rates were obtained by comparing the nucleotide sequences of germline DNA and rearranged DNA in myeloma and hybridoma.

also Bentley and Rabbitts, 1981). Although it is not clear how this high rate of somatic mutation is attained, it is certainly an effective way of generating a high degree of antibody diversity.

Influenza A viruses

Influenza remains an uncontrolled disease in man mainly because of antigenic variation of the two surface proteins, hemagglutinin and neuraminidase. From year to year the influenza A virus becomes progressively more resistant to antibodies made against older viruses. This is caused by the accumulation of amino acid changes in the surface proteins (antigenic drift). Every so often, however, a new type of virus with a major change of antigenicity appears and causes a pandemic (antigenic shift). Mutations at the four antigenic sites of the hemagglutinin seem to be responsible for antigenic drift and shift (Wiley et al., 1981). There are 12 subtypes of influenza A viruses identified at the present time. The influenza A virus is an RNA virus, but cDNA can be produced for each gene of the virus. Air (1981) sequenced the nucleotides of the cDNA from a portion of the hemagglutinin gene (a total of approximately 320 nucleotides for the signal peptide and the HA1 polypeptide) for the 12 subtypes. Some parts of his results are presented in Figure 4. It is clear from this figure that the subtypes are quite different from each other at the nucleotide level and at the amino acid level. Particularly in the signal peptide region, sequence variation is extremely high. However, there are a number of amino acid sites at which the same amino acid exists for all subtypes (two of them are shown in Figure 4). This is apparently due to the functional constraint of the protein. It is noted that although amino acids remain the same at these sites, silent substitutions have occurred.

Table 4 gives estimates of nucleotide diversities for the three nucleotide positions of codons in the hemagglutinin gene among the 12 subtypes. It is seen that the nucleotide diversity in the signal peptide region is extremely high; at the third position it is close to the maximum value of 0.75. At the second position π is considerably lower than at the third position, apparently reflecting the functional constraint of the protein. The π value for the HA1 polypeptide region of the hemagglutinin gene is smaller than that for the signal peptide at all three nucleotide positions, yet it is much higher than the value for most eukaryotic genes.

The high values of π are apparently due to an extremely high rate of mutation in RNA viruses (Holland et al., 1982). In the influenza A virus, an approximate mutation rate can be obtained by using virus strains that have been kept in refrigerators. For example, the Asian flu (subtype H2) was first isolated in 1957 and later variant strains

180

```
H1   ATG AAG GCA AAC CTA CTG GTC CTG TTA TGT GCA CTT GCA GCT GCA GAT GCA GAC ACA ATA|TGT|ATA|GGC|TAC CAT GCG AAC
     Met Lys Ala Asn Leu Leu Val Leu Leu Cys Ala Leu Ala Ala Ala Asp Ala Asp Thr Ile|Cys|Ile|Gly|Tyr His Ala Asn

H2           ATG GCC ATC ATT TAT CTC ATT CTC CTG TTC ACA GCA GTG AGA GGG GAC CAG ATA|TGC|ATT|GGA|TAC CAT GCC AAT
             Met Ala Ile Ile Tyr Leu Ile Leu Leu Phe Thr Ala Val Arg Gly Asp Gln Ile|Cys|Ile|Gly|Tyr His Ala Asn

H5       ATG GAG AGA GTA GTG CTT CTT CTT GCA ATG ATC AGT CTT GTC AAA AGT GAC CAG ATT|TGC|ATT|GGT|TAC CAT GCA AAC
         Met Glu Arg Val Val Leu Leu Leu Ala Met Ile Ser Leu Val Lys Ser Asp Gln Ile|Cys|Ile|Gly|Tyr His Ala Asn

H11      ATG AAG AAA GTA CTG CTT TTT GCA GCA ATC ATC ATC TGT ATT CGA GCA GAC GAA ATC|TGC|ATT|GGA|TAC CTG AGC AAC
         Met Lys Lys Val Leu Leu Phe Ala Ala Ile Ile Ile Cys Ile Arg Ala Asp Glu Ile|Cys|Ile|Gly|Tyr Leu Ser Asn

H6       ATG ATT GCA ATC ATT GTA GTA GCG ATA CTG GCA ACA GCC GGA AGG TCT GAC AAG ATC|TGC|ATT|GGA|TAT CAT GCC AAC
         Met Ile Ala Ile Ile Val Val Ala Ile Leu Ala Thr Ala Gly Arg Ser Asp Lys Ile|Cys|Ile|Gly|Tyr His Ala Asn

H8   ATG GAG AAA TTC ATC GCA ATA GCA ATG CTC TTG GCG AGC ACA AAT GCA TAC GAT AGG ATA|TGC|ATT|GGG|TAC CAA TCA AAC
     Met Glu Lys Phe Ile Ala Ile Ala Met Leu Leu Ala Ser Thr Asn Ala Tyr Asp Arg Ile|Cys|Ile|Gly|Tyr Gln Ser Asn
```

FIGURE 4. Partial nucleotide and amino acid sequences for the hemagglutinin gene in the six subtypes (H1, H2, H5, H11, H6, H8) of influenza A viruses. The hemagglutinin gene codes for the signal peptide and the HA1 and HA2 polypeptides. This figure shows the sequences for the signal peptide and the first ten amino acids of HA1. Air (1981) has sequenced approximately 320 nucleotides for the 5′ region of the gene for all of the 12 subtypes. The arrow sign indicates the end of the signal peptide. The amino acids in boxes are identical for all the 12 subtypes (including the six subtypes shown here). (After Air, 1981.)

TABLE 4. Estimates of nucleotide diversities (π) for the three nucleotide positions of codons for the signal peptide and HA1 polypeptide of hemagglutinin of the influenza A virus.*

Position in codon	Signal peptide (14 codons)	HA1 (81 codons)	Total (95 codons)
First	0.705	0.440	0.480
Second	0.588	0.351	0.386
Third	0.732	0.651	0.663
TOTAL	0.675	0.481	0.510

* These results were obtained from Air's (1981) sequence data (12 subtypes). Only the codons shared by all subtypes were used, excluding the initiation codon.

were isolated in 1967, 1968, 1972, 1977, and so forth. Therefore, by comparing the nucleotide sequences of these strains, one can determine the mutation rate. Figure 5 shows the patterns of accumulation of mutations with time for the genes for the signal peptide and HA1 polypeptide of hemagglutinin, the nonstructural protein (NS), and the

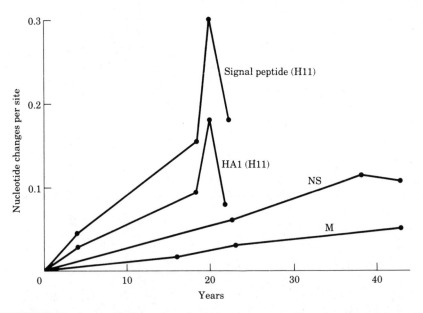

FIGURE 5. Accumulation of nucleotide changes in the signal peptide and HA1 regions of the hemagglutinin gene and in the nonstructural gene (NS) and the matrix gene (M) in the influenza A virus. (Data from Air, 1981; Air and Hall, 1981.)

matrix protein (M) for subtype H11 (Air, 1981; Air and Hall, 1981). The rate of accumulation of mutations for the hemagglutinin gene is extremely high, the average rates for the signal peptide and HA1 being 0.010 and 0.006 per site per year, respectively. The accumulation rates for the NS gene (0.0027 per site per year) and the M gene (0.0013 per site per year) are lower than the rate for the hemagglutinin gene but still extremely high compared with the rates for eukaryotic genes. The difference in the accumulation rate between the hemagglutinin gene and the other genes would not be due to the difference in the mutation rate but probably would be due to stronger purifying selection operating for the latter genes than for the former. If this is correct, the intrinsic mutation rate will be at least 1% per site per year. With this mutation rate, π would reach 0.5 in several hundred years, even if it is initially 0. The high mutation rate in this and other RNA viruses is believed to be due to the absence of replication–proofreading enzymes.

Of course, the maintenance of the polymorphism of antigenic sites must be aided by some kind of frequency-dependent selection, because viruses with low-frequency antigen are expected to have a higher fitness compared with viruses with high-frequency antigens. However, the number of amino acids involved in antigenic sites of a protein is very small (Wiley et al., 1981), and most polymorphisms are seen at other amino acid sites. Therefore, a large part of protein or DNA (RNA) polymorphisms in this virus seem to be maintained by mutation pressure.

DNA length polymorphism

In addition to the polymorphism due to nucleotide substitution, there is variation in the number of nucleotides (DNA length) for a given DNA region. One important class of these DNA length polymorphisms is those due to deletion and insertion of relatively small numbers of nucleotides. These polymorphisms are usually observed in the noncoding regions of DNA (e.g., Langley et al., 1982) but sometimes occur also in the coding regions, as seen in Figure 4. The second class of DNA length polymorphisms are those with respect to duplicate genes, which are apparently caused by unequal crossing over (concerted evolution in clustered genes; Chapter 3 by Arnheim). In this case, the number of copies of a particular gene or a particular DNA sequence may vary from individual to individual. For example, Coen et al. (1982) have shown that the length and copy number of ribosomal RNA genes vary extensively among individuals of *Drosophila melanogaster*.

183

An extensive polymorphism of this kind also exists in a DNA region approximately 500 nucleotides upstream from the human insulin gene on chromosome 11 (Bell et al., 1982). It has been shown that more than 60% of individuals in man are heterozygous with respect to this DNA length polymorphism. The third class of DNA length polymorphisms are those that are apparently caused by gene transposition. The mechanism of this gene transposition is still under investigation (Chapter 13 by Campbell), but the prevalence of repeated DNAs such as the *Alu* family genes in different chromosomes of man suggests that transpositions of DNA pieces occur quite frequently in eukaryotic genomes (Singer, 1982). Although this type of polymorphism has not been quantified, the amount seems to be extensive.

At any rate, these polymorphisms suggest that in eukaryotic genomes a large number of DNA length mutations occur every generation. A large proportion of these mutations are presumably deleterious and quickly eliminated from the population. However, there are still many mutations that become frequent and fixed in the population. At the present time little is known about the evolutionary significance of these mutations. As Orgel and Crick (1980) and Doolittle and Sapienza (1980) claim, they may simply represent neutral or nearly neutral variations.

IMPLICATIONS FOR EVOLUTIONARY THEORY

Adaptive and nonadaptive evolution

Let us now consider the implications of the above findings for the general theory of evolution. The observations we have made about protein and DNA polymorphism can be summarized in the following way. (1) The extent of protein polymorphism is nearly equal to or lower than the level expected under the equilibrium theory of neutral mutations. The difference between the observed and expected levels can be explained either by the bottleneck effect or by diversity-reducing selection. (2) The patterns of distributions of allele frequencies, single locus heterozygosity, genetic distance, and so forth are in rough agreement with the expectations from the neutral theory but are not consistent with those from several models of balancing selection. (3) Functionally important parts of genes (DNA) are generally less polymorphic than unimportant parts and evolve more slowly. (4) A large amount of genetic variation may be generated by mutation alone, as in the case of immunoglobulins and the influenza virus hemagglutinin.

The above observations indicate that mutation and genetic drift are the major factors in the maintenance of molecular polymorphism and that natural selection operates mainly for reducing genetic variability. The studies mentioned earlier certainly do not rule out all

184

types of adaptive selection, but the patterns of protein and DNA polymorphisms can be accommodated with the neutral theory without much difficulty.

Of course, this does not mean that no allelic differences produce large fitness differences. There must be some mutations that produce a definite selective advantage; otherwise, no adaptive evolution can occur. Indeed, there are a number of experimental data suggesting that some allelic differences identified by electrophoresis are associated with adaptation to different environments (see Chapter 6 by Koehn et al.). A well-known example is a pair of alleles ($Es\text{-}I^a$ and $Es\text{-}I^b$) at an esterase locus of the freshwater fish *Catostomus clarkii* (Koehn, 1969). The enzymatic activities of these alleles seem to be directly related to temperature. DiMichele and Powers (1982b) also showed that the alleles B^a and B^b at the *lactate dehydrogenase-B* locus apparently control the adenosine triphosphate level and swimming speed in the fish *Fundulus heteroclitus*. The antigenic differences in some surface proteins, such as the influenza virus hemagglutinin, are also directly related to the adaptive differences. Nevertheless, the proportion of mutations that cause significant adaptive changes seem to be generally very small at the molecular level. However, because most organisms have a large number of genes and each gene is subject to mutation in every generation, a small proportion of advantageous mutations seem to be sufficient for adaptive evolution in most organisms (Nei, 1975).

We note that most features of molecular polymorphisms mentioned above are incompatible with the balance theory of evolution. In this theory the amount of genetic variability in populations is determined mainly by natural selection, as mentioned earlier. Data on molecular polymorphism, however, suggests that the amount of genetic variability depends largely on the mutation rate and effective population size, if we exclude the effect of purifying selection that eliminates deleterious mutations. Data on the pattern of nucleotide substitution in long-term evolution also suggests that mutation rather than selection is the primary factor of molecular evolution (Chapter 11 by Kimura). Indeed, as far as proteins and DNAs are concerned, the observed pattern of evolution does not seem to deviate far from the neutral pattern.

How does this relate to morphological or physiological evolution? Obviously, any morphological or physiological character would be controlled by many structural genes as well as by regulatory genes, and at least in structural genes, silent nucleotide substitution should occur more or less in the same fashion as that for globin or insulin genes.

185

Amino acid substitutions that do not affect the function of enzymes or proteins would also be subject to random evolutionary change. In a substantial number of cases, however, allelic differences are known to affect enzyme activity. For example, the level of acid phosphatase activity in human red cells is approximately 50% greater in one homozygote (*BB*) than in another homozygote (*AA*) (Harris, 1966). Harris (1971) reported that significant biochemical differences between alleles were observed at 16 out of the 23 enzyme loci examined in man. Similar observations have been reported in a number of enzymes in different organisms. These differences are likely to be reflected in some morphological or physiological characters, but it is not clear whether they really affect fitness or not (Nei, 1975). This is particularly so when a morphological character is controlled by many loci.

In nature a considerable amount of variation in a quantitative character seems to be tolerated by the environment in which the organism lives. For example, bristle number in *Drosophila* does not seem to be directly related to fitness except at the ends of the distribution (Robertson, 1967). It is possible that the individuals for which the bristle number is not far from the mean have a nearly identical fitness, and the genetic variation within this range is maintained largely by the balance among mutation, genetic drift, and weak selection, if any (Nei, 1975, 1980). This seems to be true even if there is centripetal selection (Kimura, 1981d; Milkman, 1982; also see Latter, 1972). If a character is controlled by a large number of loci and is subject to centripetal selection, the selection coefficient for individual genes becomes very small and the gene frequency change is largely determined by genetic drift. Note also that centripetal selection is a form of purifying selection that operates at the phenotypic level; it is not a diversity-enhancing selection (Kimura, 1981d).

Even the genetic variability between populations may not always be adaptive. The Ainu living in northern Japan are morphologically different from the Japanese, the male body being quite hairy. However, it is not clear what kind of selective advantage this confers. It is possible that this character was brought about by mutation in their ancestral population and that the individuals with this mutant gene or genes later formed a new group, which later became the Ainu population. In the past several decades anthropologists, evolutionists, and (recently) sociobiologists have attempted to explain every detail of morphological, physiological, and behavioral differences between races in terms of natural selection. However, some of these differences may well be due to nonadaptive changes. We should be cautious about the adaptationist program, as was emphasized by Gould and Lewontin (1979) and Gould (1982).

In the above I have emphasized the possibility that even the genes controlling morphological characters may be subject to random evolutionary changes. However, if we note that in many organisms morphological and physiological characters are amazingly well adapted to the environment in which they live, it is clear that the evolution of these characters is controlled by natural selection to a much greater extent than molecular variation. Is it then unchallengeable that adaptive evolution of morphological characters has occurred in the way described by the balance school of evolution? In my view, this is questionable, and mutation seems to have played a vital role even in morphological evolution.

Role of mutation in evolution

In the balance theory of evolution, or in some other schools of neo-Darwinism, only a minor role in evolution is given to mutation, as mentioned earlier. This view was formed apparently because whenever artificial selection was applied to a quantitative character in a random mating population a quick and significant response was observed in many organisms. This gave the early population geneticists the impression that a random mating population contains almost all kinds of genetic variation and that the only force necessary for achieving a particular evolutionary change is natural selection. However, artificial selection is quite different from natural selection. The response to artificial selection is usually large in the early generations but gradually declines as generations proceed; and without further input of new genetic variability, the mean value of a character under selection usually reaches a plateau within a few dozen generations. Furthermore, it is noted that the initial response to selection or the heritability of a quantitative character is usually greater in those characters that are remotely related to fitness than in those that are closely related (e.g., Falconer, 1960). This suggests that the former characters have not been subjected to strong natural selection, so that the amount of genetic variability accumulated in the population is large. Therefore, if artificial selection is applied to these characters, a rapid response occurs. In the latter characters, however, natural selection seems to have reduced the genetic variation and, thus, artificial selection is less effective. If this interpretation is correct, artificial selection does not provide an accurate picture of long-term evolution by natural selection.

   Wright (1931, 1970) and Mayr (1963) have argued that the fitness of a genotype is determined by the entire set of genes in the individual, and it is difficult to isolate the effect of a single gene because of the gene interaction among loci. They have then viewed adaptive evolution as a process for a population to climb up the adaptive surface in a multidimensional space. In this view the role of mutation is minor because each locus is assumed to contain many polymorphic alleles maintained by some type of balancing selection, and a higher peak can be attained by natural selection (including the effect of environmental change) and genetic drift alone. In other words, evolution is assumed to occur mainly by gene frequency shift (without fixation of alleles) rather than by gene substitution at each locus (Wright, 1970). This shifting balance theory has been accepted by many evolutionists, but actually it lacks both theoretical and experimental proofs (Nei, 1980). At the molecular level evolution occurs mainly by gene substitution rather than by gene frequency shift, and the substitution at one locus is almost independent of that at another locus. It is also noted that natural populations do not have all kinds of alleles; different species usually having different sets of alleles. This raises a question about the validity of the shifting balance theory (see Nei, 1980, for a more detailed discussion). Furthermore, even when epistatic gene interaction is important (there are many such examples), evolution can occur by gene substitution rather than by gene frequency shift with changing balanced polymorphisms.

It is often said that genetic polymorphism is beneficial to the population, because in the presence of genetic variability the population can adapt easily to new environments (Dobzhansky, 1970). Thus, any mechanism that increases genetic variability is selected for (Ford, 1964). I am skeptical about this teleolistic explanation. In my view the genetic variability of a population at present is simply a product of evolution in the past. It may happen to be useful for future evolution, as in the case of industrial melanism, but I doubt that genetic variability is stored for future use. In many cases a population may not have the genetic variability that is required for new adaptation. In this case, the population may stay unevolved until new mutations occur or simply become extinct. It seems to me that in the evolution of unique characters, such as the camel's hump or human brain, mutation was the key factor. The mutation for human-level intelligence probably occurred in the human lineage with an exceedingly small probability. It is also possible that the primary process of improvement of brain function in man was caused by only a few regulatory mutations at the molecular level, as argued by Ohno (1970).

In my view, natural selection is a consequence of the existence of two or more functionally different genotypes in the same environment. For example, if one genotype is more efficient in obtaining food or more resistant to a certain disease compared with others, it will have a higher survival value. However, the functional efficiency of a genotype is determined by the genes possessed by the individual. Therefore, the most important process of adaptive evolution is the creation of better (functionally more efficient) genotypes by mutation or gene recombination. [Here mutation includes all sorts of genetic changes, such as nucleotide substititution and gene duplication.] Needless to say, the functional efficiency of a genotype depends on the environment in which the organism lives, but it should be noted that evolution generally occurs in the direction for specialization to a particular environment.

The above argument is similar to Morgan's mutationism or Muller's classical theory but includes some new elements. This new form of mutationism or neoclassicism may be characterized as follows:

1. At the nucleotide level many mutations are deleterious but a substantial proportion of them are neutral or nearly neutral. Only a small proportion of mutations are advantageous, and that is sufficient for adaptive evolution. Under certain circumstances, morphological characters may be subject to nonadaptive changes in evolution.
2. Natural selection is primarily a process to save beneficial mutations or eliminate unfit genotypes.
3. New mutations spread through the population either by selection or by genetic drift, but a large proportion of them are eliminated by chance.
4. Populations do not necessarily have the genetic variability needed for new adaptation, though the variability at the molecular level is usually very large. When there is not enough genetic variability needed, the population stays unchanged until new mutations occur or the population becomes extinct.

At this moment, it should be noted that Morgan's (1925, 1932) view of evolution was not typological as claimed by some authors (e.g., Mayr, 1982). In his later years he was aware of the basic concept of population genetics developed by Fisher, Haldane, and Wright; and, unlike de Vries (1901), he believed in gradual (not continuous) evolution due to mutation and selection. Indeed, his view is not much different from Muller's (1929, 1950) classical theory, in which little importance is given to balanced polymorphism, and evolution is believed to occur by successive fixation of advantageous mutations in the population. It should also be mentioned that the Morgan-Muller theory is a modern version of Darwin's theory of gradual evolution, as noted by Lewontin (1974). Lewontin coined the words *neoclassical theory* to designate the extension of the classical theory to include the case of neutral evolution. The neoclassicism I have envisioned here is somewhat different from Lewontin's (1974) definition. In my view adaptive evolution does not necessarily occur gradually, though it may. As mentioned earlier, some adaptive mutations seem to occur very infrequently (almost uniquely) but have a drastic phenotypic effect. Therefore, in some cases an organism may undergo a rapid phenotypic change, whereas in others it may stay unchanged for a long period of time. Eldredge and Gould (1972) and Gould (1982) have argued that in geologic time scale morphological evolution occurs as a discrete process rather than a continuous process. If we accept the above neoclassicism, this discontinuous evolution can be explained at least partially.

Finally, I should emphasize that my discussion on adaptive evolution is quite speculative, though it is in no way inconsistent with the molecular data available now. I presented this speculation in order to stimulate the study of the mechanism of adaptive evolution at the molecular level. Recombinant DNA technology makes it possible to study gene regulation by inserting a gene isolated from an organism into the genome of another organism (e.g., Palmiter et al., 1982). Therefore, in the near future we will probably be able to know more about the evolution of genome organization as well as the validity of neoclassicism or the balance theory of evolution.

CHAPTER 10

EVOLUTION OF THE AMINO ACID CODE

Thomas H. Jukes

The origin and evolution of the amino acid code or genetic code is an unsolved problem that is of much significance to the understanding of life from its earliest beginnings and to tracing subsequent evolution. Until recently the amino acid code was thought to be universal for all genes in all terrestrial life forms, and the lack of exceptions or variations in the genetic code made the study of its evolution highly speculative (e.g., Jukes, 1966; Crick, 1968). The discovery of variant forms of the amino acid code in mitochondrial DNA (Anderson et al., 1981, Barrell et al., 1979; Bibb et al., 1981; Bonitz et al., 1980; Heckman et al., 1980; Köchel et al., 1981) has made it possible to speculate on the origin and evolution of the amino acid code by referring to observations made with mitochondrial DNA. In this chapter I shall discuss the problem by using recent information on the amino acid code and anticodons in tRNAs. Specifically, I would like to show that available data support the basic scheme of evolution of the code I put forward in 1966 (Jukes, 1966).

THE ARCHETYPAL AMINO ACID CODE

For reasons of convenience and tradition, the amino acid code is usually written as shown in Table 1. This tabulation shows the code in the form that is used for examining DNA and RNA sequences to locate protein genes. However, Table 1 is misleading. The amino acid code is actually a tRNA code. Amino acids are attached to tRNA molecules that have anticodons rather than codons. The tRNA molecules seek out codons to translate. Messenger RNA molecules have codons, and

191

TABLE 1. The amino acid code.

Codon	Amino Acid	Codon	Amino Acid	Codon	Amino Acid	Codon	Amino Acid
UUU	Phe	UCU	Ser	UAU	Tyr	UGU	Cys
UUC	Phe	UCC	Ser	UAC	Tyr	UGC	Cys
UUA	Leu	UCA	Ser	UAA	Ter	UGA	Ter
UUG	Leu	UCG	Ser	UAG	Ter	UGG	Trp
CUU	Leu	CCU	Pro	CAU	His	CGU	Arg
CUC	Leu	CCC	Pro	CAC	His	CGC	Arg
CUA	Leu	CCA	Pro	CAA	Gln	CGA	Arg
CUG	Leu	CCG	Pro	CAG	Gln	CGG	Arg
AUU	Ile	ACU	Thr	AAU	Asn	AGU	Ser
AUC	Ile	ACC	Thr	AAC	Asn	AGC	Ser
AUA	Ile	ACA	Thr	AAA	Lys	AGA	Arg
AUG	Met	ACG	Thr	AAG	Lys	AGG	Arg
GUU	Val	GCU	Ala	GAU	Asp	GGU	Gly
GUC	Val	GCC	Ala	GAC	Asp	GGC	Gly
GUA	Val	GCA	Ala	GAA	Glu	GGA	Gly
GUG	Val	GCG	Ala	GAG	Glu	GGG	Gly

there are always 64 codons. Only 61 of them are translated. The other three do not pair with anticodons and hence are terminator or stop codons.

Table 2 lists both codons and anticodons in the universal code. Some anticodons pair with two or three codons, so there are fewer anticodons than there are codons. Anticodons are found only in tRNA molecules, and not all the anticodons have been described experimentally. Those presently missing are shown in parentheses. Codon–anticodon pairing takes place during the translation of mRNA strands into polypeptides. The pairing follows the wobble rules. These are described later.

It is a general rule of evolution that it proceeds from the simple to the complex. In living organisms, the process is achieved by increasing the amount of DNA, often by the duplication of genes (Chapter 2 by Li). With this in mind, I proposed that the universal code has evolved from an archetypal code in which only 15 amino acids were used instead of the present 20 as follows (Jukes, 1966):

The archetypal...code...is outlined as consisting of 16 quartets of triplets AAd, ACd, etc., in which the third base ["d"] is A, C, G, or U used synonymously and interchangeably. . . . It is also presumed that there could have been only 16 transfer RNAs—one for each quartet. Eight quartets have retained their original assignments; seven of the others have each kept either a purine-

192

TABLE 2. Codons (a), anticodons (b)[†], and amino acids (c) in the universal amino acid code.

a	b	c	a	b	c	a	b	c	a	b	c
UUU	GAA	Phe	UCU	IGA	Ser	UAU	GUA	Tyr	UGU	GCA	Cys
UUC		Phe	UCC	(GGA)	Ser	UAC		Tyr	UGC		Cys
UUA	UAA	Leu	UCA	UGA	Ser	UAA	–	Ter	UGA	–	Ter
UUG	CAA	Leu	UCG	CGA	Ser	UAG	–	Ter	UGG	CCA	Trp
CUU	(IAG)	Leu	CCU	IGG	Pro	CAU	GUG	His	CGU	ICG	Arg
CUC	GAG	Leu	CCC	(GGG)	Pro	CAC		His	CGC	GCG	Arg
CUA	UAG	Leu	CCA	UGG	Pro	CAA	UUG	Gln	CGA	(UCG)	Arg
CUG	CAG	Leu	CCG	(CGG)	Pro	CAG	CUG	Gln	CGG	(CCG)	Arg
AUU	IAU	Ile	ACU	IGU	Thr	AAU	GUU	Asn	AGU	GCU	Ser
AUC	GAU	Ile	ACC	GGU	Thr	AAC		Asn	AGC		Ser
AUA	*CAU	Ile	ACA	UGU	Thr	AAA	UUU	Lys	AGA	UCU	Arg
AUG	CAU	Met	ACG	(CGU)	Thr	AAG	CUU	Lys	AGG	(CCU)	Arg
GUU	IAC	Val	GCU	IGU	Ala	GAU	GUC	Asp	GGU	(ICC)	Gly
GUC	GAC	Val	GCC	(GGC)	Ala	GAC		Asp	GGC	GCC	Gly
GUA	UAC	Val	GCA	UGC	Ala	GAA	UUC	Glu	GGA	UCC	Gly
GUG	CAC	Val	GCG	(CGC)	Ala	GAG	CUC	Glu	GGG	CCC	Gly

† Anticodons in parentheses are possible anticodons but have not been identified.
* C is modified to pair only with A in *E. coli.*

terminated or pyrimidine-terminated pair of triplets for the original amino acid and have lost the other pair to another amino acid during evolution. In this way, two codes for each of the four "new" amino acids have been provided. The sixteenth quartet appears to consist of three codes for isoleucine and one for methionine, which is perhaps the newest of the amino acids that participate in the synthesis of polypeptides. . . .

Changes in the genetic code should presumably come about by changes in transfer RNA. Let us suppose that CAA was at one time one of the codes for histidine and that the corresponding transfer RNA containing a complementary coding triplet (anticodon) UUG underwent a mutational change in the region of its amino acid recognition site, so that it became charged with glutamine rather than histidine, glutamine being an amino acid that previously had no recognition site in any transfer RNA. The result would be the introduction of glutamine in proteins replacing histidine at all messenger RNA sites containing the triplet CAA. Other sites, such as CAC and CAU, would retain histidine. If an organism could survive this change, it would have acquired a "new" amino acid, which might be an evolutionary advantage. We must assume that the organism would be quite primitive and simple to endure such a disruption.

The nature of the code and its evolutionary development become much clearer when it is presented as a table of anticodons. The archetypal form of such a table is Figure 1. This shows 15 anticodons with U in the first position. Each of these anticodons pairs with four codons. U in the first anticodon position pairs with U, C, A, and G in the third codon position, and the other two positions pair conventionally. The sixteenth box contains GUA, which pairs with UAU and UAC, leaving UAA and UAG unpaired.

From this beginning, we can examine the (mammalian) mitochondrial code, shown in Figure 2. It is similar in many respects to the archetypal code in Figure 1. The mitochondrial code contains eight quartets, or family boxes, that is, boxes with one anticodon that pairs with four codons for the same amino acid in the same manner as in the archetypal code. However, the mitochondrial code also contains

UAA	UGA	GUA	UCA
UAG	UGG	UUG	UCG
UAU	UGU	UUU	UCU
UAC	UGC	UUC	UCC

FIGURE 1. Archetypal anticodons. Underlined anticodons occur also in mitochondrial code in family boxes.

194

six boxes with a pair of anticodons (nonfamily boxes). Each of these boxes contains two different amino acids, one for each of the two codons. Remarkably, each pair of anticodons consists of one starting with G and the other with U. There are no anticodons starting with either A or C, because wobble pairing is used (Crick, 1966). Anticodons with G in the first position pair with U and C in the third codon position. Anticodons with U in the first position pair with A and G in the third codon position. This is classic wobble pairing as described by Crick in 1966. Notice that in these six boxes, U does not pair with all four third-position codon bases. Notice also that this difference between the archetypal code and the mitochondrial and universal codes is accompanied by an increase in the number of amino acids from 15 to 20. The number of amino acids became frozen at 20. U may be modified in the mammalian mitochondrial code to pair only with A and G (Heckman et al., 1980; Anderson et al., 1981) in nonfamily boxes.

Mitochondria are thought to have evolved from purple photosynthetic bacteria (Woese, 1981), which as prokaryotes (eubacteria) presumably used the universal code. Subsequent evolution of mitochondria resulted in a reduction of their DNA content. The mammalian mitochondrial code has been considered as an evolutionary simplification directed toward lessening the amount of DNA assigned to genes for tRNAs and thus helping to minimize the size of the mitochondrial genome (Bonitz et al., 1980; Heckman et al., 1980; Anderson et al., 1981; Bibb et al., 1981; Jukes, 1981). The mammalian mitochondrial code (MMC), according to this theory, represents a "pruning back" of the universal code to a bare minimum of 22 anticodons (and hence 22 tRNAs) needed for translation of the 60 codons for 20 amino acids,

GAA Phe	UGA Ser	GUA Tyr	GCA Cys
UAA Leu		—	UCA Trp
UAG Leu	UGG Pro	GUG His	UCG Arg
		UUG Glu	
GAU Ile	UGU Thr	GUU Asn	GCU Ser
UAU Met		UUU Lys	—
UAC Val	UGC Ala	GUC Asp	UCC Gly
		UUC Glu	

FIGURE 2. Anticodons and amino acid assignments in mammalian mitochondrial code.

including six codons apiece (and hence two anticodons apiece) for leucine (codons UUR and CUN) and serine (codons UCN and AGY), where N = U, C, A, or G; Y = U or C; R = A or G. Such a pruning back is still in evidence, for the MMC has eliminated anticodon UCU, which is used by yeast mitochondrial code for arginine. The yeast mitochondrial code also contains two methionine tRNAs, but MMC contains only one. The elimination of redundant tRNAs could lead to a reversion toward the archetypal coding system.

EVOLUTION OF THE UNIVERSAL CODE

Let us now consider the feasible steps in evolution of the amino acid code as shown by actual examples of such steps. Mitochondrial codes provide the following examples:

1. Four-way pairing between U in anticodons and U, C, A, and G in codons (Anderson et al., 1981, 1982; Bibb et al., 1981).
2. A change in aminoacylation of a leucine tRNA from leucine to threonine while retaining anticodon UAG, which paired with codons CUN (Bonitz et al., 1980). This change was presumably a single mutational event in the amino acid recognition site of the leucine tRNA, perhaps a nucleotide substitution. This is an important clue to the method of addition of "new" amino acids during the evolution of the code.
3. The loss of anticodon UCU by deletion, so that codons AGA and AGG have become chain terminators (Anderson et al., 1981, 1982). By the same token, anticodon UUA may never have been present, even in the archetypal tRNAs, so that codons UAA and UAG were the original chain terminators because they were not paired with an anticodon in any tRNA.

The first step proposed in evolution, starting with the archetypal code (Figure 1), is tRNA gene duplication (except for GUA) without change in aminoacylation. One duplicate could undergo anticodon change from UNN to GNN. This would convert the code to classic wobble (Crick, 1966) in codon–anticodon pairing and at the same time would provide two tRNAs for each amino acid (except tyrosine). The stage would then be set for the second step: a change in aminoacylation of one of the two tRNAs of one amino acid. Apparently this occurred in seven quartets, thus introducing new amino acids. Just which were the new ones and which the original ones is a matter of conjecture that can be made at the preference of the reader. Lagerkvist (1981) has proposed that partitioning of family boxes was favored by weakness of hydrogen bonding between codon and anticodon and that, therefore, partitioning did not take place in boxes containing the

196

family codons CCN, GCN, CGN, and GGN. However, other possibilities exist. Partitioning might also depend upon the degree of ease with which an aminoacylation site could be changed to accept a different amino acid by mutation in a tRNA molecule.

The introduction of new amino acids proceeded until freezing of the code took place at its present number of 20 amino acids. This occurred when living systems became too complex to tolerate further changes in the amino acid assignments of codons except in mitochondria, or to tolerate introduction of new amino acids. Requirement for increased amounts of an amino acid, such as lysine, was henceforth accommodated by selection of existing codons. Figure 3 shows the universal code at this stage.

However, the code continued to evolve. Anticodons with C in the first position were introduced. Modifications of the first anticodon base took place. As an example, valine had the anticodon UAC (Figures 1 and 2). The tRNA with this anticodon underwent duplication, and one of the duplicates underwent a nucleotide substitution of G for U. This anticodon, GAC, pairs with codons GUU and GUC. The other anticodon, UAC, pairs with codons GUA and GUG. UAC underwent another duplication to produce anticodons UAC and CAC. The latter anticodon pairs only with codon GUG. A third duplication took place in the GAC anticodon and AAC was produced by a mutation in one of the duplicates. This would normally pair with codon GUU. However, the first base (A) in this anticodon is deaminated to hypoxanthine by the enzyme anticodon deaminase (Kammen and Spengler, 1970). The nucleoside form of hypoxanthine is called inosine (I), and the anticodon

GAA	Phe	GGA	Ser	GUA	Tyr	GCA	Cys
UAA	Leu	UGA	Ser	—		CCA	Trp
GAG	Leu	GGG	Pro	GUG	His	GCG	Arg
UAG	Leu	UGG	Pro	UUG	Glu	UCG	Arg
GAU	Ile	GGU	Thr	GUU	Asn	GCU	Ser
*CAU	Ile	UGU	Thr	UUU	Lys	UCU	Arg
CAU	Met						
GAC	Val	GGC	Ala	GUC	Asp	GCC	Gly
UAC	Val	UGC	Ala	UUC	Glu	UCC	Gly

FIGURE 3. Anticodons in generic universal code and amino acid assignments. *C is modified to pair only with A in *E. coli*.

197

IAC pairs with codons GUU, GUC, and GUA, as explained by Crick (1966). This series of changes is shown in Figure 4. The same evolutionary pathway was followed in the cases of leucine (codons CUN), serine (codons UCN), proline, threonine, alanine, arginine (codons CGN), and glycine. Not all the anticodons for these amino acids have been discovered (Table 2). Their discovery awaits the isolation and sequencing of more cytoplasmic tRNAs. All these amino acids have four codons apiece in the universal code, corresponding to the case of valine. Serine, leucine, and arginine each have two additional codons. All the anticodons in Figure 2 have been described (Anderson et al., 1981, 1982; Bibb et al., 1981).

The second example is a case that follows the same first step as that of the first example, except that one of the duplicates acquires a different amino acid, following a change in the aminoacylation site of its tRNA. This is explained in Figure 5. This presents the case for anticodon UAA, postulated as being assigned to phenylalanine (possibility A), or leucine (possibility B) in the archetypal code. UAA duplicates, and one of the duplicates undergoes a mutation to GAA, which persists as the anticodon for phenylalanine. The other duplicate (UAA) is in a tRNA that becomes a leucine tRNA by mutation in its aminoacylation site. In the mitochondrial code, UAA is the anticodon for the leucine codons UUA and UUG. The same is true in the universal code. In addition, a third anticodon (CAA) exists for leucine; it

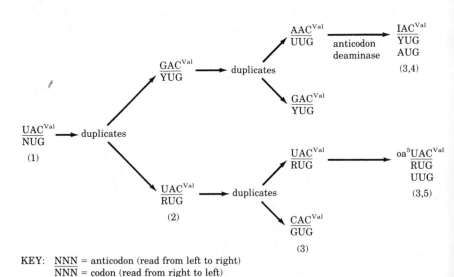

KEY: $\underline{\text{NNN}}$ = anticodon (read from left to right)
$\overline{\text{NNN}}$ = codon (read from right to left)

FIGURE 4. Evolution of valine codons. U = uridine; C = cytidine; A = adenosine; G = guanosine; N = U, C, A, or G; Y = U or C; R = A or G; oa⁵U = uridine-5-oxyacetic acid. 1, Archetypal and mitochondrial codes; 2, generic universal code; 3, present universal code; 4, eukaryotes; 5, prokaryotes.

198

was formed by a mutation from UAA and pairs with leucine codon UUG. In possibility B, the tRNA with anticodon GAA becomes acquired by phenylalanine following a mutation in the aminoacylation site, and UAA continues as an anticodon for leucine.

This pathway of evolution has been followed for the following pairs of amino acids and their anticodons; phenylalanine–leucine (UUR), histidine–glutamine, asparagine–lysine, aspartic acid–glutamic acid, serine–arginine (AGN). Two possibilities in each case exist. Each will depend upon which of the two amino acids was present in the archetypal code.

The third case is that of tyrosine. In the archetypal code, tyrosine had anticodon GUA pairing with codons UAU and UAC. Codons UAA and UAG had no anticodon in a tRNA and, hence, no amino acid assignments. They became chain terminators or stop codons, a role that they play in all versions of the mitochondrial code and in the

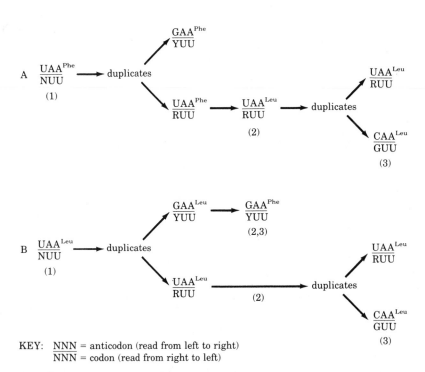

KEY: $\underline{NNN}$ = anticodon (read from left to right)
$\underline{NNN}$ = codon (read from right to left)

FIGURE 5. Evolution of anticodon UAA and codons for phenylalanine and leucine (UUR). 1, Archetypal code; 2, mitochondrial code and universal code; 3, present universal code. See Figure 4 legend for base abbreviations.

199

universal code. This role secured for these codons the recondite appellations of *amber* and *ochre*.

The fourth example is that of arginine with anticodon UCU in the ascomycete mitochondrial code. In the case of yeast, its use far exceeds that of the other arginine anticodon, UCG. The tRNA with anticodon UCU has been deleted from the human, bovine, and mouse mitochondrial genomes. As a result, codons AGA and AGG have become chain terminators in human and bovine mitochondrial codes, although they are used but sparsely (Figure 6).

The fifth example is that of cysteine, with archetypal anticodon UCA. This duplicated and mutated to produce anticodon GCA in the mitochondrial and universal codes, with the cysteine codons UGU and UGC. Anticodon UCA pairs with tryptophan codons UGA and UGG in the mitochondrial code. The anticodon UCA was replaced by CCA, presumably by a mutation, in the universal code. This left the codon UGA without an anticodon, so that it became a stop codon. Its resemblance to the existing stop codons UAA and UAG facilitated this step (Figure 7).

The next example is that of isoleucine, with archetypal anticodon UAU. This underwent duplication to produce two tRNAs with anticodons GAU and UAU. GAU pairs with isoleucine codons AUU and AUC in the mitochondrial and universal codes. The other anticodon,

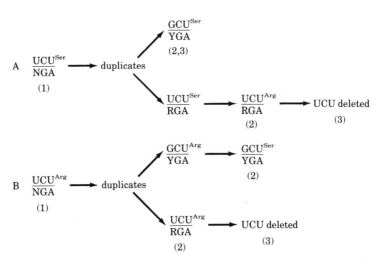

KEY: $\underline{\text{NNN}}$ = anticodon (read from left to right)
      $\overline{\text{NNN}}$ = codon (read from right to left)

FIGURE 6. Evolution of anticodon UCU and codons for serine (AGY), arginine (AGR), and terminators. 1, Archetypal code; 2, mitochondrial and universal codes; 3, human and bovine mitochondrial codes; AGA and AGG have become terminators. See Figure 4 legend for base abbreviations.

200

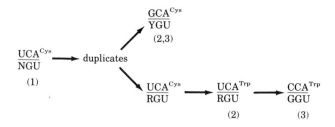

KEY: NNN = anticodon (read from left to right)
       NNN = codon (read from right to left)

FIGURE 7. Evolution of anticodon UCA and codons for cysteine (UGY), tryptophan (UGG), and terminator (UGA). Anticodon UCA disappeared and codon UGA became a terminator. 1, Archetypal code; 2, mitochondrial code; 3, universal code. See Figure 4 legend for base abbreviations.

UAU, should pair with codons AUA and AUG. However, methionine tRNA in the mitochondrial codes has anticodon CAU, which normally pairs with AUG. AUA is a methionine codon in mammalian mitochondria, but in *Neurospora crassa*, AUA is an isoleucine codon (Bonitz et al., 1980). In prokaryotes, isoleucine codon pairs with *CAU, in which C has been modified (*C) to pair with A (in codon AUA), but not with G (in codon AUG) (Kuchino et al., 1980; Fukada and Abelson, 1980). A different route was followed in eukaryotes. This was brought about by the enzyme anticodon deaminase, which changed the fourth anticodon in the series, namely AAU, into IAU, pairing with isoleucine codons AUU, AUC, and AUA, as noted by Crick (1966). Because IAU pairs with AUA, but not with AUG, distinction between these two codons took place, so that one is an isoleucine codon and the other is the sole methionine codon in the universal code (Figure 8).

The next question to be discussed is the matter of the first base in anticodons. This obviously could be either U, C, A, or G. However, A and C were not used in the first anticodon position in primitive codes, such as the archetypal and mitochondrial codes. This is because A pairs only with U, and C only with G. Therefore, A and C in the first position of anticodons would not participate in the multiple pairing that is needed for economizing in the number of tRNA genes in primitive systems that have limited amounts of DNA. As evolution proceeded and duplications of DNA took place, A and C were added to the anticodon–codon pairing system. The introduction of C increased the strength of pairing with G and, hence, increased fidelity of trans-

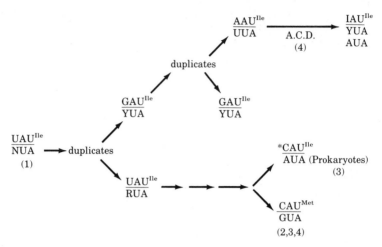

KEY: NNN = anticodon (read from left to right)
      NNN = codon (read from right to left)

FIGURE 8. Evolution of isoleucine (AUA) and methionine (AUG) codons and anticodons. 1, Archetypal code; 2, mammalian mitochondrial code; 3, universal code (*Escherichia coli*); 4, universal code (yeast). A.C.D., Anticodon deaminase. See Figure 4 legend for base abbreviations.

lation. A is found only very rarely in the first position of anticodons, because it is converted to I by a specific enzyme, anticodon deaminase, as previously explained. A is present in the first position of anticodons ACG (arginine) and ACC (glycine) in yeast and *Aspergillus* mitochondria, respectively (Bonitz et al., 1980; Köchel et al., 1981); possibly these mitochondria do not contain anticodon deaminase.

Transfer RNA molecules are in an ongoing state of evolution as shown by the nucleotide substitutions seen in comparisons of their sequences. Changes in aminoacylation may occur. We postulated this in the case of *E. coli* $tRNA_{Val}$ changing to $tRNA_{Gly}$ (or vice versa), also in the case of yeast $tRNA_{Arg}$ changing to $tRNA_{Lys}$ (or vice versa) when their genes were in a temporarily nonfunctional, nontranscribed state (Holmquist et al., 1973). The findings with yeast mitochondrial $tRNA_{Leu(CUN)}$ changing to $tRNA_{Thr(CUN)}$ and possibly maize mitochondria $tRNA_{Arg(CGG)}$ changing to $tRNA_{Trp(CGG)}$ (Fox and Leaver, 1981) are other examples. Except in the case of mitochondria, it is not likely that mutations causing such changes can persist. However, these findings are a serious challenge to theories that postulate that a stereochemical relationship between an anticodon and an amino acid determine specificity in the amino acid code. Table 3 shows variations among mitochondrial codes.

202

THE TWO-LETTER CODE

The first suggestion of a possible two-letter code in which two nucleotides specify an amino acid was made by Roberts (1962a,b) on the basis of "shared" doublets (Speyer et al., 1962). Such two-letter codes actually consist of three nucleotides, the third of which is nonspecific, or degenerate, so that each quartet of codons with a nonspecific third letter would be assigned to one amino acid, as in the archetypal code. A code consisting of only two nucleotides per amino acid could conceivably exist, but there could be no continuity or transition between organisms using such a code and organisms using a three-letter code. All the genetic information would be lost by switching from a two-letter code to a three-letter code.

A nagging question remains. If there was a code for only 15 or 16 amino acids, with the third letter degenerate or silent, why was it a three-letter code rather than a two-letter code? The obvious advantage of a three-letter code is to expand the resources of protein synthesis by adding more amino acids, but nature is more likely to discard an old method and displace it with an entirely new one rather than to act in anticipation of a future event.

It may be concluded that the three-letter code with the degenerate third letter was necessary for protein synthesis because three bases are needed for effectively binding the adapter to the messenger, an idea that has been expressed by Eigen (1971).

ORIGIN OF THE AMINO ACID CODE

There are two different ways of considering the origin of the universal amino acid code. The first approach says that the code is a "frozen accident," a term used by Crick (1968). The essence of this is that the present universal code cannot be changed without tolerable conse-

TABLE 3. Differences from the generic mitochondrial code in other mitochondrial codes.

Species	Codons	Anticodons	Amino Acids
Yeast	CUN	UAG	Thr
Yeast	AUA	GAU	Ile
Yeast	AUG	CAU	Met
Human, bovine	AGR	–	Ter
Maize	CGG	CCG	Trp

203

quences to living organisms, because any change would be dispersed throughout all proteins in the cell, and "at the present time, any change would be lethal, or at least very strongly selected against . . . because the code determines the amino acid sequences of so many highly evolved protein molecules" (Crick, 1968).

The idea of the "frozen accident" is that all living organisms have evolved from an ancient single ancestor, or a small group of ancestors, that displaced all contemporaries with other codes. After the evolutionary expansion of the descendants of this founder group had started, changes in the amino acid assignments of codons were not possible.

In support of the "frozen accident" model, we have pointed out that the code is not even the best code for presently existing organisms (Jukes et al., 1975). Arginine has six codons when it actually needs only two or three in terms of the extent to which it is used, on the average, in proteins. Lysine has only two codons, but the extent of its use demands that it should really have four. Aspartic and glutamic acids should have more than two codons apiece. These conclusions follow when the distribution of codons among amino acids is compared with the percentage occurrence of amino acids in proteins.

The second theory is that the code is a product of evolution and its composition was inevitable because it is the "best possible code" or because of a specific fit or affinity between each amino acid and its codon or anticodon. This is discussed as the "stereochemical theory" by Crick (1968). The number of codons for an amino acid, in this theory, depends upon the extent to which the amino acid is needed in proteins. For example, the simple amino acid alanine, with four codons, is used four times as frequently as tryptophan, with one codon. It can be argued, in reverse, that the fact that alanine is used four times as often as tryptophan is because it has four times as many codons.

The "stereochemical theory" is opposed by the finding that codons CUN can change from leucine to threonine in the yeast mitochondrial code (Bonitz et al., 1980) and codon CGG can possibly change from arginine to tryptophan in maize mitochondrial code (Fox and Leaver, 1981).

The recent observations with mitochondria show that other codes can successfully exist and that there is not just one mitochondrial code; it can vary in mitochondria of different species. Therefore, the code can change if evolutionary pressures are sufficiently strong and if the genome is very small and contains only a few proteins.

DISCUSSION

Leucine, serine, and arginine each have six codons in the universal code, so that two tRNAs with different anticodons must in each case

204

be aminoacylated by the same amino acid. The simplest suggestion for leucine is that it originally had eight codons, UUN and CUN, and two similar anticodons, UAA and UAG. In the case of serine, its AGY codons have long been perceived as anomalous because of the obvious difference between UCN and AGY. The original code for serine was probably UCN, with anticodon UGA. The arginine anticodons UCG and UCU are similar to each other; therefore, arginine may have had the archetypal anticodon UCU. This suggestion runs contrary to a proposal that arginine is an "evolutionary intruder" into protein synthesis (Jukes, 1973); this proposal was made because arginine is represented with such paucity in proteins in comparison with its large proportion of codons in the amino acid code.

In contrast, lysine, with only two codons, occurs in proteins at an average of 6.6%, so that it really "needs" four codons. This could fit the case if lysine entered the code as a "new" amino acid by acquiring two codons through anticodon UUU and if its role in proteins subsequently expanded at the expense of that of arginine, which is the other strongly basic amino acid.

Evolution of the universal code has resulted in a wide variety of modified bases occurring at the first anticodon position. These are in Table 4.

The origin and evolution of the amino acid code have long been topics for speculative articles, and it is easy to become carried away by the temptation to theorize on this interesting subject. In this communication, I have endeavored to limit the proposals to those that have some support from observations with mitochondrial codes.

The use of U in anticodons for four-way pairing may well represent a return to an earlier, simpler coding system. Perhaps the use of uridine-5-oxyacetic acid for three-way pairing with U, A, and G (Nishimura, 1972) is a vestige of this.

CONCLUSION

A proposal is made for evolution of the amino acid code. Variations arise principally by duplication of tRNAs followed by substitutional changes in the first bases of anticodons. Variations also arise by changes in aminoacylation, so that a different amino acid becomes attached to a tRNA molecule without a change in the anticodon. Loss of anticodons may occur by mutation of the first base to a different anticodon or by deletion. These steps would provide an origin for chain termination codons.

The proposal is compatible with differences between the universal

TABLE 4. First base of anticodons in sequenced tRNAs compared with third base of codons with which the anticodon pairs.

First base in anticodon*	Pairing with	Organisms‡	Present in tRNAs for:§	Missing in tRNAs for:
H (I)	U, C, A	E, P	Ile, Val, Ser (UCN), Ala, Arg (CGN), Thr, Pro	Gly
G	C, U	E, P	Phe, Leu (CUN), Ile, Val, Tyr, Asp, Arg, (CGN), Gly, Ser (AGY), Thr, Cys	Ser (UCN), Pro, Ala
Gm	C, U	E	Phe	
G† ("Q")	C, U	P	Tyr, His, Asn, Asp	
U	A, G	P	Gly, Leu (CUN)	Arg (CGR)
s^2U	A	E, P	Gln, Lys, Glu	
oa^5U	U, A, G	P	Val, Ser (UCN), Ala	
mO^5U	A	E	Thr	
U†	A, G(?)	E, P	Leu (UUR), Arg (AGR), Val, Pro	
C	G	E, P	Leu (CUN, UUR), fMet, Gln, Lys, Trp, Gly, iMet, Met, Ser (UCN), Val	Pro, Thr, Ala, Glu, Arg (CUG), Arg (AGG)
ac^4C	G	P	Met	
Cm	G	E	Trp, Leu	

* H, Hypoxanthine; I, inosine; oa⁵U, uridine-5-oxyacetic acid; Gm, 2′-O-methylguanosine; G† ("Q-base"), (7-[4,5-dihydroxyl-1-cyclopentenyl-aminomethyl]-7-deazaguanosine); s²U, 2-thiouridine; mO⁵U, 5-methoxyuridine; U†, unidentified uridine derivative; ac⁴C, 4-acetylcytidine; Cm, 2′-O-methylcytidine.
‡ E, Eukaryotes; P, prokaryotes.
§ N: A, C, G or U; R, purine; Y, pyrimidine; fMet, N-formylmethionine; iMet, initiator methionine. First base in prokaryotic anticodon for Ile (AUA) is modified C.

code and mitochondrial codes and with the principle that gene duplication, followed by modifications, is largely responsible for molecular
evolution. The list of 20 amino acids partaking in protein synthesis is
"frozen," but evolution of the amino acid code has continued and is
characterized by substitutions and modifications of nucleotides that
occur in the first position of anticodons.

The freezing of the code for 20 amino acids was a crucial event.
These amino acids are responsible for the properties of the vast pan-
oply of living organisms that inhabit the biosphere. A two-letter code
could have specified only 15 amino acids and might have given rise to
a far more meager biota.

CHAPTER 11

THE NEUTRAL THEORY

OF MOLECULAR EVOLUTION

Motoo Kimura

The neutral theory asserts that the great majority of evolutionary changes at the molecular level are caused not by Darwinian selection but by random fixation of selectively neutral or nearly neutral mutants in the species. The theory also asserts that much of the intraspecific genetic variability at the molecular level is essentially neutral, so that most polymorphic alleles are maintained in the species by the balance between mutational input and random extinction (or fixation). In other words, it regards protein and DNA polymorphisms as a transient phase of molecular evolution and rejects the notion that the majority of such polymorphisms are adaptive and actively maintained by some form of balancing selection.

The essential aspect of the neutral theory is not that the alleles involved are selectively neutral in the strict sense. Rather, the emphasis is on mutation and random drift as explanatory factors in molecular evolution, because the selection intensity involved must in general be exceedingly small. In this respect, the designation "the neutral mutation–random drift hypothesis" may be more appropriate. However, the term "the neutral theory" is simpler and now widely used, so I shall adhere to this term. One of the characteristics of the neutral theory is that it is quantitative; it proposes to treat evolution and variation at the molecular level by using various models of population genetics and examining if they agree with quantitative data.

The neutral theory does not deny the occurrence of deleterious mutations. In fact, selective constraint due to such negative selection

208

is a very important part of the neutralist explanation of some prominent features of molecular evolution. The neutral theory does not deny
the possibility that some changes are adaptive. Thus, it is by no means
antagonistic to the Darwinian theory of evolution by natural selection.
However, because of its emphasis on mutation and random drift, and
also because of its accent on negative selection rather than positive
Darwinian selection, the neutral theory clearly differs in its theoretical framework from the traditional neo-Darwinian or "synthetic" theory of evolution.

The neutral theory has two roots. One is the stochastic theory of
population genetics, the foundation of which traces back to the great
work of R. A. Fisher, J. B. S. Haldane, and Sewall Wright early in
the 1930s and is mathematical in nature. Extension of their work in
terms of diffusion models (Kimura, 1964; Kimura and Crow, 1964;
Kimura, 1968b; Kimura and Ohta, 1969a,b; Kimura, 1969b) was very
useful for the subsequent development of "molecular population genetics" (Kimura, 1971; Nei, 1975). The other is molecular genetics,
which has revolutionized our concept of life and the impact of which
we are still feeling.

The proposal of the neutral theory (Kimura, 1968a), followed immediately by strong support from King and Jukes (1969) with their
provocative title "non-Darwinian evolution" and emphasis and extension of the theory by our group (Kimura, 1969a; Kimura and Ohta,
1971a,b; for review, see Kimura, 1979b and 1982a), led to a great deal
of controversy. This is often referred to as "the neutralist-selectionist
controversy" and has been documented by many authors, particularly
by Crow (1972, 1981), Calder (1973), and Lewontin (1974). Such a
controversy is not surprising because evolutionary biology has been
dominated for more than half a century by the neo-Darwinian view
that organisms become progressively adapted to their environments
by accumulating beneficial mutants, and evolutionists naturally expected this principle to extend to the molecular level.

I believe that this controversy stimulated much research, not only
in molecular evolution but also in population genetics of protein polymorphism. I also believe that it triggered reexamination of the orthodox synthetic theory of evolution, which at one time appeared to be
so firmly established. This orthodox theory was dominated by a pan-
selectionist view that the speed and direction of evolution are determined almost exclusively by positive natural selection with mutation
and random genetic drift playing only a minor and subsidiary role.
However, the traditional synthetic theory is no longer as firm as it
was in the late 1950s and early 1960s.

NEUTRAL EVOLUTION BY RANDOM GENETIC DRIFT

Before I present detailed data pertinent to the preceding points, I would like to enumerate some important properties of the behavior of neutral mutations in a finite population upon which the neutral theory is based.

1. The probability that a selectively neutral mutant eventually spreads through the whole population is equal to its initial frequency. Thus, in a population of N diploid individuals, if a mutant allele is represented only once at the moment of appearance, the probability of its eventual fixation is $1/(2N)$.

2. The rate of decrease of the heterozygosity by random drift is $1/(2N_e)$ per generation, where N_e is the effective population size (Wright, 1931). Note that N_e is roughly equal to the number of breeding individuals in one generation. Usually, N_e is considerably smaller than N.

3. If a new allele is produced at a locus with the rate v per generation, then the average length of time between consecutive substitutions of alleles in the population is $1/v$ generations (Crow and Kimura, 1970).

4. For each mutant allele destined to reach fixation, it takes on the average $4N_e$ generations from its first appearance until fixation, where N_e is the effective size of the population (Kimura and Ohta, 1969a). On the other hand, for a mutant allele destined to eventual loss, it takes on the average $(2N_e/N) \log_e(2N)$ generations until loss (Kimura and Ohta, 1969b).

5. If we assume that every mutation is unique and leads to a new allele (i.e., not preexisting), then the expected frequency of homozygotes (or sum of squares of allelic frequencies) under mutation-random drift equilibrium is $\bar{H}_0 = 1/(4N_e v + 1)$, where v is the mutation rate and N_e is the effective population size. The reciprocal of this quantity is called the effective number of alleles (n_e) so that $n_e = 4N_e v + 1$ for this model of mutation. The average heterozygosity at equilibrium, i.e., $1 - \bar{H}_0$, is then

$$\bar{H}_e = \frac{4N_e v}{4N_e v + 1} \tag{1}$$

(Kimura and Crow, 1964).

6. Let us call a population monomorphic at a given locus if the frequency of the most common allele happens to be $1 - q$ or higher, where q is a small quantity such as 0.01. Then, the probability of

a population being monomorphic is

$$P_{mono} = q^{4N_e a} + v \tag{2}$$

(Kimura and Ohta, 1971a).

7. Consider the process by which molecular mutants are substituted one after another, as shown in Figure 1. Let k be the rate of evolution in terms of mutant substitutions. This is defined as the long-term average of the number of molecular mutants that are substituted at a given locus or site in the species per unit time. If we restrict our consideration to only selectively neutral mutations, then

$$k = v \tag{3}$$

where v is the mutation rate per unit time (Kimura, 1968a). In other words, the rate of evolution in terms of mutant substitutions in the population (species) is equal to the mutation rate per gamete and is independent of population size. This remarkable property is only valid for neutral alleles. If the mutant substitution is due to positive Darwinian selection acting on definitely advantageous mutants, the corresponding formula for the rate of evolution is

$$k = 4N_e s v \tag{4}$$

(for derivation, see Kimura and Ohta, 1971b). In this formula s is the selective advantage of the mutant alleles (assumed to be

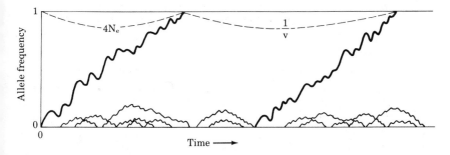

FIGURE 1. Behavior of mutant alleles in a finite population. Courses of change in the frequency of mutant alleles destined to fixation are shown by thick paths. For selectively neutral mutations, the average time until fixation of a mutant (excluding the cases in which mutants are lost) is $4N_e$ generations, and the average interval of consecutive fixations is $1/v$, where v is the mutation rate for neutral alleles, and N_e is the effective population size.

$4N_es \gg 1$) and v is the mutation rate for such advantageous alleles. Thus, in this case the rate of evolution depends on the effective population size (N_e) and on selective advantage (s) as well as the rate (v) at which mutants having such selective advantage are produced in each generation. However, if mutant alleles are nearly neutral such that their selective advantage or disadvantage (as measured by s) is much smaller than $1/(2N_e)$, Equation (3) holds approximately.

CONSTANT RATE
OF MOLECULAR EVOLUTION

Among the features that have been revealed by recent molecular evolutionary studies, the following two are particularly noteworthy. (1) For each protein, the rate of evolution in terms of amino acid substitutions is approximately constant per amino acid site per year for various lineages. (2) Molecules or parts of molecules that are subject to less functional constraint evolve faster (in terms of mutant substitutions) than those that are subject to stronger constraint. These two features show that the patterns of molecular evolution are quite different from those of phenotypic evolution and suggest that the laws governing them are also different.

The constancy (or uniformity) of the evolutionary rate is most apparent in hemoglobin (see Kimura, 1979b). The hemoglobin molecule in bony fishes and higher vertebrates is a tetramer, consisting of two identical α chains and two identical β chains. In mammals, amino acid substitutions in the α chain, consisting of 141 amino acids, occur roughly at the rate of one substitution per 7 million years. This corresponds to approximately 10^{-9} substitution per year per amino acid site, and, surprisingly, this does not seem to depend on such factors as generation time, living conditions, and population size. In Figure 2, the number of amino acid differences between the α chains of several vertebrates, including man, are given together with their phylogenetic tree and a list of geologic epochs to show the times of their divergence.

I once stated (Kimura, 1969a) that if the hemoglobins and other molecules of "living fossils" were shown to have undergone as many DNA base (and therefore amino acid) substitutions as the corresponding genes (protein) in more rapidly evolving species, this would support the neutral theory. Since then, the amino acid sequences of the β chain and the α chain of the principal hemoglobin of the Port Jackson shark have been determined (Fisher et al., 1977). According to Romer (1968), this shark is a relict survivor of a type of ancestral shark, which had numerous representatives in the late Paleozoic days, notably in the Carboniferous period (270–350 million years ago). So, this shark is well entitled to be called a living fossil. In Table 1, I present a result of comparison between the α and β chains of the Port

212

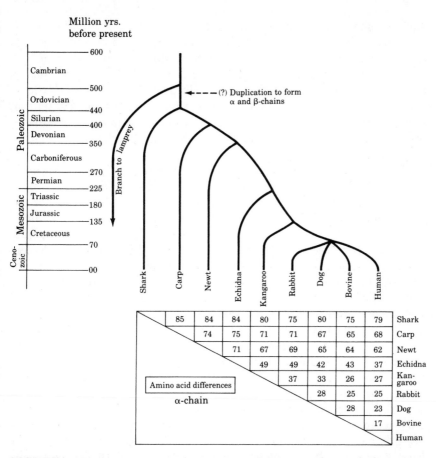

FIGURE 2. Phylogenetic tree of several vertebrates together with their times of divergence. The numbers of amino acid differences between the hemoglobin α chains of each pair of vertebrates are also given. Note that the numbers of differences are approximately 20 among the four mammals, whereas it is approximately 70 when the carp is compared with any one of the four mammals. (From Kimura, 1982a.)

TABLE 1. Comparison of amino acid differences between α and β hemoglobin chains.

Type of change*	0	1	2	3	Gap	Total
Sharp α vs. β	50	56	32	1	11	150
Human α vs. β	62	55	21	0	9	147

*In each comparison, the number of amino acid sites that can be interpreted from the code table as due to a minimum of 0, 1, 2, and 3 nucleotide substitutions are given together with the number of gaps (expressed as equivalents of the number of amino acid sites).

213

Jackson shark together with a similar comparison of α and β chains of humans. In each comparison, the numbers of amino acid sites that can be interpreted from the code table as due to a minimum of 0, 1, 2, and 3 nucleotide substitutions are listed. From the two sets of comparisons, it is clear that genes coding for the α and β chains of hemoglobin in the shark have diverged roughly to the same extent (or slightly more) as have the corresponding two genes in humans by accumulating random mutations since the origin of the α and β globin genes by duplication.

The constancy of the evolutionary rate, however, is not as uniform as the rate of radioactive decay. In fact, as shown by Ohta and Kimura (1971b), the observed variance (squared standard deviation) of the evolutionary rates for hemoglobins and cytochrome c among mammalian lines is roughly 1.5 to 2.5 times as large as the variance theoretically expected, if the variation is purely due to chance. Essentially the same result was obtained by Langley and Fitch (1974), who carried out more elaborate statistical analyses. Using data for α and β hemoglobins, cytochrome c, and fibrinopeptide A, they found that variation of the evolutionary rate in terms of mutant substitutions among branches of a phylogenetic tree is approximately 2.5 times as large as that expected from chance fluctuations. They took this as evidence against the neutral theory. At the same time they showed that when the estimated numbers of mutant substitutions are plotted against the corresponding geologic dates (in years), they are on a straight line, suggesting the uniformity of the evolutionary rates! The only exceptions are three observed points for primates for which there are other uncertainties. In my opinion, it is misguided to emphasize local fluctuations as evidence against the neutral theory while neglecting the inquiry of why the rate is intrinsically so regular or constant. The constancy of evolutionary rates per year at the molecular level has also been shown by Wilson and his group, using an immunological method involving albumin over a wide range of vertebraté species (Wilson et al., 1977).

The near-constancy of the amino acid substitution rate in evolution was termed "a molecular evolutionary clock" by Zuckerkandl and Pauling (1965) and has since been one of the most controversial subjects in molecular evolutionary studies. Particularly, it has been debated in relation to the date of human–ape divergence (Wilson et al., 1977), and as evidence for and against the neutral theory (see Kimura, 1979b).

A strong claim of nonconstancy has been made by Goodman and his associates in their study of globin evolution (Goodman et al., 1974, 1975). They maintain that mutant substitutions occurred at a very high rate in the early stage of globin evolution, soon after gene duplication to form myoglobin and α and β hemoglobins and that this

214

was followed by a markedly reduced rate during the last 300 million years from the ancestral amniote to the present. They make use of an extensive series of computer programs to assign mutations to various branches of the phylogenetic tree, to determine ancestral and contemporary nucleotide sequences. Together with their plausible explanation that, when new functions emerged in duplicated globin genes, rapid mutant substitutions occurred by positive Darwinian selection, the claim of Goodman et al. (1975) of nonconstancy has been quoted widely as evidence against the neutral theory.

I have pointed out (Kimura, 1981b) that the relatively high evolutionary rates obtained by Goodman et al. (1974, 1975) are not the result of their claimed superior statistical method but rather of the wrong assignment of geologic dates to duplication events in the early history of globin evolution, and that when this is corrected, there is no basis for their claim. I also pointed out that the great majority of globin codons determined by Goodman et al. (1974) using their maximum parsimony method are wrong when compared with the actual nucleotide sequences of globins determined by direct sequencing.

If we examine the data used by Goodman et al. (1974, 1975) carefully, we find that their estimate of the high evolutionary rate in the early stage of globin evolution comes almost entirely from their assumption that the gene duplication leading to lamprey globin on the one hand and, on the other, to the ancestral globin that later produced myoglobin and, still later, the α and β globins, occurred 500 million years ago (Kimura, 1981b). I pointed out that their positioning of gene-duplication points in time contains some errors, of which the most serious is their assumption that the duplication responsible for the myoglobin versus hemoglobin divergence occurred approximately 470 million years ago, after the ancestor of jawed fishes diverged from the ancestor of the lamprey. In other words, they assumed that the history of myoglobin is younger than that of the jawless fish (agnatha). The recent finding by Romero-Herrera et al. (1979) that true myoglobin exists in the red muscle of the lamprey heart seems to invalidate the assumption of Goodman et al. that myoglobin originated only approximately 470 million years ago.

In addition, the fallacy of maximum parsimony codons determined by Goodman and his associates has been revealed in a dramatic way by a recent study of Holmquist (1979), as discussed by Kimura (1981c). Furthermore, the "augmentation" procedure that has been used extensively by Goodman and his associates contains various problems as pointed out by Tateno and Nei (1978) and by Kimura (1981c).

All these findings clearly show that the study of Goodman et al.

(1974, 1975) gives no evidence that amino acid substitution was very rapid in the early stage of vertebrate globin evolution. Their claim that the high evolutionary rate is due to positive Darwinian evolution, although plausible, has no real supporting case in molecular evolution. It is much more likely, as I shall explain later, that the high rate is usually caused by removal of preexisting functional constraint, allowing a large fraction of mutations to become selectively neutral (not harmful) so that they can become fixed by random drift. Very high evolutionary rates recently revealed for globin pseudogenes are particularly suitable examples.

This, of course, does not mean that no adaptive mutant substitutions have occurred in the course of globin evolution. In my opinion, however, such definitely adaptive substitutions are much less frequent than selectively neutral or nearly neutral substitutions caused by random drift.

SELECTIVE CONSTRAINT AND THE NEUTRAL SUBSTITUTIONS

The second feature of molecular evolution is even more remarkable than the first. It has become increasingly clear that the weaker the functional constraint to which a molecule or a part of one molecule is subject, the higher the evolutionary rate of mutant substitutions. Among the proteins so far investigated, fibrinopeptides show the highest evolutionary rate, 8.3×10^{-9} per amino acid site per year (Ohta and Kimura, 1971a), although a lower estimate 4.5×10^{-9} has also been reported (Barnard et al., 1972). Fibrinopeptides have little known function after they become separated from fibrinogen during blood clotting. A similar example is the C peptide of the proinsulin molecule, which is removed when active insulin is formed. As shown in Figure 3, it evolves at a rate approximately six times as fast as that of insulin (for data, see Jukes, 1979). More examples of this sort will be forthcoming because tailoring of a precursor molecule to produce a functionally active form appears to be a common phenomenon. Dickerson (1971) explained the relationship between the functional importance (or more strictly, functional constraint) and the evolutionary rate using four molecules as follows. In fibrinopeptides, virtually all amino acid changes (mutant substitution) that permit the peptides to be removed are "acceptable" to the species. Thus, the rate of amino acid substitution may be close to the actual mutation rate. Hemoglobins, because of their definite function of carrying oxygen, are subject to more restrictive specifications than fibrinopeptides and thus have a lower evolutionary rate. Cytochrome c interacts with cytochrome oxidase and reductase, both of which are much larger than cytochrome c itself. Therefore, there is more functional constraint in cytochrome

216

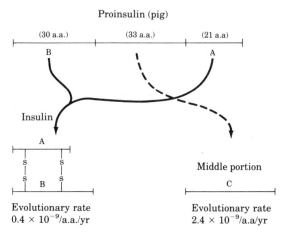

Proinsulin (pig)

FIGURE 3. Comparison of evolutionary rates of segments of insulin. When the active insulin molecule is formed from its precursor molecule, proinsulin, the middle segment (C peptide) is cut off and discarded. The rate of evolution is several times higher in the C peptide than in peptides A and B, which by joining through disulfide bridges form the insulin. (From Kimura, 1982a.)

c than in hemoglobins. Thus, cytochrome c has a lower evolutionary rate than hemoglobins. Histone H4 binds to DNA in the nucleus, and is believed to control the expression of genetic information. Thus, Dickerson considered it quite probable that a protein so close to the genetic information storage system is highly specified, with little evolutionary change over a billion years.

All these examples strongly suggest that natural selection (mostly in the form of negative selection) acts through the function of the folded protein but not the amino acid sequence itself. A similar relationship between functional constraint and evolutionary rate also holds in different parts within hemoglobin molecules. It is known (Perutz and Lehman, 1968) that in hemoglobins the surface part generally has much fewer constraints in maintaining the structure and function of the molecules than the heme pocket, which is vitally important. We have shown (Kimura and Ohta, 1973) that in both α and β hemoglobins the surface part evolves approximately ten times as fast as the heme pocket (Table 2). An additional example is afforded by the report of Barnard et al. (1972). According to them, sequence 15–24 of pancreatic ribonucleases evolves at a very high rate comparable to rapidly evolving parts of fibrinopeptides, and this "hypervari-

217

TABLE 2. Comparison of evolutionary rates between the surface and the heme pocket in the α and β hemoglobin chains. (From Kimura and Ohta, 1973.)*

Region	Hemoglobin α	Hemoglobin β
Surface	1.35	2.73
Heme pocket	0.165	0.236

* Values listed are in units of 10^{-9} substitutions per amino acid site per year.

ability" can be correlated with a lack of any contribution of this part either to the enzymatic activity or to the maintenance of structure required for the activity. Incidentally, their Table III (listing frequencies of amino acids in hypervariable segments) suggests that in such regions there might still exist some selective constraint in amino acid substitutions, so that not all of the mutations are tolerated.

As I mentioned already, there is growing evidence suggesting that the rate of nucleotide substitution is very high at the codon's third position where a large fraction (approximately 70%) of random nucleotide substitutions are synonymous (i.e., do not lead to amino acid changes). A remarkable finding was reported by Grunstein et al. (1976), who compared the histone H4 messenger RNA sequences from two sea urchin species, *Strongylocentrotus purpuratus* and *Lytechinus pictus*. Despite the fact that this protein has kept its amino acid sequence practically unchanged over billions of years, a number of synonymous nucleotide changes are found even between these two sea urchin species (see Figure 4). Using their data, together with paleontological knowledge on the time of divergence of these two species, I

L. pictus	mRNA	GAU AAC AUC CAA GG**A** AU**A** AC**U** AA**A** CC**G** GCA
S. purpuratus	mRNA	GAU AAC AUC CAA GG**C** AU**C** AC**C** AA**G** CC**U** GCA
Histone H4	Amino acid sequence	Asp Asn Ile Gln Gly Ile Thr Lys Pro Ala
		24 25 26 27 28 29 30 31 32 33

FIGURE 4. A part of the comparison between histone H4 messenger RNAs of two sea urchin species *Lytechinus pictus* and *Strongylocentrotus purpuratus*. Although the amino acid sequences of histone H4 of these two species are identical, there are five "synonymous" differences at the third positions of the codon (as indicated by small boxes) even in this short stretch of the messenger RNA. This diagram is constructed using data from Grunstein et al. (1976) and Grunstein and Grunstein (1978).

218

have estimated (Kimura, 1977) that the rate of nucleotide substitution is $(3.7 \pm 1.4) \times 10^{-9}$ per year at the third position of the codon. This is a very high rate, because the rate of nucleotide substitution for fibrinopeptides converted from the amino acid substitution rate amounts to approximately 4.0×10^{-9}. It is remarkable that synonymous mutant substitutions have occurred in the H4 gene almost at the highest known rate.

The preceding observations can be explained very simply and consistently by the neutral theory. Suppose that a certain fraction f_0 of the molecular mutants are selectively neutral and that the rest of the mutants are definitely deleterious. Although advantageous mutations will no doubt occur from time to time, we assume that their frequency is negligible in determining the evolutionary rates as estimated by comparative studies of amino acid and nucleotide sequences. If we denote by v_T the total mutation rate, then the mutation rate for neutral alleles is $v = v_T f_0$, so that from Equation (3) we obtain

$$k = v_T f_0 \tag{5}$$

We now assume that the probability of a mutational change being neutral (i.e., not harmful) depends strongly on functional constraints. The weaker the functional constraint, the larger the probability, f_0, of a random change being neutral, with the result that k in Equation (5) increases. According to this explanation, the maximum evolutionary rate is attained when $f_0 = 1$, that is, when all the mutations are neutral (assuming that v_T per site is the same among different molecules). In my opinion the high evolutionary rates observed at the third position of the codon are rather near this limit, although there may still exist a weak selective constraint shown by "nonrandom" synonymous codon usage, which I shall discuss later.

SYNONYMOUS AND OTHER SILENT SUBSTITUTIONS

Since the high evolutionary rate of synonymous nucleotide substitutions is of much interest for the neutral theory, I shall examine such substitutions more systematically in this section.

Table 3 lists the evolutionary distances in terms of the number of mutant substitutions per nucleotide site at the first, second, and third positions of the codon for pregrowth hormones and preproinsulins (both with respect to human versus rat comparison), β globins (chicken versus rabbit comparison), α tubulin (chicken versus rat), and histones H2B and H3 (both involving the comparison between the same pair of sea urchin species). The table also lists K_s, i.e., the synonymous

TABLE 3. Evolutionary distance (and standard errors) per site.

Comparison	Evolutionary distance per nucleotide site*			
	K_1	K_2	K_3	K_S
Human vs. rat pregrowth hormones	0.26 ± 0.04	0.18 ± 0.03	0.53 ± 0.07	0.44 ± 0.07
Human vs. rat I preproinsulins				
A + B chains (insulin)	0.04 ± 0.03	$0.00†$	0.46 ± 0.12	0.38 ± 0.12
C peptide	0.18 ± 0.06	0.27 ± 0.10	0.95 ± 0.46	0.77 ± 0.51
Chicken vs. rabbit β globins	0.30 ± 0.04	0.19 ± 0.03	0.64 ± 0.10	0.54 ± 0.10
Chicken vs. rat α tubulins	0.025 ± 0.008	0.005 ± 0.003	0.58 ± 0.05	0.47 ± 0.05
S. purpuratas vs. *P. miliaris*				
Histone H2B	0.09 ± 0.03	0.02 ± 0.01	0.48 ± 0.10	0.43 ± 0.10
Histone H3	0.008 ± 0.008	0.008 ± 0.008	0.47 ± 0.08	0.41 ± 0.08

*K_1, K_2, and K_3 stand for the number of base substitutions at the first, second and the third positions of the codon, whereas K'_S denotes the synonymous component of the distance at the third position of the codon.
† No observed changes among 51 codons.

component of the distance at the third position of the codon. Note that nucleotide substitutions at the third position of codons contain both synonymous and amino acid-altering substitutions. These estimates were obtained by using a model that I termed the "three substitution type" (3ST) model (Kimura, 1981a). The evolutionary rates per year may then be obtained by dividing these estimates by $2T$, where T is the divergence time for the two lineages compared. Throughout this chapter, I shall use the lower case letter k to denote the evolutionary rate and the capital letter K to denote the evolutionary distance.

The first line of Table 3 shows the estimates of distances obtained by comparing human pregrowth hormone (presomatotropin) (data from Martial et al., 1979) with rat pregrowth hormone (Seeburg et al., 1977). Because the human and the rat probably diverged late in the Mesozoic, some 80 million years ago, we may take $T = 8 \times 10^7$ for the divergence time. Then the evolutionary rates per year at these three codon positions are $k_1 = 1.6 \times 10^{-9}$, $k_2 = 1.1 \times 10^{-9}$, and $k_3 = 3.3 \times 10^{-9}$, and $k'_S = 2.8 \times 10^{-9}$ for the synonymous component. Comparison of human preproinsulin (Bell et al., 1980; Sures et al., 1980) with rat preproinsulin gene I (Cordell et al., 1979; Lomedico et al., 1979) presented in the second and third lines of Table 3 show very large differences of the evolutionary rates between $A + B$ chains (insulin) and C peptide with respect to the first two codon positions. In other words, with respect to amino acid substitutions, insulin evolves only approximately one-tenth as rapidly as proinsulin C peptide. Yet with respect to the synonymous component (K'_S), the evolutionary rate of insulin is not significantly lower as compared with that of C peptide, particularly if we note the large statistical error involved. Furthermore, K'_S value of insulin is practically equal to that of pregrowth hormone (both regarding human–rat divergence). The fourth line gives the estimates of the evolutionary distances between the nucleotide sequence of the chicken β globin (Richards et al., 1979) and that of the rabbit β globin (Efstratiadis et al., 1977), and the fifth line gives the distances obtained by comparing the sequences of the chicken α tubulin (Valenzuela et al., 1981) with the rat α tubulin (Lemischka et al., 1981). In both comparisons, the divergence times may be assumed roughly as $T = 3 \times 10^8$ years (see Romer, 1968). As far as the amino acid substitution rate is concerned, these two proteins have quite different evolutionary rates: β globin shows a standard evolutionary rate, whereas α tubulin is one of the most conservative proteins. In fact, the data reported by Lemischka et al. (1981) show that the amino acid substitution rate for α tubulin is only approximately 1/76 of that of β globin. Nevertheless, the genes for these two proteins

221

show essentially the same evolutionary rate with respect to synonymous substitutions. Such a remarkable agreement of synonymous substitution rates is also shown in the last two lines of Table 3, where the estimates of evolutionary distances between two sea urchin species *S. purpuratus* and *P. miliaris* are listed regarding histone H2B and H3 sequences (data from Sures et al., 1978 and Schaffner et al., 1978). These two species belong to different families, which diverged (6 ~ 16) × 10^7 years ago (Kedes, 1979). These two histones, particularly H3, are quite conservative; H3 has an amino acid substitution rate only 1/7 of that of H2B. However, with respect to the synonymous substitution at the third position of codons, these two histone genes are very similar. Furthermore, their synonymous substitution rates per year are not very different from the corresponding rates for pregrowth hormone ($k'_S = 2.8 \times 10^{-9}$), because, if we tentatively assume $T = 10^8$ years for the divergence time for these two sea urchin species, we get $k'_S = 2.2 \times 10^{-9}$ for histone H2B and $k'_S = 2.1 \times 10^{-9}$ for histone H3. Considering the uncertainty of the divergence time and the error of statistical estimation involved, such an agreement of the synonymous substitution rates is remarkable.

The fact that the rates of synonymous substitutions are not only higher than those of amino acid-altering substitutions but that they are also approximately equal among different genes have also been observed by Miyata et al. (1980).

In this connection, the evolutionary rate of nucleotide substitutions in introns is of interest, because the selective constraint in these parts is expected to be very weak (except for short stretches on either side at the junctions, which appear to have an important function as recognition sites in "splicing"). Using data presented by van Ooyen et al. (1979), who investigated similarity between the nucleotide sequences of rabbit and mouse β globin genes, I estimated the substitution rate in introns (Kimura, 1980). There are two introns in the β globin gene. The evolutionary distance per site between the small introns of rabbit and mouse turned out to be $K = 0.60 \pm 0.12$ (excluding five gaps that amount to six nucleotides). On the other hand, the corresponding value for the synonymous component of the distance at the third position of the codon between the two β globins was $K'_3 = 0.36 \pm 0.07$. Noting that 2/3 of random base substitutions at the third position of the codon are synonymous, we may estimate the true synonymous distance by $K_S = K'_S/(2/3)$, which gives $K_S = 0.54 \pm 0.11$. This means that the rate of mutant substitutions in the small intron is approximately equal to that of synonymous substitutions. On the other hand, the large introns of rabbit and mouse β globin genes differ considerably in length, being separated from each other by 14 gaps (determined by optimization of alignment of the sequences) that amount of 109 nucleotides. Excluding these parts, we get $K = 0.90 \pm 0.07$. This value

222

is significantly larger than K_S. It is likely, as pointed out by van Ooyen et al. (1979), that insertions and deletions occur rather frequently in this part in addition to point mutations and that they inflate the estimated value of the "nucleotide substitution rate," because a majority of these changes may also be selectively neutral and subject to fixation by random genetic drift.

RAPID EVOLUTIONARY CHANGE OF PSEUDOGENES

A pseudogene is defined as a region of DNA that shows definite homology with a known functional gene but has lost ability to produce a functional product due to mutational changes (Proudfoot, 1980). Pseudogenes are of unusual interest from the standpoint of the neutral theory for the following reason. If they are "dead genes," as they are sometimes called, being liberated from the constraint of negative selection, then there is a possibility that they have been accumulating various mutational changes at the maximum speed as predicted by the neutral theory. Indeed, comparison of pseudogenes with their normal counterparts shows that not only do base substitutions occur at a very high rate but also deletions and addition of nucleotides occur quite frequently. Unlike the "conservative" mode of change that characterizes the evolution of many normal genes, base substitutions at the first and the second positions of codons in the pseudogenes occur just as frequently as those at the third position.

Although some casual calculations were reported earlier (Kimura, 1980; Proudfoot and Maniatis, 1980) showing that pseudogenes evolved faster (in terms of base substitutions) than their normal counterparts, it was Miyata and Yasunaga (1981) who first made meticulous analyses to estimate the evolutionary rate of a pseudogene. They made use of the data reported by Vanin et al. (1980) on the nucleotide sequence of a mouse α globin pseudogene, which is called $\psi\alpha30.5$ and which is essentially equivalent to the mouse $\alpha3$ gene of Nishioka et al. (1980). This gene has a remarkable feature that it completely lacks the two introns present in all the functional α and β globin genes. In this sense, it resembles an α globin mature mRNA. Miyata and Yasunaga (1981) compared the sequence of this pseudogene with those of productive α globins from the mouse and the rabbit, as well as the latter two sequences. The result suggests that the pseudogene was created from a normal α globin gene by gene duplication in the mouse genome after the mouse line diverged from the rabbit line.

In order to reconstruct the evolutionary history of this α pseudogene, they assumed a molecular phylogenetic tree similar to Figure 8 in Chapter 2 by Li. In carrying out their analysis, Miyata and Yasunaga (1981) carefully noted that the third exon of the mouse α pseudogene ($M\psi\alpha$) differs from the other exons in that it is highly conserved. This suggests the possibility that this part of $M\psi\alpha$ was derived from one of the closely related α globin genes ($M\alpha$) by unequal crossing-over relatively recently, so they excluded this part in their analysis. They assumed that the gene duplication that led to $M\psi\alpha$ occurred in the mouse line T_d years ago and that the duplicated gene evolved at the same rate as the normal α globin gene until T_n years ago when it became nonfunctional (see Figure 8 in Chapter 2). Let us designate the amino acid-altering substitution rate by k_A and the synonymous substitution rate by k_S. We also designate by k_0 the rate of nucleotide substitution in pseudogenes, which is assumed to be the same for all three positions of codons.

Assuming that the divergence time of the mouse and the rabbit (T in Figure 8 in Chapter 2) is 75 million years ago, they obtained $T_d = 24$ million years and $T_n = 17$ million years. They also obtained $k_A = 0.82 \times 10^{-9}$, $k_S = 6.6 \times 10^{-9}$, and $k_0 = 12.6 \times 10^{-9}$ per site per year, so that $k_0/k_S = 1.91$. This means that the rate of mutant substitutions in $M\psi\alpha$ is 1.9 times as high as the synonymous substitution rate for the functional α genes. This suggests that, in the normal functional gene, there exists some selective constraint, so that not all synonymous mutations are selectively neutral. Miyata and Hayashida (1981) consider that the evolutionary rate of the pseudogene may correspond to the maximum rate predicted by the neutral theory (Kimura, 1977), namely, the case of $f_0 = 1$ in Equation (5). Whether the rate of evolution in terms of mutant substitutions in this case is really equal to the rate of point mutations, i.e., $k = v$ (both being measured with the unit of one year), is an extremely important problem for the neutral theory. Eventually, it will be possible to estimate the mutation rate (v) directly thereby testing the validity of this prediction.

A similar analysis was carried out by Takahata and Kimura (1981) assuming a rather general model of mutation. We followed Miyata and Yasunaga (1981) in excluding the third exon in the analysis. The main difference is that we assumed that the pseudogene $M\psi\alpha$ lost its function immediately after its birth by duplication, that is, $T_n = T_d$. We calculated the evolutionary rate separately for the three codon positions, denoting by k_i the rate at position i ($i = 1,2,$ or 3) for the normal gene. We assumed that duplication occurred T_d years ago, and immediately thereafter a duplicated gene became "dead" and started to evolve at the rate k_i' instead of k_i. Then the following ratios were obtained for the first, second, and the third positions of codons: $k_1'/k_1 = 11.5$, $k_2'/k_2 = 13.9$, and $k_3'/k_3 = 0.9$. Also, the ratio T_d/T for the three codon positions turned out to be 0.26, 0.42, and 0.43, giving the average of 0.37. Thus, if we assume $T = 8.0 \times 10^7$ years for the diver-

224

gence time of the mouse and the rabbit, we get $T_d = 3.0 \times 10^7$ years, suggesting that this dead gene was created approximately 30 million years ago by gene duplication. Although these values are tentative, it is likely that base substitution rates increased some tenfold in the first two codon positions after the gene lost its function, with the result that selective constraint disappeared entirely.

An elegant statistical analysis of the evolutionary rates of pseudogenes has been carried out by Li et al. (1981). They used sequence data for the human globin pseudogene $\psi\alpha1$ (Proudfoot and Maniatis, 1980), rabbit globin pseudogene $\psi\beta2$ (Lacy and Maniatis, 1980) and mouse $\psi\alpha3$ (Nishioka et al., 1980). They estimated that the average substitution rates at the first, second, and third positions of codons for the normal globin genes used are $k_1 = 0.71 \times 10^{-9}$, $k_2 = 0.62 \times 10^{-9}$, and $k_3 = 2.64 \times 10^{-9}$, and the average substitution rate per site for the pseudogenes during their nonfunctional periods is $k_0 = 4.6 \times 10^{-9}$ (per years). They noted that 4.6×10^{-9} is one of the highest rates of the nucleotide substitutions so far estimated.

Although the estimate of Li et al. for k_0 is less than one-half the corresponding estimate by Miyata and Yasunaga, the ratio $k_0/k_3 = 1.74$ does not differ much from the value of $k_0/k_S = 1.91$ obtained by Miyata and Yasunaga. These results suggest that the synonymous substitution rate, even if high, is not really the maximum rate predicted by the neutral theory, and therefore the synonymous mutations are subject to negative selection, though the intensity of selection involved must be very weak. This is reflected in the "nonrandom" usage of synonymous codons, as I shall discuss in the next section.

It is likely that the very high evolutionary rate observed for globin pseudogenes in mammalian evolution (roughly 5×10^{-9} per nucleotide per year) is very close to the maximum rate corresponding to $f_0 = 1$ in Equation (5) (see also Chapter 2 by Li). It is also estimated that this maximum rate is approximately ten times as high as the amino acid-altering substitution rate in the evolution of normal globin genes, and it is nearly twice as high as the synonymous substitution rate.

"NONRANDOM" USAGE OF SYNONYMOUS CODONS

It has become increasingly evident that synonymous or "degenerate" codons are used quite unequally or in "nonrandom" fashion in many genes of various organisms (e.g., Grantham et al., 1980b). Furthermore, as will be mentioned later, there is a consistent pattern of usage for various genes within the genome of a species, and there is good evidence suggesting that choice of synonymous codon is largely constrained by tRNA availability. This leads to a hypothesis that the preferential codon usage represents the optimum state in which the population of synonymous codons matches that of cognate tRNA available in the cell. I have shown (Kimura, 1981d) that the concept of stabilizing selection can be applied to treat the problem using the diffusion equation method and that the observed pattern of nonrandom codon usage and its effect on retarding the fixation of synonymous mutants can be explained in quantitative terms by this model. Thus, the universal phenomenon of nonrandom codon usage can be explained satisfactorily within the framework of the neutral theory in terms of selective constraint. For me, this is most reassuring, because this phenomenon has often been mentioned as evidence against the neutral theory.

One of the best examples showing nonrandom usage of synonymous codons is given by the codons for the amino acid leucine (Leu). There are six codons that code for Leu, namely, UUA, UUG, CUU, CUC, CUA, and CUG. Of these six codons, CUG is used predominantly by the genes in bacteria (Grantham et al., 1981). In fact, the frequency of usage of CUG amounts to approximately 60% of all the six codons coding for leucine.

A quite different pattern of codon usage is apparent in yeast (*Saccharomyces*). In this organism, among the six codons coding for leucine, UUG is by far the most frequently used codon (amounting to approximately 80%). It is also known that as far as the usage pattern of leucine codons is concerned, man is nearer to bacteria than to yeast.

The most thorough analysis of the pattern of codon usage that has been made so far is that of Grantham and his associates (Grantham et al., 1980a,b; 1981), who compiled extensive tables of codon usage from reported messenger RNA sequences of various organisms. They applied "correspondence analysis," a multivariate statistical method, to the distances between messenger RNAs based on differences in the usage of the 61 codons; each messenger was treated as a point in a multidimensional space. Then the data were projected onto a plane whose horizontal and vertical axes correspond to the first and second most important factors. Grouping was done by an automatic classification. The results of analysis of 119 mRNAs (Grantham et al., 1980a) showed that most genes in a genome have the same "coding strategy" with respect to choices among synonymous codons. Thus, mammalian, bacterial, virus, mitochondrial and yeast plus slime mold genes fall in different classes. In other words, there is a consistent choice of degenerate bases, and this confirms "the genome hypothesis" of Grantham et al. (1980b), which states that each gene in a genome tends to conform to its species' usage of the codon catalog. On the other hand, application of correspondence analysis to amino acid frequencies of 119 proteins shows that no grouping of proteins by genome type is

226

evident; viral, bacterial, mammalian, and other proteins lie in the same class.

Grantham (1980) claims, based on these observations, that messenger RNA is an evolutionary structure in its own right and that the work by his group reveals protein-independent molecular evolution of a "nonneutral character." In my opinion, however, what he calls "nonneutral" is an expression for selective constraint rather than that of the adaptive mutant substitutions that Grantham seems to have in mind.

Then, the problem is, what is the main cause of selective constraint that brings about such a choice of degenerate codons characteristic of each genome type? An important step forward in understanding this cause has been made by Ikemura (1980, 1981a,b). Using two-dimensional polyacrylamide gel electrophoresis, he separated the 26 known tRNAs of *Escherichia coli* and measured their relative abundance in terms of molecular numbers in the cells. He then investigated the relationship between the tRNA abundance and the frequency of usage of the corresponding codons in *E. coli* genes, such as tryptophan synthetase A protein (*trp* A), lac operon repressor (*lac* I), and ribosomal protein (r-protein). A very strong correlation was found between the tRNA abundance and choice of codons among synonymous codons and also codons corresponding to different amino acids. Of particular interest, in the present context, is the finding that among synonymous codons for an amino acid the most frequently used codon invariably corresponds to the most abundant isoaccepting tRNA species. Figure 5 illustrates this relationship for the six synonymous codons coding for leucine. The solid columns represent the relative frequencies of codon usage and the hatched columns represent the abundance of the cognate tRNA species. It is clear that the agreement between these two is excellent. Note that codons CUC and CUU are recognized by a single tRNA species (tRNA$_2^{Leu}$), and, so also are UUG and UUA.

Ikemura's proposal that codon usage is mainly constrained by the availability of the corresponding tRNA species has been strengthened by his finding that such a correlation also exists in yeast, which has a very different pattern of codon usage from that of *E. coli* (Ikemura, 1982; see also Bennetzen and Hall, 1982). For example, among six synonymous codons coding for leucine, the most frequently used one is UUG in yeast. Ikemura found that, in this case, the most abundant cognate tRNA species is exactly the one that recognizes UUG (see Figure 6). Incidentally, these findings cast doubt on the validity of the suggestion made by Modiano et al. (1981) that UUA and UUG are "pretermination codons" (that can mutate to termination codons by a

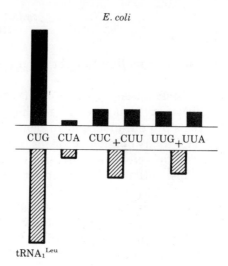

E. coli

CUG   CUA   CUC ₊CUU   UUG₊UUA

tRNA₁^{Leu}

FIGURE 5. Comparison of the relative frequencies of use of codons for leucine (solid columns) and the relative abundance of the corresponding cognate tRNA species (hatched columns) in *Escherichia coli*. The frequencies of use are based on the data compiled by Grantham et al. (1981), and the data on the abundance of tRNAs are from Ikemura (1981a,b). The plus sign between codons CUC and CUU means that these two codons are recognized by a single tRNA species (i.e., tRNA$_2^{Leu}$), the relative abundance of which (taking that of tRNA$_1^{Leu}$ as unity) is given by the hatched column. There is a similar meaning for the plus sign between UUG and UUA. The relative amount of tRNA corresponding to CUA, which is a minor species, is 0.1 (Ikemura, pers. comm.).

single nucleotide substitution) and because there are four other non-pretermination codons available for leucine, codons UUA and UUG are never used in the normal α and β globin genes. These authors claim that usage of pretermination codons is avoided whenever possible as an evolutionary strategy for reducing the rate of mutation with drastic effects. It seems to me that the selective advantage coming from such a strategy is too small (presumably the order of the mutation rate) to be effective in the actual course of evolution.

Previously, Fitch (1980) examined the β globin mRNA sequences for human, mouse, and rabbit and observed that there is a significant bias against the use of codons differing from terminating codons by only one base. Rarity of usage of UUA and UUG is also evident in other human genes, and, it is likely, if Ikemura's hypothesis is valid, that this is due to the paucity of the tRNA species that recognize these two codons rather than to their "pretermination" codons.

We may then assume that, in general, the choice of synonymous codon is largely constrained by tRNA availability and that this is

228

related to translational efficiency as suggested by Post et al. (1979) and Ikemura (1981b). As I already mentioned, this leads to a hypothesis that preferential codon usage represents the optimum state in which the population of synonymous codons matches that of cognate tRNA available in the cell. This will help in carrying out more efficient cell function leading to higher Darwinian fitness. Then, the concept of stabilizing selection can be applied to treat the problem. Because the mathematical treatment involved is rather intricate (Kimura, 1981d), I shall not repeat it here. Instead, I would like to mention a few observations that are pertinent to this theoretical treatment. First, this is compatible with the genome hypothesis of Grantham et al. (1980b), which states that a surprising consistency in the choice of degenerate bases exists among genes of the same or similar genomes. The observation that synonymous substitution rates are not only high but are similar among different genes in a genome can also be understood by noting that they are subject to the same cause of constraint coming from the availability of tRNA species in the cell. Furthermore,

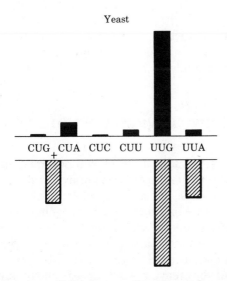

FIGURE 6. Comparison of the frequencies of use of synonymous codons for leucine and the abundance of the corresponding cognate tRNA species in yeast. The solid columns represent the relative frequencies of use and the hatched columns the relative tRNA abundance. The frequencies of use are based on the data compiled by Grantham et al. (1981), and the data on the relative abundance of tRNAs are from Ikemura (pers. comm.).

the observation suggesting that strongly expressed genes (corresponding to abundant proteins) are subject to stronger constraint than weakly expressed ones (Ikemura, 1981a,b,; Bennetzen and Hall, 1982) can readily be understood from the relationship between the phenotypic effect and selection coefficient in quantitative characters (see Crow and Kimura, 1970, p. 228). Second, this approach enables us to estimate the extent by which the rate of evolution in terms of mutant substitutions is retarded by unequal codon usage. The result suggests that the observed pattern of unequal usage and observed rate of retardation are indeed compatible.

Probably the most significant fact revealed from the various studies mentioned above is that evolution (in terms of base substitutions) is *slowed down rather than accelerated* by nonrandom choice among synonymous codons. It supports the neutralist interpretation that the selection involved is negative (coming from stabilizing selection) rather than positive Darwinian selection usually envisaged by the selectionists.

Although much remains to be clarified, I find it pleasing that such diverse observations as the rapid change of pseudogenes and nonrandom usage of synonymous codons can be understood in a unified and quantitative way under the framework of the neutral theory by incorporating the concept of stabilizing selection.

Existence of a maximum evolutionary rate set by the total mutation rate in accordance with the formula $k \leq v_T$, I believe, is one of the very successful predictions (Kimura, 1977) of the neutral theory; this prediction is now vindicated by comparative studies of DNA sequences, which have suddenly become possible on a large scale during the last few years.

EVOLUTION AT TWO LEVELS, PHENOTYPIC AND MOLECULAR

Classical evolutionary studies have shown beyond doubt that positive Darwinian selection is the major cause of evolutionary change at the phenotypic level, that is, at the level of form and function. On the other hand, as I have explained in the foregoing sections, mutation pressure and random genetic drift prevail in evolutionary change at the molecular level. As more and more data on nucleotide sequences accumulate, I am convinced that the neutral theory will prove to be increasingly useful for treating such data from the standpoint of evolution and intraspecific variability studies.

One final question is why natural selection is so prevalent at the phenotypic level and yet random fixation of selectively neutral or nearly neutral alleles prevails at the molecular level. The answer to this question, I believe, comes from the fact that the most common type of natural selection at the phenotypic level is "stabilizing" selection. It eliminates phenotypically extreme individuals and preserves those that are near the population mean (Mather, 1953; Haldane, 1959).

Unlike the type of natural selection that Darwin had in mind when he tried to explain evolutionary change, stabilizing selection is a conservative force acting to keep status quo. Since the early work of H. C. Bumpus on the house sparrow and W. F. R. Weldon on the land snail, many examples of stabilizing selection have been reported (see Parkin, 1979, for review). Probably the best example in man is the relationship between the birth weights of babies and their neonatal mortality studied by Karn and Penrose (1951). These authors found that babies whose weight is very near the mean have the lowest mortality. This optimum weight is slightly heavier than the mean, and the mortality increases progressively as the weight deviates from this optimum.

Let us consider a quantitative character, such as height, weight, concentration of some substance, or a more abstract quantity that represents Darwinian fitness in an important way. We assume that the character is determined by a large number of loci or sites, each with a very small effect in addition to being subjected to environmental effects. Let us also assume that genes are additive with respect to the character and that under the continued action of stabilizing selection the mean has been brought very near the optimum or even to coincide with it. Under such conditions, it can be shown (for details, see Kimura, 1981d) that the intensity of natural selection involved between alleles at an individual locus or site can be exceedingly small and that every mutation is slightly deleterious but nearly neutral. In fact, the selection is frequency dependent and is equivalent to negative overdominance. Furthermore, it can be shown that negatively overdominant alleles are far more susceptible to random drift than unconditionally deleterious alleles having the same magnitude of selection coefficient. Thus, the mutant substitutions are mainly controlled by random drift, although the rate of neutral evolution is lower than when all the mutations are strictly neutral.

I have already mentioned one application of this theory to the problem of nonrandom or unequal usage of synonymous codons, and that this model gives a satisfactory solution of the problem within the framework of the neutral theory.

More generally, we can think of a large number of quantitative characters that collectively constitute the entire phenotype of an individual. During its lifetime, an individual is subject to natural selection through such characters. Many of these characters are mutually correlated, so let us suppose that we can choose a certain number of independent characters that represent, as a first approximation, the total pattern of selection. This collection may be called the total phenotype.

Under a simplifying assumption that the component quantitative characters are equivalent, we can estimate the average selection intensity per segregating nucleotide site. The selection intensity per site may be expressed by a coefficient s_s, which may be called the selection coefficient for stabilizing selection. Then, we obtain (Kimura, 1981d)

$$s_s = - [\ln (1 - L_T)]\rho^2/(n_{nuc}\bar{h}_e) \tag{6}$$

where L_T is the total selection intensity or the load; ρ^2 is the fraction of genotypic (hereditary) variance in the total phenotypic variance; n_{nuc} is the total number of nucleotide sites concerned; and $\bar{h}_e$ is the average heterozygosity per site. For mammals, a typical set of values of these parameters may be as follows: $L_T = 0.5$ (50% selective death), $\rho^2 = 0.5$ (50% broad sense heritability), and $n_{nuc}\bar{h}_e = 10^6$ (one million heterozygous nucleotide sites per individual). Substitution of these values in Equation (6) yields $s_s = 3.5 \times 10^{-7}$, which is an exceedingly small selection coefficient. This means that the great majority of mutations at the molecular level are nearly neutral but very slightly deleterious. This agrees with Ohta's hypothesis of very slightly deleterious mutations (Ohta, 1973, 1974), but the fitness of the species does not drift downward in this view as it does in Ohta's hypothesis. Also, in this view, those genes that are substituted by random drift and those that are responsible for phenotypic variability of quantitative traits belong to the same class.

This approach has the merit of showing that neutral molecular evolution is an inevitable process under stabilizing phenotypic selection when a very large number of nucleotide sites are involved. This approach was called "a unified selection theory" by Milkman (1982), who conceived essentially the same idea independently and who presented his motivation and rationale in his article.

I would like to add here that there is a possibility that a certain fraction of nucleotide sites (presumably a large fraction) produce no phenotypic effects at all and therefore are completely neutral with respect to natural selection. Mutational changes in such sites are selectively neutral in the strict sense of the word. On the other hand, there is an additional, biologically much more important possibility, namely, some of the "neutral alleles" that are functionally equivalent or nearly so under a prevailing set of environmental conditions of a species become selected when a new environmental condition is imposed. Experiments strongly suggesting this possibility have been reported by Dykhuizen and Hartl (1980) and Hartl and Dykhuizen

232

(1981). Indeed, neutral mutants can be the raw material for adaptive evolution.

Thus, we are led to a general picture of evolution as follows. From time to time, the position of the optimum of a phenotypic character shifts as a result of change of environment, and the species tracks such a change rapidly by altering its mean. During this short period of change, extensive shift of gene frequencies is expected to occur at many loci, but this process itself will seldom cause gene substitutions. But, most of the time, stabilizing selection predominates, under which neutral evolution or random fixation of mutant alleles occurs extensively, transforming all genes including those of living fossils.

CHAPTER 12

EVOLUTION OF NEW
METABOLIC FUNCTIONS
IN LABORATORY ORGANISMS

Barry G. Hall

Studies of electrophoretic variants in a wide variety of proteins from an almost unending number of species have shown us that there is an enormous amount of genetic variability in natural populations. Theoretical considerations have led to the realization that much of the observed variability must have resulted from the fixation of neutral or nearly neutral mutations by stochastic processes. A good deal of effort has gone into trying to estimate the fraction of observed genetic diversity that is neutral, as opposed to that which is "adaptive." Nearly ten years ago Lewontin (1974) pointed out that to a large extent we had become sidetracked from considering the fundamental question: "How much genetic variation is there that can be the basis of adaptive evolution?" He pointed out that it will be necessary to evaluate the potential for adaptive change that is conferred by presently neutral or nearly neutral mutations.

It has been a major source of frustration in the field of molecular evolution that while we can observe evolutionary changes in protein sequences and enormous genetic variabilities in natural populations, we have not been able to directly relate those changes or variabilities to fitness differences. This is mainly due to the fact that we have been forced to observe, primarily, a single point (the present population) of a long dynamic process of evolution. No matter how detailed the observation is, it is virtually impossible to understand any long-term process by observing a single time point.

234

In the study of electrophoretic variation in natural populations or amino acid substitutions in a set of homologous proteins, it is difficult to know whether or not a particular electrophoretic variant is selectively neutral or whether or not a particular amino acid substitution occurred by genetic drift alone (see Chapter 6 by Koehn et al.). If we cannot evaluate the adaptive differences associated with specific alleles, how can we begin to approach the question of "potentiality for adaptive evolution in genetic variation that may currently be nonadaptive" (Lewontin, 1974)?

To answer such questions we are forced to turn from studies of natural populations to those of model systems in which the adaptive differences of specific alleles can be measured and manipulated. Experimental evolutionists have attempted to directly establish causal relationships between fitness differences and molecular variation by applying strong selective pressures to populations of microorganisms and monitoring the molecular changes that occur as the populations adapt. We are thus studying evolution by observing it as a dynamic process. The selection pressure is usually for the acquisition of a new catabolic function, that is, for the ability to utilize a novel resource as a source of carbon, nitrogen, or energy. In this way the evolution of new metabolic functions is used as a model system to study general questions of the relationships between molecular variation and fitness differences. As is always the case, model systems represent simplified situations that permit us to pose questions in ways that are amenable to experimental analysis. Several problems have been experimentally studied with these model systems. How many ways are there to evolve a particular new metabolic function? What are the relative roles of regulatory versus structural gene mutations in evolution of new functions? How many new, related functions can evolve from a single ancestral gene? How are newly evolved genes integrated into the preexisting regulatory system of the cell? How many mutational steps are required to evolve a particular new function? Must all of these steps be advantageous mutations, or can some of them be neutral mutations? Can we identify any neutral mutations that have the potential for being adaptive?

There are legitimate concerns about the degree to which we can extrapolate the results from laboratory experiments to natural populations. One special concern is the applicability of models based upon unicellular organisms to multicellular organisms, where specialized tissues and the isolation of the germ line from somatic cell lines may provide a strong buffer between the environment and selection acting on transmission of genetic information. Nevertheless, these studies have for the first time allowed us to make direct connections between adaptive fitness and the evolution of single genes.

EVOLUTION OF NEW FUNCTIONS PRIMARILY BY MUTATIONS IN REGULATORY GENES

Evolution of xylitol utilization

How do cells acquire new metabolic functions? Studies with a variety of organisms in numerous laboratories have shown that the most common way for a microorganism to acquire a new metabolic capability is to synthesize constitutively an enzyme that already has a specificity for the novel resource, but for which the novel resource is not an inducer. E. C. C. Lin and his students carried out some of the earliest experiments deliberately designed to study the evolution of new metabolic functions by applying specific selective pressures to populations of microorganisms. *Aerobacter* (now *Klebsiella*) *aerogenes* strain 1033 cannot use xylitol as a sole source of carbon and energy. Because xylitol can freely diffuse into the cell, Lin's group chose to study the mutations that would enable strain 1033 to utilize xylitol (Learner et al., 1964). Wild-type cells were plated onto xylitol minimal medium, and a spontaneous xylitol-utilizing mutant was obtained. The mutant, designated strain X1, expressed the ribitol operon constitutively (see Figure 1). Several lines of evidence showed that ribitol dehydrogenase (RDH) activity was responsible for xylitol metabolism and that constitutive synthesis of RDH was both a necessary and a sufficient condition for the evolution of xylitol utilization. First, an independent RDH constitutive mutant, selected by other means, grew on xylitol at exactly the same rate as did strain X1. Second, a mutant of X1 that had lost its RDH activity had also lost the ability to grow on xylitol. Third, RDH from extracts of strain X1 was able to oxidize both ribitol and xylitol; and during thermal inactivation experiments the activity toward ribitol was lost at exactly the same rate as activity toward xylitol. Finally, the RDH from strain X1 was indistinguishable from the RDH synthesized by the parental strain when it was grown on ribitol.

RDH oxidizes xylitol to xylulose, a normal intermediate in the pathway for D-arabitol metabolism. In the parental strain, xylitol does not induce expression of the D-arabitol operon, but in strain X1 (and in other RDH constitutives) the D-arabitol operon is induced by growth on xylitol. It is thought that the intermediate xylulose may be the normal inducer of the D-arabitol operon. Thus, a single regulatory mutation was sufficient for the evolution of xylitol utilization. Note, however, that this simple alteration was adequate only because the product formed by the action of RDH on xylitol (xylulose) was able to gain entry to the central metabolic network via the preexisting D-arabitol pathway. Mortlock and his students obtained similar results when they selected xylitol-utilizing mutants of *Aerobacter aerogenes* strain PRL-R3 (now designated *Klebsiella pneumoniae* variety *oxytoca*) (Mortlock and Wood, 1964a,b; Fossitt et al., 1964; Mortlock et al., 1965).

When Learner et al. (1964) treated strain X1 with nitrosoguanidine and selected for faster growth on xylitol, they obtained strain X2, which grew 2.5-fold faster on xylitol than did strain X1. The mutation in strain X2 was in the structural gene for RDH, and the RDH in strain X2 showed a 2.5-fold increase in activity toward xylitol relative to ribitol. In a later study (Wu et al., 1968), mutant X3 was isolated following another round of nitrosoguanidine treatment and selection for rapid growth on xylitol. Strain X3 grew twice as fast as X2 on xylitol, and the rapid growth was shown to result from the constitutive expression of the D-arabitol permease, which already had low transport activity toward xylitol. This third mutation was thus another regulatory mutation that led to constitutive expression of a protein with a weak activity toward the novel substrate.

It is important to note that the changes in fitness conferred by the individual mutations depended strongly upon the genetic background of the mutation. The RDH constitutive mutation in strain X1 would not have increased fitness (allowed growth on xylitol) if the resulting xylulose could not have been metabolized rapidly via another pathway. The mutation in X2 that improved the activity of RDH toward xylitol would not have been advantageous in a nonconstitutive background because xylitol is not an inducer of the ribitol operon. The mutation in strain X3 was advantageous only because the greatly increased activity of the improved RDH had made xylitol permeation a rate-limiting step. Thus, the individual mutations increased the fitness of the organism with respect to a particular environment (xylitol as the only available nutrient) and with respect to a particular set of biochemical information that we call "genetic background."

If we are concerned with relating particular biochemical changes to changes in fitness, then we must consider the effects of the mutations in the *absence* of the specific selective pressure that was applied. This is an important consideration, because in nature organisms do not exist in constant environments, and the amount of a novel resource in the environment may vary in either a random or a cyclical manner. A mutation that creates a new metabolic function must be advantageous when its fitness is averaged over all the environments that the host organism encounters. Thus, if the mutation were disadvantageous in the absence of the novel resource and if the novel resource were in the environment only rarely, the new mutation would tend to be lost rapidly during periods of absence of the novel resource. There is an intuitive feeling that, in the absence of the substrate for the enzyme, constitutive enzyme synthesis must be wasteful and hence disadvantageous. There is little direct evidence to substantiate that intuitive feeling; but constitutive synthesis may be disadvantageous for reasons other than waste of resources.

Scangos and Reiner (1978) have shown that constitutive expression of the ribitol operon in *Escherichia coli* C allows utilization of xylitol, just as it does in *Klebsiella*. However, these xylitol-positive mutants are inhibited by other pentitols to which the parent strain is resistant. The parental strain is unable to utilize L-arabitol or galactitol, but these polyols do not inhibit the growth of the parent on other resources. The xylitol-positive mutants are strongly inhibited by these or other nonmetabolizable polyols. The inhibition is caused by the activity of another ribitol operon enzyme, D-ribulokinase (DRK). Constitutive expression of the ribitol operon leads to high levels of DRK, as well as RDH, and the DRK apparently phosphorylates the nonmetabolizable polyols, thereby generating toxic polyol phosphates. Mutants that lose their DRK activity are no longer inhibited by nonmetabolizable polyols.

These examples point out that evolution of a new metabolic function involves more than simple synthesis of the necessary degradative enzymes; it also requires integration of the new enzyme system into the overall metabolic network of the cell.

Evolution of D-arabinose utilization

Neither *K. pneumoniae* nor *E. coli* normally utilizes D-arabinose. Mutants of *Klebsiella* selected to grow on D-arabinose are found to be constitutive for the synthesis of L-fucose isomerase, which converts D-arabinose to D-ribulose (Oliver and Mortlock, 1971). D-Ribulose is a normal intermediate in the ribitol pathway and is the internal inducer for expression of the ribitol operon. Thus, in *Klebsiella*, the evolutionary strategy for acquiring D-arabinose utilization is virtually identical to that for xylitol utilization; it is to synthesize an enzyme constitutively to convert the novel resource into an intermediate that is normally metabolized by another pathway.

Constitutive synthesis, however, is not the only way to exploit an existing pathway to metabolize a novel resource. In a mutant of *E. coli* K12 that could grow on D-arabinose, the L-fucose pathway had become inducible by D-arabinose (LeBlanc and Mortlock, 1971). Unlike *E. coli* C, *E. coli* K12 does not have a ribitol operon, so the enzymes of the ribitol pathway are unavailable for further metabolism of the D-ribulose formed by the D-arabinose-inducible L-fucose isomerase. In-

238

stead, in *E. coli* K12, the D-arabinose was metabolized entirely by the enzymes of the L-fucose pathway (see Figure 1). In this case, a regulatory mutation allowed a single existing pathway to be exploited for the metabolism of the novel resource. In *E. coli* K12 the nature of the regulatory mutation allowed the new capability to be immediately integrated into the existing metabolic network.

Evolution of propanediol utilization

The evolution of L-1, 2-propanediol utilization in *E. coli* K12 provides a particularly striking example of the disruption of a normal regula-

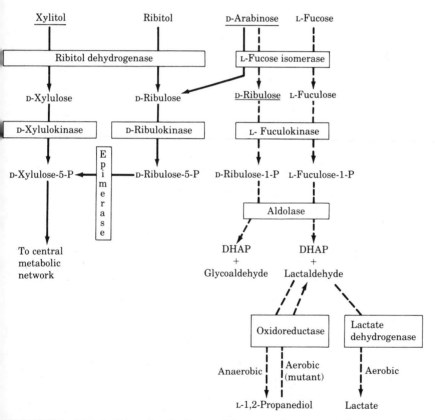

FIGURE 1. Metabolism of xylitol, D-arabinose, and propanediol in *Klebsiella* and *E. coli*. Novel substrates and intermediates are underlined. Enzymes are enclosed in boxes. Solid arrows show *Klebsiella* and *E. coli* C pathways and dashed arrows show *E. coli* K12 pathways.

tory network by mutations that lead to a new metabolic capability. During anaerobic growth on L-fucose, the lactaldehyde formed by the aldolase (see Figure 1) is converted to L-1, 2-propanediol by an oxidoreductase, and the propanediol is excreted. The oxidoreductase is not synthesized aerobically, and during aerobic growth on L-fucose the lactaldehyde is oxidized to lactate by lactate dehydrogenase.

Mutants that were selected to grow aerobically on propanediol were shown to synthesize the oxidoreductase constitutively. This was a major regulatory change in that it led to the constitutive *aerobic* expression of an enzyme that was normally synthesized only anaerobically (Cocks et al., 1974). The propanediol-positive mutant was unable to utilize L-fucose either aerobically or anaerobically. This disruption in the fucose pathway was shown to result from the L-fucose permease, the fucose isomerase, and the fuculokinase having become uninducible. The aldolase, on the other hand, was synthesized constitutively (Hacking and Lin, 1977). Subsequent studies showed that constitutive synthesis of the oxidoreductase was always required for growth on propanediol but that the mutation(s) leading to disruption of the L-fucose pathway were not required. When cultures were examined soon after selection for propanediol utilization, in replicate experiments, several of the propanediol-positive isolates were still fucose positive. Continuous growth of these mutants on propanediol, however, always led to their being displaced by a population in which the fucose pathway was disrupted in exactly the same manner as described above (Hacking et al., 1978). This is a particularly good example of a situation in which increased fitness resulted from the *loss* of a function (L-fucose utilization). The basis of the selective advantage to fucose-negative cells growing on propanediol is not yet known.

EVOLUTION OF NEW FUNCTIONS BY MUTATIONS IN BOTH STRUCTURAL AND REGULATORY GENES

Evolution of the aliphatic amidase of *Pseudomonas aeruginosa*

The aliphatic amidase of *Pseudomonas aeruginosa* has been exploited very successfully as a model system by P.H. Clarke and her associates. *Pseudomonas aeruginosa* grows well on acetamide and proprionamide, both of which are good inducers of amidase enzyme synthesis and good substrates for the amidase (Clarke, 1978). The aliphatic amidase is the product of the *amiE* gene, which is subject to positive control by the product of the *amiR* gene (Farin and Clarke, 1978). A variety of amides have been utilized to isolate mutants with altered regulatory and enzymatic properties. These studies have been thoroughly re-

240

viewed (Clarke, 1978), and I shall only touch upon some of the highlights of those studies here.

Formamide is both a poor inducer and a poor substrate for the amidase. Formate is not utilized as a carbon source, but mutants able to utilize formamide as a nitrogen source were isolated (Brammer et al., 1967). These mutants synthesize an unaltered amidase constitutively. Butyramide is also a poor substrate and a poor inducer of the amidase. One of the constitutive mutants, strain C11, was unable to utilize butyramide despite a high level of synthesis of the wild-type amidase. When a butyramide-utilizing mutant, strain B6, was isolated from that strain, it synthesized an altered amidase with increased activity toward butyramide (Brown et al., 1969). Phenylacetamide is not a substrate of the wild-type amidase, and neither strains C11 nor B6 grew on phenylacetamide as a nitrogen source. A phenylacetamide-utilizing mutant was isolated from strain B6, and the new strain, designated PhB3, had a further altered amidase with substrate specificities different from both wild-type and strain B6 enzyme (Betz and Clarke, 1972). A series of three mutations, one in *amiR* and two in *amiE*, had occurred. The *amiR* mutation increases the range of amides used to include formamide. The first *amiE* mutation further increases the range to include butyramide. The second *amiE* mutation enables utilization of phenylacetamide and valeramide, but results in a loss of the ability to utilize acetamide (Betz et al., 1974). Working with these and other amide analogs, Clarke and her co-workers have demonstrated multistep evolutionary pathways in the laboratory. An important finding from their work is that both regulatory and structural gene mutations are required for the evolution of these new functions.

Clarke has isolated a *regulated* mutant that can utilize butyramide (Turberville and Clarke, 1981). The wild-type regulatory protein is not only insensitive to butyramide as an inducer, but butyramide is an anti-inducer. A recombinant was constructed that carried the structural gene mutation present in strain B6 and also carried a wild-type *amiR* gene. Because of the anti-inducer effect, the recombinant strain was unable to utilize butyramide despite the fact that its amidase was active toward butyramide as a substrate. Nineteen butyramide-utilizing mutants were isolated from the recombinant strain, and all but one of them proved to be constitutive (i.e., equivalent to strain B6). One mutant, however, was found to be inducible by butyramide. The properties of the amidase were unaltered. The regulatory properties, on the other hand, were dramatically altered. Butyramide had become a more powerful inducer than acetamide, and the sensitivity to lactamide as an inducer had increased tenfold.

Evolution of a new β-galactosidase function in *E. coli* K12

The previous studies have all adopted the strategy of forcing populations to utilize resources that are truly novel in the sense that the wild-type organism cannot utilize the resource. In choosing such a novel resource, the experimenter runs the risk that *several* new functions (such as permease, hydrolases) may be required for metabolism of the new resource. In addition, it may be necessary to evolve functions to deal with toxic metabolic by-products. For these reasons, xenobiotic compounds may not represent good choices for the development of model systems. An alternative strategy is to remove a specific normal function irreversibly by deletion of the appropriate gene and then apply selective pressure for the re-evolution of the same function.

In order to avoid problems of simultaneous evolution of multiple functions and to focus on a single clearly defined new function, we have chosen to direct the evolution of lactose utilization in strains of *E. coli* in which the *lacZ* gene has been deleted. Lactose utilization requires (1) a lactose permease to transport lactose across the plasma membrane; (2) a β-galactosidase (or more specifically a lactase) enzyme to hydrolyze lactose; and (3) existing pathways to metabolize the hydrolysis products glucose and galactose. If we begin with a strain that carries a large internal deletion in the *lacZ* gene (the gene for β-galactosidase) but that has an intact *lacY* gene (the gene for lactose permease), we have an organism that cannot hydrolyze lactose but that can transport lactose and can metabolize glucose and galactose. For such a strain, lactose is truly a novel resource, and we can apply selective pressure for the evolution of an enzyme that can hydrolyze the β-1,4 bond of lactose—a very specifically defined new function. I shall refer to such a *lacZ* deletion strain as the "ancestral" or "unevolved" strain.

Selection of mutants that have evolved lactose utilization. A model system should, as much as possible, mimic nature. We have, therefore, chosen to utilize only spontaneous mutations in our study of the evolution of lactose utilization. In nature we might expect to find a population of microorganisms in an environment in which there is some primary resource that is readily utilized and a secondary resource that is not utilized. As the primary resource is exhausted, mutants capable of utilizing the secondary resource would be expected to be favored. To mimic that sort of situation, the *lacZ* deletion strain is streaked onto medium that contains broth as a primary metabolizable resource and lactose as a secondary nonmetabolizable resource. In addition, the medium includes triphenyltetrazolium chloride as a fermentation indicator and isopropyl-thio-galactoside (IPTG) to induce-synthesis of the *lac* permease.

242

Colonies arise by growing at the expense of the broth. Those colonies are deep red, indicative of a failure to ferment lactose. As the colonies grow to a size of approximately 10^{10} cells, they exhaust the broth in their vicinity. Rare, spontaneous, lactose-utilizing mutants that arise grow at the expense of the lactose and eventually form outgrowths (papillae) on the surface of the colony. The papillae are white and are thus easily identified as lactose-fermenting cells. This selection scheme enables isolation of large numbers of independent mutants by the simple expedient of picking only one papilla from each colony.

The EBG operon. The mutations that enable *lacZ* deletion strains of *E. coli* to utilize lactose are all in the genes of the EBG (evolved β-galactosidase) operon. All attempts to isolate lactose-utilizing mutants from a strain deleted for both *lacZ* and the EBG operon have failed (B.G. Hall, unpublished). The EBG operon is located at 66 minutes on the *E. coli* map (Hall and Hartl, 1975), far away from the Lac operon. Three genes have been identified in the EBG operon (see Figure 2). *ebgA* is a structural gene specifying the EBG β-galactosidase enzyme. EBG enzyme is a protein of 720,000 MW and consists of six identical 120,000 MW subunits (Hall, 1976). The EBG and *lacZ* β-galactosidases are serologically unrelated, as anti-*lacZ* enzyme does not react with EBG enzyme (Campbell et al., 1973) and anti-EBG enzyme does not react with *lacZ* β-galactosidase (Arraj and Campbell, 1975; Hall, unpublished). The wild-type allele of *ebgA* is designated *ebgA*^{O} ("O" for original), *ebgA*$^+$ being reserved for mutant alleles specifying enzyme with improved activity toward lactose (Hall and Hartl, 1975). Much of our work has focused upon the mutations that convert *ebgA*^{O} to *ebgA*$^+$ alleles.

The second structural gene in the operon is *ebgB* (Hall and Zuzel, 1980b). *ebgB* specifies a 79,000 MW protein with no known function and has been of little importance to our evolutionary studies.

Expression of the EBG operon is subject to negative control by a repressor that is the product of the tightly linked *ebgR* gene (Hall and Hartl, 1975). *ebgR*$^-$ mutants are constitutive, synthesize approximately 5% of their soluble protein as EBG enzyme (Hall, 1976) and are recessive to *ebgR*$^+$ (Hall and Hartl, 1975).

An important aspect of this approach to studying evolution is that the experimenter can study the *unevolved* strain directly, then compare it with a strain that has evolved the new function in response to known selective pressures. It must be emphasized that the EBG operon is present and functional in wild-type *E. coli* K12 strains. The activity

243

of EBG enzyme is, however, so low that it is detectable only in the absence of *lacZ* β-galactosidase activity.

The wild-type or unevolved EBG enzyme is designated EBG^O, and it has detectable activity toward the synthetic chromogenic substrate O-nitrophenyl-β-galactoside (ONPG) (see Table 1). The basal level of EBG enzyme is approximately 0.3 units/mg of cell protein, corresponding to 3–5 molecules per cell (Hall and Clarke, 1977; Hall, 1981). Powerful inducers of the *lac* operon, such as TMG or IPTG, do not increase this basal level of synthesis nor do most other galactosides

TABLE 1. Induction of EBG^O enzyme in strain DS4680A (*ebgA^O ebgR$^+$*). (Data from Hall and Clarke, 1977.)

Additions	Crude extract specific activity†
NONINDUCERS	
None	0.28 ± 0.058
Isopropyl-thio-galactoside, 0.2 mM*	0.30 ± 0.110
Melibiose, 0.1%*	0.20 ± 0.045
Galactose, 0.1%	0.20 ± 0.052
Lactobionate, 0.1%	0.20 ± 0.077
Glycerol-β-D-galactoside, 1.0 mM*	0.20 ± 0.059
β-Methyl-thio-galactoside, 1.0 mM*	0.34 ± 0.047
Phenyl-β-galactoside, 0.1%	0.13 ± 0.080
Galacturonic acid, 0.1%	0.22 ± 0.060
WEAK INDUCERS	
Methyl-β-galactoside, 0.1%	1.4 ± 0.3
Thiodigalactoside, 1.0 mM	1.3 ± 0.6
Lactulose, 0.1%	3 ± 1.9
Galactosyl-β-1,3-D-arabinose, 0.1%	2.3 ± 0.6
STRONG INDUCERS	
α-Lactose, 0.1%	29 ± 5.4
β-Lactose, 0.1%	33 ± 3.1
STRAIN 1B1 (*ebgA^O ebgR$^-$*) CONSTITUTIVE	
None	536 ± 43

* Indicates 0.2 mM IPTG included in growth medium.
† Units, nanomole ONPG hydrolyzed per minute (Units mg^{-1} ± 95% confidence interval).

244

tested (see Table 2). Lactose is the best inducer of EBG enzyme synthesis yet found. The ancestral strain cannot utilize lactose. However, if it is grown on another carbon source such as glycerol or succinate in the presence of lactose, EBG^O enzyme synthesis is induced approximately 100-fold (Hartl and Hall, 1974; Hall and Clarke, 1977). Still, lactose is not a very good inducer of the unevolved EBG operon, because the induced level of synthesis is only approximately 5% of the maximal level of expression as defined by $ebgR^-$ strains, which synthesize EBG enzyme 2000-fold above the basal level.

TABLE 2. Properties of EBG enzymes and growth of strains synthesizing these enzymes constitutively. (Data from Hall, 1978a and 1981, and Hall and Zuzel, 1980a.)

		Substrate			
Class*	Property†	Lactose	Lactulose	Galactosyl arabinose	Lactobionate
EBG^O	V_{max}	620	270	52	No
(n=1)	K_m	150	180	64	detectable
	Specificity	4.0	1.5	0.81	activity
	Growth rate	0	0	0	0
Class I	V_{max}	3566	69	185	No
(n=3)	K_m	22	34	14	detectable
	Specificity	160	2.1	12.7	activity
	Growth rate	0.45	0	0.03	0
Class II	V_{max}	2353	1887	356	No
(n=3)	K_m	59	26	25	detectable
	Specificity	40	73	14.4	activity
	Growth rate	0.19	0.26	0.02	0
Class IV	V_{max}	1461	430	737	105
(n=9)	K_m	0.82	7.9	3.0	15.4
	Specificity	1800	55	244	6.7
	Growth rate	0.37	0.18	0.13	0
Class V	V_{max}	590	215	349	370
(n=1)	K_m	0.69	6.5	4.9	3.0
	Specificity	850	33	70	123
	Growth rate	0.18	0.10	0.07	0.20

* n, Number of purified enzymes from independent mutants.
† V_{max} in units mg^{-1} of pure enzyme; K_m in mM substrate; growth rates in reciprocal hours. Standard errors of these mean values are less than 10% of the values.

Evolution of lactose utilization. Campbell et al. (1973) isolated the first lactose-utilizing mutant of a *lacZ* deletion strain following five rounds of selection similar to that described above. The lactose-utilizing strain was designated ebg-5, and it was assumed to have resulted from five mutations. That assumption was later shown to be in error (Hall, 1977), and it was determined that exactly two mutations are required to evolve lactose utilization (Hall and Clarke, 1977). One mutation must be in *ebgA* in order to increase the lactase activity of the EBG enzyme. These mutations occur spontaneously at the rate of 2×10^{-9} per cell division (Hall, 1977). The other mutation must be in *ebgR*, in order to increase the level of EBG enzyme synthesis above that permitted by the wild-type repressor (Hall and Clarke, 1977). We shall, for convenience, consider mutations in *ebgA* and *ebgR* separately.

Evolution of the EBG enzyme. We shall consider the fitness of various strains on a variety of β-galactoside sugars: lactose, lactulose, galactosyl-D-arabinose, and lactobionic acid. Of these, only lactose is an effective inducer of EBG enzyme synthesis; hence we shall compare *ebgR⁻* (constitutive) strains with each other. We can take growth rate as a measure of fitness and compare growth rates with the properties of purified EBG enzymes from each of the strains being considered (Table 2).

The unevolved enzyme (EBG^O) does not permit growth on any of the β-galactoside sugars being considered. Table 2 shows that activity of EBG^O enzyme is low toward all of those sugars; however, the sugars can be ranked with respect to their effectiveness as substrates: lactose > lactulose > galactosyl-arabinose >> lactobionic acid.

When lactose-utilizing mutants are selected, the majority exhibit the pattern of growth rates shown for Class I mutants: rapid growth on lactose, very slow growth on galactosyl-arabinose, and no growth on either lactulose or lactobionate. These growth rates are consistent within a class and vary less than ±10% from the mean value.

Table 2 also shows that the properties of EBG enzymes from Class I strains have altered dramatically from the properties of unevolved enzyme. The specificities (V_{max}/K_m) for lactose and for galactosyl-arabinose increased 40-fold and 17-fold, respectively, whereas the specificities for lactulose and lactobionate did not increase at all. It is clear, then, that Class I enzymes represent a specific adaptation to lactose: the specificity for lactose is vastly improved, a change that results in good growth on lactose. The reason for considering the parameter "specificity" is that at substrate concentrations much below the K_m value the ratio V_{max}/K_m determines the rate of substrate hydrolysis. Although this is a convenient parameter to consider, the important parameter for the cell is the rate of substrate hydrolysis at physiological substrate concentrations. For lactose, the physiological

246

concentration is on the order of 25 mM; however, we do not have good estimates for the physiological concentrations of the other substrates. Because these Class I mutants arise at the rate of 10^{-9} per cell division (Hall, 1977), we can conclude that they are the result of single point mutations and that the altered enzyme properties are the result of a single amino acid substitution. Although these substitutions clearly affect the active site, M. L. Sinnott (pers. comm.) has obtained evidence that they are not actually within the active site.

When lactose utilization is selected, approximately 10% of the mutants have an entirely different pattern of growth rates. These mutants are designated Class II (Table 2). If, instead of selecting for lactose utilization, one selects directly for lactulose utilization, then Class II mutants are obtained 100% of the time. These again arise at a frequency of approximately 10^{-9}. In terms of growth, Class II mutants differ from Class I in two respects: they grow only 40% as fast on lactose, and they grow well on lactulose. Like Class I mutants they show only marginal growth on galactosyl-arabinose. Enzyme from such Class II mutants is as distinct from Class I enzyme as it is from the unevolved enzyme. Its specificity for lactose has increased only 10-fold compared with unevolved enzyme, but its specificity for lactulose has increased 50-fold. An even more dramatic difference can be seen when we examine the activities of the enzymes toward the synthetic substrate ONPG. The K_m of Class II enzyme for ONPG is 125-fold lower than that of unevolved enzyme and 6-fold lower than that of Class I enzyme. Figure 3 shows how Class I and Class II enzymes represent alternative evolutionary descendants of the common ancestral enzyme EBG^O.

It was earlier apparent that one of the major differences between Class I and II strains was the failure of Class I strains to utilize lactulose (Hall, 1978a). Several Class I strains were subjected to selection for lactulose utilization, and lactulose-utilizing mutants were readily obtained. Although these double mutants (designated Class IV) could utilize both lactose and lactulose, they were not at all like Class II strains (Table 2). First, the Class IV strains grew faster on lactose than they did on lactulose. Second, and more surprisingly, Class IV strains grew at a respectable rate on galactosyl-arabinose. Characterization of Class IV EBG enzymes served to confirm that this was indeed a new class. Class IV enzymes exhibit a very low K_m for lactose, leading to a specificity for lactose more than 10-fold above that of Class I enzyme. Similarly, the specificity for lactulose was improved 25-fold compared with that of Class I enzyme. A 20-fold increase in the specificity for galactosyl-arabinose resulted in an enzyme that hydrolyzes that substrate at a rate sufficient to support good growth. As before, the changes in the properties of EBG enzyme correlated well with changes in the growth characteristics of strains synthesizing the enzyme. The strong correlation between growth rate on a substrate and the specificity of an enzyme for the substrate suggests that hydrolysis is the rate-limiting step in growth on these β-galactosides. The growth rate of Class IV strains on galactosyl-arabinose appears to be lower than would be predicted from the high specificity of Class IV enzyme for galactosyl-arabinose. This is because the D-arabinose moiety of galactosyl-arabinose is not used by *E. coli*; thus, each hydrolysis yields only one metabolizable hexose sugar. When this is taken into account, growth rate on galactosyl-arabinose correlates well with specificity.

At this point it could be argued that no new enzymatic activities had been evolved and that the only changes had been to improve the efficiencies of previously existing enzyme activities. Class IV enzyme, however, did exhibit a *completely new* activity: lactobionate hydrolysis. The specificity of Class IV EBG enzymes for lactobionate is not high enough to enable utilization of lactobionate as a sole carbon and energy source, but the activity is detectable *in vitro*.

Class IV strains were clearly double mutants and were directly descended from Class I strains. With the realization that galactosyl-arabinose-positive mutants existed, we wondered whether galactosyl-arabinose positive mutants could also be selected from Class II strains and, if so, what their phenotype would be. A series of galactosyl-arabinose-utilizing mutants were selected from previously characterized Class II strains. These new mutants exhibited an increase in growth rate on lactose and a decrease in growth rate on lactulose relative to their Class II parents (Hall and Zuzel, 1980a).

More striking, however, was the observation that the growth rates of those mutants on all substrates tested was indistinguishable from that of Class IV strains. This observation led to the suggestion that the new mutants were indeed Class IV strains, which in turn led to the testable hypothesis that *Class IV strains are the result of a Class I mutation and a Class II mutation being present in the same ebgA gene*. The hypothesis makes two predictions. (1) A cross between an unevolved strain and any Class IV strain should lead to the recovery of both Class I and Class II recombinant progeny independent of the route by which the Class IV strain was selected. (2) Class IV strains should be recovered from crosses between Class I and Class II strains but not from crosses between two Class I or two Class II strains. Both of these predictions were verified, and in doing so data was obtained that permitted mapping of the Class I and Class II sites within the *ebgA* gene (Figure 2).

Studies of the enzymes from some of the recombinant strains also supported the hypothesis. A cross between one of the original Class IV strains and an unevolved strain led to the recovery of a recombinant that was apparently Class II. Enzyme from that strain was clearly Class II EBG enzyme based upon its kinetic parameters on ten different substrates. Similarly, Class IV strains generated by crosses between Class I and Class II strains yielded enzyme that was indistinguishable from Class IV enzyme by all criteria. It was therefore concluded that the Class IV alleles were the result of two mutations (one in the Class I site and the other in the Class II site) being present in the same *ebgA* gene at the same time. Merodiploid strains that carried a Class I *ebgA*$^+$ allele and a Class II *ebgA*$^+$ allele simultaneously were constructed. Those strains were not able to utilize galactosyl-arabinose, an observation that leads to the conclusion that Class IV enzyme could not arise by complementation and required instead two amino acid substitutions in the same polypeptide chain (Hall and Zuzel, 1980a).

Class IV strains cannot utilize lactobionate, although there is detectable activity of Class IV EBG enzyme toward lactobionate *in vitro*. Attempts were made to select lactobionate-utilizing mutants from the unevolved strain and from Class I, Class II, and Class IV strains (Hall, 1978a). Only Class IV strains yielded lactobionate-utilizing mutants, which were designated Class V. Class V mutants arose from Class IV at a frequency of 10^{-9}, exactly the frequency expected for single point mutations within the *ebgA* gene. It was therefore concluded that lactobionate utilization required three mutations within *ebgA*. Enzyme from one Class V strain was purified and characterized (Table 2). As expected, it exhibited an increased V_{max} and a decreased K_m for lactobionate. That change was accompanied by a roughly twofold decrease in the V_{max} for the other disaccharides tested.

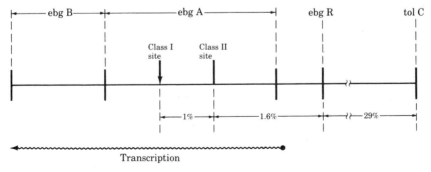

FIGURE 2. Map of EBG operon. Distances given in percent recombination.

The growth properties of constitutive strains correlate strongly with the *in vitro* properties of the EBG enzymes that they synthesize. This suggests that we are, indeed, applying selective pressure directly on the β-galactosidase function specified by EBG enzyme. We have demonstrated the existence of an evolutionary pathway in the laboratory (Figure 3). We have also provided direct evidence for the evolution of new functions by *intragenic* recombination. In a cross between a Class I and a Class II strain, a recombination event between the two sites generates a Class IV allele that specifies functions possessed by neither parent, nor by the grandparent. At the level of growth, Class IV strains have the new function "galactosyl-arabinose utilization." At the level of enzyme activity, EBG enzyme possesses a new activity: lactobionate hydrolysis. At another level, Class IV strains possess the easily realized potential for evolving lactobionate utilization.

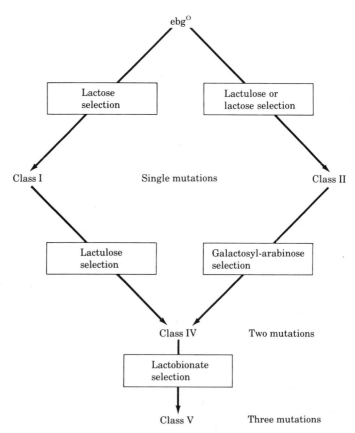

FIGURE 3. Pathway for directed evolution of EBG enzyme.

Evolution of regulatory functions. The first EBG$^+$ mutant isolated by Campbell et al. (1973) synthesized EBG enzyme constitutively. In contrast, 90% of the EBG$^+$ mutants I have isolated are inducible by lactose (Hall and Hartl, 1974). We initially believed that these inducible mutants were altered only in their structural genes for EBG enzyme. However, careful comparisons of the levels of induction in the ancestral strain DS4680A with the level in the inducible evolved strain A4 showed that the evolved strain was fourfold more inducible by lactose (Hall and Clarke, 1977). This hyperinducibility was shown to result from a mutation in the *ebgR* gene, a mutation that resulted in a repressor that was more sensitive to lactose than the wild-type repressor (Table 3). By utilizing published information on the properties of the lactose permease, we were able to estimate the physiological concentration of lactose under our conditions. From our kinetic studies, we could estimate the lactase activities of various EBG enzymes at that concentration and thus calculate the *in vivo* lactase activities of EBG enzyme in our various evolved strains. There was a linear relation between *in vivo* lactase activity and growth rate on lactose, but the line did not pass through the origin (Figure 4). The threshold for growth was estimated as 5.8 units/mg of EBG lactase activity. We showed that strains synthesizing the most active EBG enzyme could not produce that threshold level of activity if the enzyme was under control of the wild-type repressor but that the increased inducibility of the mutant repressor enabled synthesis of sufficient enzyme for growth. We designated the evolved repressor *ebgR$^+$(U)*, the "U" indicating increased sensitivity toward lactose. We concluded

TABLE 3. Inducers of strains DS4680A and A4. (Data from Hall and Clarke, 1977.)

Inducer	Specific synthesis × 10^2 ± 95% CI*	
	DS4680A (ebgR$^+$)	A4 (ebgR$^+$U)
None	0.0022 ± .0003	0.0030 ± .0007
Thiodigalactoside	0.012 ± .0054	0.015 ± .0036
Methyl-β-galactoside	0.012 ± .0025	0.014 ± .0013
Lactulose	0.027 ± .017	0.056 ± .028
Lactose	0.26 ± .05	1.23 ± .25

* Specific synthesis = the fraction of total soluble protein that is EBG enzyme.

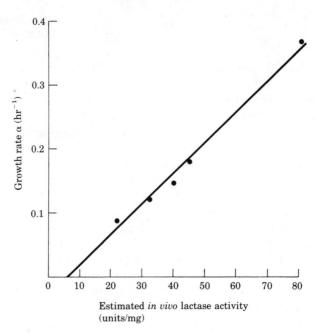

Estimated *in vivo* lactase activity
(units/mg)

FIGURE 4. Growth rate on lactose versus *in vivo* EBG lactase activity for five *ebgA*⁺ strains.

that *all* of the regulated evolved strains carried such "U" mutations in the *ebgR* gene in addition to a mutation in the *ebgA* gene.

Directed evolution of the EBG repressor. Neither the unevolved repressor nor the type U evolved repressor is sensitive to lactulose as an inducer. To determine whether a regulatory protein could be subjected to the same sort of directed evolution as a catalytic protein, I decided to attempt to direct the evolution of the EBG repressor so that it responded to lactulose as an inducer (Hall, 1978b). I began with strain 5A1, which has a class II evolved EBG enzyme under control of the wild-type repressor. Although the class II enzyme can hydrolyze lactulose efficiently, strain 5A1 can not grow on lactulose, which is not an inducer of the wild-type repressor. Following selection for lactulose utilization, I obtained approximately 3000 lactulose-positive mutants, of which all except 9 were constitutive. The nine exceptions turned out to have fully functioning EBG repressors that had become sensitive to lactulose as an inducer. These mutants were designated "L" for lactulose inducible; Table 4 shows the properties of one such mutant.

This study demonstrated that regulatory proteins could be subjected to the same sorts of selective pressures as could enzymes, but

252

more importantly, it provided the tools for deliberately directing the evolution of the EBG system so that the new functions were integrated into the normal cellular regulatory network. Before discussing that evolutionary pathway, it is necessary to consider another function required for lactose utilization.

Regulation of synthesis of the lactose permease by EBG enzyme. Lactose utilization depends upon the functioning of the entire Lac pathway, that is, both lactose permease and lactose hydrolase functions are required. In this system, the permease function is supplied by the *lacY* gene product. Lactose, however, does not induce expression of the Lac operon. In wild-type cells, lactose is converted to allolactose by the activity of the *lacZ* β-galactosidase, and allolactose is the true inducer of the Lac operon (Jobe and Bourgeois, 1972). Early in these studies it was realized that EBG β-galactosidase did not convert lactose into a *lac* operon inducer and that it was therefore necessary to include the artificial inducer IPTG in the medium in order to turn on synthesis of the *lac* permease (Campbell et al., 1973; Hall and Hartl, 1974). The inclusion of this synthetic inducer creates a somewhat artificial situation because strains that require IPTG for growth on lactose have clearly not evolved a fully functional lactose utilization system.

We have isolated a mutant EBG enzyme that is capable of converting lactose into an inducer of the *lac* operon (Rolseth et al., 1980). The parent strain synthesized a Class II EBG enzyme constitutively. Following selection for rapid growth on lactose, a mutant was isolated

ABLE 4. Induction of *ebg* enzyme by various β-galactosides.* (Data from Hall, 978b.)

Strain	*ebgR* allele	None	Lactulose	IPTG	Galactosyl-arabinose	Methyl galactoside	Lactose
		Galactoside added					
)S4680A	*ebgR*⁺	0.28	3	0.30	2.3	1.4	29
A1	*ebgR*⁺	0.23	4.1	0.26	1.2	0.8	23
A101	*ebgR103*⁺ᴸ	0.24	58	0.47	64	14	75
A103	*ebgR105*⁺ᴸ	0.27	126	0.48	211	55	230
A108	*ebgR110*⁺ᴸ	0.30	84	0.45	215	51	199

Values shown are units of *ebg* enzyme activity per milligram of protein in crude extracts.

that not only grew faster than the parent on lactose, but also grew on lactose without the presence of IPTG in the medium. It was shown that the IPTG-independent lactose utilization was the result of a mutation of *ebgA*. The mutant also grew on galactosyl-arabinose, and characterization of the EBG enzyme from the mutant showed that it was a typical Class IV enzyme (Hall, 1981). It has been shown that all Class IV EBG enzymes are capable of converting lactose into a *lac* operon inducer and that the inducer is allolactose (Hall, 1982a).

This finding represents still another new function for Class IV EBG enzymes: allolactose synthesis. Class IV strains are thus free from the artificial constraint of requiring IPTG for lactose utilization, that is, they are capable of expressing all of the genes necessary for lactose utilization without an artificial inducer.

Integration of the EBG system into the normal regulatory network. A fully evolved metabolic system that is integrated into the normal regulatory network of the cell should (1) synthesize a sufficient amount of each protein required for the pathway to function, and (2) synthesize those proteins in direct response to appropriate environmental signals that indicate the need for the pathway to function. As described above, the EBG system does not meet the criteria of being a fully evolved system for lactose utilization. The unevolved strain synthesizes EBG enzyme in response to lactose in the environment, but neither the amount nor the activity of that enzyme is sufficient for lactose utilization. In addition, the second protein of the pathway, lactose permease, is not synthesized in response to lactose in the environment. Evolved strains carrying mutations in both *ebgR* and *ebgA* show sufficient enzyme activity for growth, but fail to synthesize lactose permease in response to lactose. Class IV strains, on the other hand, synthesize the lactose permease in response to lactose in the environment, but they synthesize EBG enzyme constitutively rather than in response to lactose induction.

As described earlier, strain 5A103 carries a Class II mutation in *ebgA* and an "L" mutation in *ebgR*. It therefore synthesizes a high level of an EBG enzyme that has good activity toward lactose (Table 5). The only protein not synthesized in response to lactose is the lactose permease.

Our earlier results had suggested that a second mutation in *ebgA*, a mutation in the Class I site, would generate a Class IV EBG enzyme capable of converting lactose to allolactose and thereby turn on expression of the lactose permease. I selected a galactosyl-arabinose utilizing mutant of strain 5A103, taking advantage of the fact that galactosyl-arabinose is a powerful inducer of the $ebgR^+L105$ repressor (Hall, 1982b). The new mutant grew on lactose without the presence of IPTG (Table 5). The failure of the parent to grow on lactose alone

254

is not due to a failure to synthesize EBG enzyme but is due to its failure to synthesize the lactose permease activity (Table 5). The mutant strain possesses a fully evolved, integrated system for lactose utilization. In the absence of lactose, both the permease and the EBG β-galactosidase are synthesized at very low repressed levels. In the presence of lactose, EBG enzyme synthesis is induced, and that enzyme in turn converts lactose into allolactose, resulting in induction of the *lac* permease, that is, both proteins of this lactose utilization pathway are made in sufficient amount in specific response to the presence of the substrate in the environment. Figure 5 shows the pathway for the evolution of this fully regulated system.

DISCUSSION

Microorganisms have enormous potential uses as model systems for exploring evolution. That potential is only beginning to be exploited, and we can look forward to more complex and interesting systems in the future.

Throughout this chapter fitness has implicitly been equated with growth rate, and adaptive mutations have been discussed in the context of improved growth rate. Growth rate is an important, but limited, aspect of fitness. In nature, organisms are not confronted with a constant supply of nutrients. Indeed, starvation is more likely the norm. Mutations that improve survival under nongrowing conditions are

TABLE 5. Evolution of the EBG system for lactose utilization. (Data from Hall, 1982b.)

Strain	Genotype*	*In vivo* lactase activity Uninduced	Induced by lactose	Permease activity induced by lactose	Growth on lactose with IPTG	w/o IPTG
DS4680A	$ebgA^O\ ebgR^+$	0.0036	0.44	<0.3	0	0
5A1	$ebgA^+\ ebgR^+$	0.021	1.92	<0.3	0	0
5A103	$ebgA^+\ \underline{ebgR^{+L}}$	0.021	21.7	<0.3	0.124	0
5A1032	$\underline{ebgA^{++}}\ ebgR^{+L}$	0.028	31.7	2.7	0.176	0.168

* The gene underlined indicates the site of the new mutation in this strain. The "$ebgA^{++}$" indicates that this allele carries two mutations, that is, it is a Class IV allele.

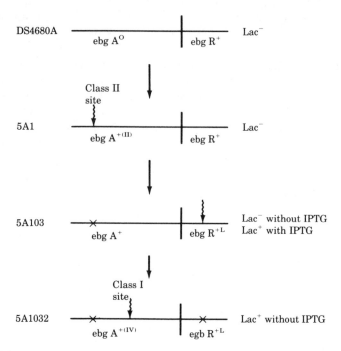

FIGURE 5. Pathway for evolution of the lactose utilization system by mutations in the EBG operon.

expected to be strongly selected. For instance, in a nonconstitutive *Klebsiella* strain, a mutation that improves the activity of RDH toward xylitol will not permit growth on xylitol because synthesis of RDH is not induced. The mutation will, however, increase the basal level of activity toward xylitol. Such an increase might well enable a cell to obtain enough energy and carbon from xylitol to survive until better conditions occur. Such a mutation, then, might be neutral under growing conditions but advantageous during starvation in the presence of xylitol.

Another important aspect of fitness could include the ability to discriminate between a utilizable substrate and a structurally similar toxic compound. If increased activity toward the substrate were accompanied by a reduced ability to discriminate, mutations causing the increased activity (and thus rapid growth) might be disadvantageous. Such considerations are important because, in nature, microorganisms must deal with fluctuating environments that may occasionally include toxic analogs of growth substrates.

Finally, it must be realized that in nature microorganisms do not exist as pure cultures. Instead, they function as members of microbial communities within which they must both compete for resources and take advantage of any synergistic opportunities available. Slater and his colleagues have been modeling such situations by studying the ways by which microbial communities evolve the ability to utilize halogenated compounds (Slater and Bull, 1982).

These examples suggest that there may be many ways to improve fitness other than by increasing growth rate. The future of experimental evolution depends on devising model systems to explore the other aspects of fitness.

CHAPTER 13

TRANSPOSONS AND THEIR EVOLUTIONARY SIGNIFICANCE

Allan Campbell

One important accomplishment of classical genetics was the demonstration that genes generally occupy fixed locations on chromosomes. New genotypes may arise through reshuffling of alleles by homologous recombination; but the gene order remains fixed except for rare rearrangements, which might result from chromosome breakage. This "classical" concept of the genome was not arrived at casually but rather summarizes a large body of valid observations. Any mechanisms for rapid or extensive genetic rearrangement must be viewed against this general constancy of genome organization.

The most conspicuous agents of genetic plasticity are collectively called "movable genetic elements." Movable elements differ from ordinary genes in that they are not always present at a given location but rather can be added onto a chromosome by intercalation. Typically, the insertion of movable elements is catalyzed by enzymes encoded by the elements themselves, which specifically recognize base sequences at the termini of the inserted element. Thus, these elements can move in an active sense (under their own direction) rather than merely being "movable" by other agents.

Movable elements can be divided into *episomes*, which can replicate independently on the chromosome, and all other elements, which are found only in the inserted state. Both types of elements are discussed in this chapter, but the latter group (sometimes called *transposons*) are emphasized.

258

Movable elements were discovered in maize by McClintock (1952). Her rather revolutionary findings had little impact on the general course of genetic research until recently, probably for three reasons: (1) The elements were harder to study than classical genes with fixed locations. (2) No good methods were available to demonstrate their widespread distribution among other organisms. (3) Their significance in developmental and evolutionary genetics was uncertain.

The recent revival of interest in movable elements has attended technical developments that eliminate the first two problems. Movable elements are clearly common in the genomes of many (perhaps all) organisms. Modern DNA technology has facilitated their detection and study. Their significance remains a live issue. The purpose of this chapter is to survey the facts that appear most relevant to that issue.

The basic question to be examined concerns the selective forces that maintain movable elements in nature. Do they provide some benefit to the organism, either in its individual biology or in its evolution, that is adequate to account for their survival? Or are they more appropriately regarded as genetic parasites, which persist because they are hard to get rid of?

Many of the elements known in prokaryotes show signs of being highly evolved, so that "survival" over a long time span is a real issue. There are examples of controlled DNA rearrangements that serve useful biological functions, but we cannot yet say whether directed DNA rearrangement constitutes a general, rather than an exceptional, mechanism of achieving stable changes in gene expression.

If movable elements provide some benefit at the population level, it should be by promoting recombination or genetic rearrangements of selective value. An important role for movable elements in chromosomal evolution would be strongly indicated if they provided a pathway for producing rearrangements of different types than those that occur anyway by other mechanisms. Current knowledge of rearrangement mechanisms, both element-directed and otherwise, will be reviewed with reference to what qualitatively new potentialities movable elements may create.

CLASSIFICATION OF MOVABLE ELEMENTS

Movable elements can be classified according to their state prior to insertion. Some elements (such as plasmids and most temperate viruses) replicate extrachromosomally but can also become inserted and passively replicated as part of the chromosome. Such elements are called *episomes*. Other elements (such as the bearers of many antibiotic resistance genes in bacteria) cannot replicate separately and are found only in the inserted state. They may transpose into a new location, either on the same molecule or on a different one. Some viruses (such as bacteriophage Mu) fall into this category. Unlike episomes, Mu replicates by transposition, not as a separate molecule.

Many authors call all such nonautonomous transposable elements *transposons*. Others distinguish two classes of such elements: true *transposons* (which include genes that determine phenotypic traits) and *insertion sequences* (the only known properties of which are transposability and promoter or terminator activity). The origin of the distinction is partly historical: in bacteria, insertion sequences were discovered and named before true transposons were found. The maize elements are also insertion sequences by this definition. True transposons are usually studied experimentally by introducing them into a cell line from which they were formerly absent, their presence being recognized by the phenotype they confer. On the other hand, insertion sequences are generally endogenous to the genome under study and have been recognized only by their effects on other genes.

MECHANISM OF TRANSPOSITION

For most elements, the biochemical mechanism of transposition or insertion is not known in detail. At least two general mechanisms can be distinguished. With some elements, such as bacteriophage λ, insertion of the element proceeds by breaking and rejoining of both strands of both parental molecules at a specific site on the DNA, a reciprocal event that entails neither degradation nor synthesis. This process has been called *conservative site-specific recombination*. The λ insertion reaction has been studied *in vitro* with purified components. Host and viral DNA are identical in a 15-base core within which the crossover event occurs (Figure 1). One virus-coded protein is required, λ integrase. Two host-coded polypeptides are also required. Integrase binds specifically to λ DNA at the crossover point. The specific recognition of this site by integrase is not confined to the core but is directed to a DNA segment that is approximately 200 base pairs long and that includes the core and four separate integrase binding sites. All four sites are necessary for a normal rate of insertion; DNA from which either end of the 200-base pair segment has been deleted is an inferior substrate *in vitro*. The nucleotide positions at which the two strands of the molecule are cut are not directly opposite one another but are seven base pairs apart (Nash, 1981).

The other mechanism, generally used by prokaryotic transposons and so far studied only *in vivo*, differs from conservative site-specific recombination in that both the element and the target site are replicated during the transposition process. Thus, the result of transposi-

260

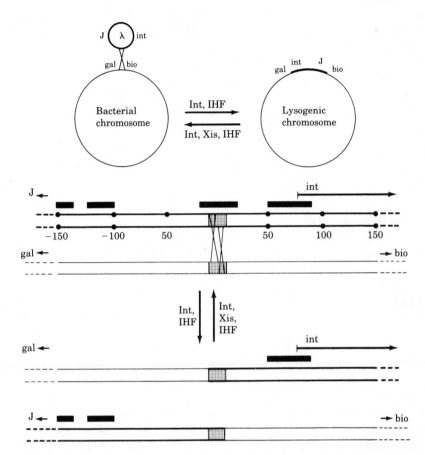

FIGURE 1. Insertion of bacteriophage λ into the *Escherichia coli* chromosome by conservative site-specific recombination. A. Overall process, showing fusion of circular viral and host DNA. B. Details of the crossover event. Crossing-over takes place within a 15-base segment (shown as stippled) that is identical in the two partners. The crossover points in the upper and lower strands are displaced by seven bases within the core. Sites of integrase binding to viral DNA are indicated by solid bars. *gal, bio*, Host genes flanking the crossover site on the bacterial chromosome; *J, int*, phage genes flanking the crossover site on the phage; Int, phage-coded integrase; Xis, phage-coded excisionase; IHF, host-coded proteins needed in the insertion reaction.

tion is a cell that carries the transposon still inserted at its original site and an additional copy inserted at a new site; and at the new site, the transposon is bracketed by a direct repeat of an oligonucleotide

261

sequence of the target DNA (Figure 2). It is not known whether replication is part of the transposition mechanism itself or whether ordinary chromosomal replication is followed by transposition, during which the chromosomal donor is destroyed. Besides transposition, transposons cause another type of rearrangement, which is most easily studied when the target site is on a different replicon from the donor. In this case, the transposon and target site are duplicated, but also the two replicons are joined together by the transposons in a structure called a *cointegrate*. It is not known whether cointegrate formation is a necessary step in transposition or an alternative to it. Some (but not all) transposons contain an internal resolution site and encode an enzyme that catalyzes conservative site-specific recombination at that site. In such elements, mutational inactivation of the enzyme or the site prevents transposition but not cointegration, as expected if transposition proceeds by forming cointegrates and then resolving them by recombination (Heffron et al., 1981).

In a typical transposon, the base sequence at one end of the element is an imperfect reverse repeat of the sequence at the other end. For example, in the insertion sequence IS*1*, 28/34 base pairs at the two ends match. This pseudosymmetry might make the element look the same to the transposase when viewed from either end. The fact that the match is imprecise suggests that base pairing between the two ends is not part of the transposition mechanism. Transposases are presumed to recognize these terminal sequences because a defective transposon whose own transposase gene has been deleted can move when complemented by a complete transposon of the same type. Transposons frequently show little specificity for the target site. There are sequence preferences, whose basis is poorly understood. Some elements, like IS*1*, can transpose to almost any site. At the other extreme, IS*4* inserts repeatedly at a unique site of the *E. coli* chromosome (Klaer et al., 1981).

Conservative site-specific recombination and duplicative transposition are unlikely to be the only mechanisms for incorporation of movable elements. In fact, the maize elements conform to neither of these prototypes: transposition can be associated with replication (Greenblatt, 1968). However, unlike bacterial transposons, the maize elements are usually lost from the donor site when they insert elsewhere (McClintock, 1957).

Elements that lack any specific insertion or transposition system can become inserted into chromosomes or other replicons at a low rate. In bacteria, such nonspecific insertion is very rare (Pogue-Beile et al., 1980). In animal cells, DNA of the tumor viruses SV40 and polyoma can become inserted in a fairly random manner. The DNA junctions do not lie at fixed points on either viral or host DNA. Frequently, the

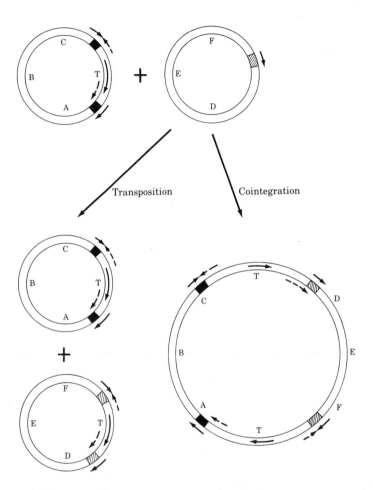

FIGURE 2. Actions of a typical bacterial transposon. The transposon (T), initially in the donor molecule ABC, has imperfect inverted repeats at its termini (dotted arrows) and is bracketed by a direct oligonucleotide repeat of host DNA (solid arrows). The target site with which the transposon will interact in molecule DEF is also shown with a solid arrow. In both transposition and cointegration, the recipient site and the transposon are replicated. In cointegration, donor and recipient molecules end up fused together in one molecule, which contains two copies of the transposon, one at each junction. Arrow within the transposon is included to indicate that the two transposons in the cointegrate are in direct orientation. Where target site and transposon are both in the same molecule, the same steps that lead to cointegration here would generate adjacent deletions or inversions instead.

263

insert is not even an integral number of viral genomes. The provirus can stimulate its own excision or amplification in a manner that is dependent on replication initiated by viral proteins at the viral replication origin and that apparently requires homologous recombination in the duplicated portions of the insert (Botchan et al., 1979; Colantuoni et al., 1980).

It is possible that SV40 insertion is mediated by endogenous transposable elements at or near the site of insertion. In this chapter, I will assume that it indicates the ability of nonspecific cellular mechanisms (as contrasted with the specific mechanisms introduced by movable elements) to cause rearrangements.

EVOLUTION AND COMPARATIVE STRUCTURE OF TRANSPOSONS AND OTHER MOVABLE ELEMENTS

One common feature of all movable elements is that under appropriate circumstances, their DNA can replicate faster than the cellular genome—either by autonomous replication of episomes or by duplicative transposition. Furthermore, at least in prokaryotes, transposition allows the elements to become associated with agents (viruses or conjugative plasmids) that can move from one cell to another. The potential for amplification and transfer gives transposons, like viruses and plasmids, an evolutionary history of their own, distinct from that of the cellular genome.

How do such subcellular genetic elements survive in nature? This question has long interested virologists. The best idea for the origin of viruses is that they consist of host genes that have broken away from their normal constraints and evolved in the direction of greater independence. One version of this hypothesis (which could equally well apply to all movable elements, not just viruses) supposes that new viruses arise frequently and generally pass quickly to extinction. Alternatively, one might imagine that new viruses originate very rarely and that extant viruses have evolved extensively in directions allowing their long range survival as viruses.

At least some viruses appear to be complex and highly evolved, hence unlikely to be recently derived from nonviral DNA. In what follows, I will discuss some evidence that other movable elements are likewise highly evolved. The evidence is far from compelling. My own outlook is colored by experience with bacteriophage λ, a large and sophisticated virus that may be atypical.

Among movable elements there is a range of size and complexity (Figure 3). The simplest bacterial elements are insertion sequences like IS*1* (768 nucleotides long), which apparently includes little more than the information for the transposase enzyme (as yet unidentified) and its recognition sites at the ends of the element.

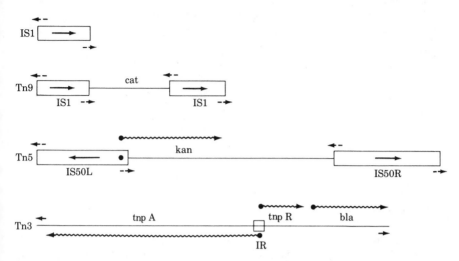

FIGURE 3. Comparative structure of some transposable elements of bacteria. IS*1* is 768 base pairs long, with an imprecise inverted repeat (arrows with dashed tails) in the terminal 34 nucleotides. Tn*9* consists of two IS*1* elements bracketing a gene for chloramphenicol acetyltransferase. In Tn*5*, two oppositely oriented, independently transposable, 1500-base pair sequences (IS*50*), which differ by a single mutational change that creates a rightward promoter in IS*50L*, bracket a gene for kanamycin resistance. Tn*3*, 4957 nucleotides in length, has a precise inverted repeat (arrows with solid tails) of 38 bases at the termini. Of the three genes shown, *tnpA* encodes the transposase (required for transposition); the *tnpR* product binds to Tn*3* DNA at the internal resolution site IR, thereby repressing transcription of both *tnpA* and *tnpR*, and also catalyzes conservative site-specific recombination at IR (a process that can convert the cointegrate structure of Figure 2 into the transposition products, or vice versa); the β lactamase gene *bla* confers ampicillin resistance.

The simplest true transposons consist of a short DNA segment (including the structural gene for some enzyme) bracketed by two insertion sequences such as IS*1*. In Tn*9*, for example, IS*1* elements in direct orientation bracket an 1102-base DNA segment that includes a gene for chloramphenicol acetyltransferase. The ability of the complex to move as a unit requires no special properties of the central DNA segment. Any DNA lying between two identical insertion sequences or transposons can be transposed along with the bracketing elements—a result that is comprehensible in that the termini of the composite element contain the same recognition sequences as the bracketing elements themselves.

265

One might expect that, where selection favors transposability of a particular trait over a long period, the composite element might further evolve so that the bracketing elements degenerate and only the complex can transpose. Tn5 and Tn3 may represent stages in such a progression.

The Tn5 element consists of two oppositely oriented 1500-base pair insertion sequences (called IS50) surrounding a 2700-base pair segment that includes a gene for kanamycin resistance. The two IS50 elements differ by a single base pair change, which simultaneously creates a strong promoter for the kanamycin resistance gene and inactivates a gene product needed for transposition (Rothstein and Reznikoff, 1981). Thus, of the two IS50 elements, only the one on the right (IS50R) can transpose on its own, although IS50L is transposable in a cell containing IS50R or Tn5. One may imagine that Tn5 evolved from an element that had two IS50R elements, through selection for high-level expression of the antibiotic resistance genes.

An element like Tn5 might further evolve in the direction of an indivisible unit by deletion of the inside ends of the bracketing insertion sequences. The Tn3 element exemplifies the expected result. The terminal inverted repeat of Tn3 is only 38 base pairs long (and 38/38 bases match at the two ends), as though the rest of an insertion sequence originally at the right end had been removed by a deletion internal to the transposon.

Both Tn3 and Tn5 encode repressors of their own transposase genes. The Tn3 repressor also catalyzes reciprocal site-specific recombination (needed for cointegrate resolution) at its binding site. Repression of transposase limits the extent to which transposons can proliferate to fill up the genome of the cell. The existence of well-organized regulatory systems suggests that the elements possessing them are products of extensive natural selection.

Even stronger evidence for a highly evolved state is seen in bacteriophage λ. λ inserts its DNA into the bacterial chromosome by the action of λ integrase. The reverse reaction—cutting the viral DNA out of the chromosome—requires two enzymes, integrase and excisionase. The two genes are so regulated that only integrase is made in those infected cells that will survive infection, neither enzyme is formed in established lysogenic cells harboring the inserted prophage, and both enzymes appear in those occasional lysogenic cells where phage reproduction is activated. Regulation is achieved by use of two different promoters, which are responsive to different effectors; this type of regulation enables either cotranscription of the genes for integrase and excisionase or transcription of the integrase gene alone. A second, posttranscriptional control prevents integrase production in those infected cells headed for lysis rather than lysogeny. A noteworthy fea-

266

ture of this posttranscriptional control is that it requires a site that is downstream from the integrase gene in phage DNA but is separated from it by the insertion event. With this arrangement, the state of insertion can feed back on the regulation of the integrase gene (Miller et al., 1981).

Natural λ-related coliphages have diverged extensively in base sequence and in specificity of insertion, replication, and regulation. All these phages are similar in size and gene order and appear to be descended from a common ancestor with the same general structure. Furthermore, many *E. coli* strains harbor DNA segments homologous to parts of the λ genome, which appear to be defective remnants of λ-related proviruses. Both facts suggest that λ has a long evolutionary history as a phage, rather than constituting a recent amalgamation of host genes derived from various sources (Campbell, 1983).

Although they have not been shown to transpose, the retroviruses of vertebrates appear to insert their DNA into the host chromosome by the same mechanism used by transposons, with the generation of flanking oligonucleotide repeats. In addition to the genes needed for viral development, highly oncogenic retroviruses carry a protein kinase gene that is closely related to a host gene. Several such host protein kinase genes (generically called *onc* genes) are found in different retroviruses. At some stage in their history, these retroviruses must have picked up the *onc* genes from their host in a manner perhaps analogous to the acquisition of host genes by λ in specialized transduction or of the incorporation of antibiotic resistance genes into transposons. One may ask whether new complexes of this type arise frequently and continually (in which case existing combination might have a brief life span) or whether they have arisen rarely but have been preserved and refined by natural selection. No definite answer can be given. There have been some claims that new combinations arise at a frequency detectable in the laboratory, but these can be questioned for technical reasons (Lee et al., 1981).

A survey of the spectrum of movable elements indicates that some elements, such as bacteriophage λ, are highly evolved and probably have been around for a long time. As we go to progressively simpler elements (such as retroviruses, transposons, and insertion sequences), the evidence for a long evolutionary history diminishes. In what follows, I shall assume that those elements that are common enough to have been found are likely to have evolved strategies that assure their long-range survival.

POSSIBLE FUNCTIONS OF TRANSPOSONS IN CELLS AND ORGANISMS

How do transposons survive in nature? Their presence constitutes at least some metabolic burden on the organisms that harbor them, which should decrease the selective potential of these organisms with respect to their transposon-free counterparts. Two hypotheses can be entertained: (1) transposons serve some positive function, either at the individual or the populational level, that counterbalances the cost of maintaining them; (2) except perhaps incidentally, transposons serve no useful purpose. Because of their ability to amplify, they are hard to get rid of; it doesn't pay for the organism to bother, so long as the burden is small. They are most appropriately regarded as "selfish DNA."

No definite evidence is available that enables one to choose between these alternatives. We can point to a few examples of controlled DNA rearrangements that serve some useful purpose: the diversification of immunoglobulins in vertebrates (controlled deletion, occurring in certain somatic cells only); the alternation of antigenic specificities in bacteria such as *Salmonella* (controlled inversion, by conservative site-specific recombination, Figure 4); the switching of mating type in yeast (controlled transposition, in this case a replacement rather than an addition); and the variation in antigenic specificity of trypanosomes (DNA rearrangement of unknown nature, remarkable for the large number of alternative states). An important question that remains unresolved is whether commitment to a differentiated state during development of higher eukaryotes generally comes about by DNA rearrangement.

So far as we know, none of the preceding examples involves a movable *element* such as a typical transposon. If movable elements play an important role in the biology of the organism (rather than simply serving their own ends), the important process they reveal, which transcends the class of movable elements themselves, is the enzymatic breakage and reunion of DNA at specific sites, which allows genomes to undergo controlled reorganization. Such reorganization requires enzymes of the types encoded by movable elements and mechanisms for regulating them, but need not require as a general rule the presence of elements that behave as self-contained units, which encode machinery for moving themselves around. The organism's purposes should best be served by moving specific segments of DNA, not necessarily under their own agency.

Movable elements could plausibly be derivatives of such cellular rearrangement systems, which have evolved (selfishly) in the direction of independence from the normal genetic program of the organisms.

268

It is equally plausible that such systems were derived from movable elements that lost some of their former independence. Probably changes have occurred in both directions. Some cellular rearrangement mechanisms are definitely related to movable elements. For example, the enzyme catalyzing specific inversion during phase variation in *Salmonella* has significant sequence homology with the internal resolution enzyme of Tn*3* (Zieg and Simon, 1980).

Do movable elements themselves ever serve a positive function? Clearly transposons carrying specific genes such as those determining antibiotic resistance can confer a selective advantage on their bearers under appropriate circumstances. However, the important question is why the genes remain associated with the transposon rather than becoming a stable part of the cellular genome. The most likely answer is that transposability promotes amplification of the transposable gene and its dissemination to hosts of various chromosomal genotypes (Campbell, 1981).

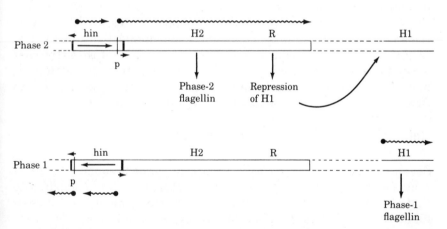

FIGURE 4. Phase variation in *Salmonella typhimurium*. In phase 2, genes *H2* and *R* are transcribed from a promoter p that lies within an invertible 995-base segment whose termini form an exact 14-base inverted repeat (arrows). Cells in phase 2 synthesize only phase 2 flagellin, because the phase 1 flagellin gene is repressed by the product of gene R. In phase 1, the invertible region is in the opposite orientation. *H2* and *R* are not transcribed, and *H1* is active. Inversion is catalyzed by the product of the *hin* gene, which lies within the invertible segment and causes conservative site specific recombination between the 14-base termini. (Based on Zieg and Simon, 1980; Iino, 1977.)

TRANSPOSONS AND SPECIFIC REARRANGEMENT MECHANISMS AS AGENTS OF CHROMOSOMAL EVOLUTION

Both recent studies with bacteria and the older work on maize show that transposons can cause various kinds of rearrangements besides transposition of the element itself. These include adjacent deletions, adjacent inversions, and cointegrate formation. How important have transposons been in shaping the course of chromosomal evolution?

There are really several separate questions to consider. First, how important are chromosomal rearrangements in evolution? Second, are a large fraction of these rearrangements caused by transposons? Third, are there mechanisms for rearrangement that do not depend on transposons but that can produce the same end result as transposon-mediated rearrangement?

Rearrangement as an evolutionary mechanism

It is no news to geneticists that chromosomal rearrangement has been frequent in nature. One may debate selectionism versus neutralism for rearrangements as for point mutations, but there is no reasonable doubt that rearrangement has had evolutionary impact, sometimes by the fusion of genes or operons to create novel genetic units, sometimes by affecting genetic recombination and reproductive isolation. Whether one considers the inversions found in different wild populations of *Drosophila*, the frequency of tandem duplications, or the distribution of repeated sequences in DNA, it is apparent that stretches of DNA of all sizes have moved to new locations many times. Molecular biologists have been overly preoccupied with missense changes in coding regions, as witnessed by the composition of this volume.

Relative frequencies of rearrangement and point mutation

Several questions may be asked concerning the frequency of rearrangement. What is the relative rate of rearrangement compared to that of point mutation? What fraction of rearrangements are successful in evolution? What fraction of rearrangements are caused by transposons?

Probably the answers to these questions vary widely depending on the system under study. It is noteworthy that, under some circumstances, the majority of recognizable mutations prove to be small rearrangements rather than point mutations (Arber et al., 1981).

Uniqueness of transposon-mediated rearrangements

The central issue concerning transposon-mediated rearrangements is whether there are cellular mechanisms, not requiring movable elements, that would have produced the same end result.

There is a striking difference in the explanatory value of transposons for students of development as compared to students of evolution. At least to some genetically oriented developmentalists, the commitment of certain somatic cells to a differentiated state that is transmitted to their cellular descendants virtually requires specific DNA rearrangements. If transposons had not been discovered, developmentalists would have found it necessary to postulate something similar to them. Thus, the idea that immunoglobulin diversity is generated by DNA rearrangement was proposed long before direct evidence was available, at a time when the only model for controlled DNA rearrangement was λ prophage insertion (Dreyer and Bennett, 1965).

Do evolutionists "need" transposons in the same sense that developmentalists need them? I think it is fair to say that they have seldom perceived such a need. This could mean that other mechanisms are adequate to account for the facts of evolution. Alternatively, it might indicate only that evolutionists do not care much about mechanisms.

Rearrangements occur during evolution. Transposons can cause rearrangements. Are transposons then needed in evolution? If there are important classes of rearrangements that can only take place through the agency of transposons, a clear case can be made that transposons are needed.

We may start with deletions. For many deletions, both termini have been localized within known genes. For some of these, the DNA sequence at the junction point is known (Farabaugh et al., 1978). Such deletions differ from the deletions typically caused by movable elements, which terminate at the element itself. As the biochemistry of deletion formation is unknown, it is of course technically possible that some of the enzymes used in the process are coded by unidentified transposons somewhere in the cellular genome; but that hypothesis is both gratuitous and currently untestable.

For other known rearrangements (duplications, inversions, reciprocal translocations), determination of the end points is seldom precise enough to preclude the presence of movable elements at the junction points. Movable elements can sometimes play a role in generating duplications; for example, specialized transducing phages produce partially diploid bacteria when they lysogenize. On the other hand, there are small duplications for which the complete nucleotide sequence has been determined with no movable elements nearby (Calos et al., 1978). It seems likely that rearrangements of all these types can take place without the sequence-specific recognition characteristic of movable elements.

Transposition and replicon fusion may fall into a different category. Transposition (or "insertional translocation") should be very rare if it

271

depended simply on random chromosome breakage, because multiple events are required. Replicon fusion (a term most applicable to the circular DNA molecules of prokaryotes) can take place by three known mechanisms: (1) by site-specific recombination, as in λ insertion; (2) by recombination between homologous DNA segments, as in the insertion of the F factor into the chromosome (where the homologous segments are insertion sequences, whose presence in both partners results from a prior transposition event); and (3) by cointegrate formation, in which a transposon is replicated as part of the Hfr fusion process.

Can replicon fusion ever take place through random breakage and reunion of DNA molecules? The answer depends on whether the joining of nonhomologous DNA sequences is ever a reciprocal event. It is

(A)

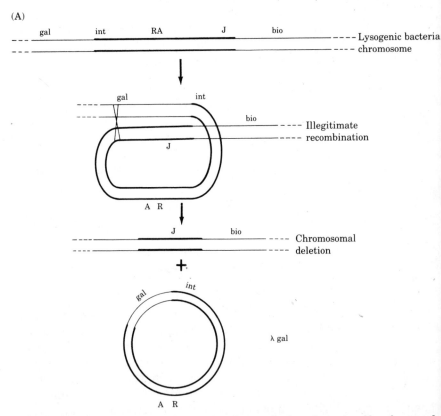

FIGURE 5. A. Generation of specialized transducing phage λ*gal* by abnormal excision from the chromosome of a lysogenic bacterium. The reciprocal product (chromosomal deletion of *gal* extending into the prophage) is known to occur. It is not known whether both products can arise in the same event, as implied here. B. Generation of F' elements by abnormal excision from Hfr chromo-

not known, for example, whether duplications and deletions (other than those generated by homologous recombination between previously duplicated or transposed segments) can arise as products of the same event.

Deletions could arise by breakage and rejoining of DNA molecules, in which case a small circle would be excised from the chromosome. Alternatively, they might result from replication errors, where some DNA in the template strand is skipped over. Specialized transducing phages have the formal structure expected of the reciprocal products of a recombinational mode of deletion formation (Figure 5). However,

(B)

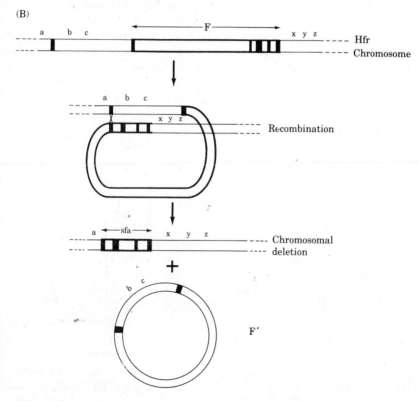

some. The reciprocal product [chromosomal deletion with retention of part of F, creating a sex-factor affinity (*sfa*) locus] can arise in the same event as F'. Where reciprocity is observed, it could result from homologous recombination between insertion sequences in F and insertion sequences preexisting in the chromosome. The figure shows one such sequence between a and b, homologous to the innermost IS3 sequence in F.

273

they too might arise by replication errors. In the analogous case of F' generation, reciprocal production of F' elements and chromosomal deletions is demonstrable (Berg and Curtiss, 1967). However, interpretation of the F' results is complicated by the fact that some F's arise, not by random breakage of bacterial DNA, but by homologous recombination within IS elements preexisting in the bacterial chromosome (Ohtsubo et al., 1974).

Precise fusion of two replicons by DNA breakage and reunion requires a reciprocal event. It may be that the only efficient mechanisms for precise fusion depend on movable elements, either to catalyze the fusion or to provide homology between the two replicons. It seems likely that imprecise fusions, in which most of one replicon becomes inserted into another, can come about by less specific mechanisms.

In higher eukaryotes, heterologous joining proceeds readily with no obvious help from movable elements. The insertion of SV40 and polyoma virus was mentioned earlier. Even more striking is the ability of cultured cells to take up and integrate heterologous transforming DNA (Robins et al., 1981). A role of endogenous movable elements in the process is not excluded (and might even be suggested by the correlation between integration and rearrangements of chromosomal segments adjacent to the insertion site). At any rate, if one interprets these observations to indicate a high rate of nonspecific DNA joining, then it is hard to see how movable elements could contribute much in the way of qualitatively new potentialities.

Specific rearrangements in evolution

The fact that nonspecific mechanisms may be capable of generating a wide spectrum of chromosomal rearrangements does not imply that specific rearrangement mechanisms have had no influence at all on the course of evolution. The progress of organisms or populations from one adaptive peak to the next generally requires many successive changes and should tend to follow channels determined by the rates of change of the individual steps. Specific rearrangements should facilitate certain pathways of change. If specific rearrangements generally occur in somatic cells as part of differentiation, we might expect that they would sometimes accidentally occur germinally as well. Those that are not dominant lethal should contribute to the genetic diversity of the population and create certain pathways for evolution that might roughly be categorized as "phylogeny recapitulates ontogeny" or perhaps "the new orthogenesis."

It is a short step from "the new orthogenesis" to "the new Lamarckianism." More than 20 years ago, I can remember being criticized by some of my colleagues for referring to lysogenization by phage λ as an example of Lamarckian inheritance. My statement was based on the fact that lysogeny is an acquired heritable trait and that it has adaptive value (immunity to infection) in the presence of the inducing agent (bacteriophage). My colleagues' objections, as I remember them, were that Lamarckianism had been disproved with respect to traits determined by real genes and that what viruses do to their hosts is irrelevant to the issue.

Although the advances of the last 20 years do not change the nature of the argument, they justify some shift in perspective. We know now that specific rearrangement is not an exclusive attribute of viruses. It can be carried out by other subcellular systems; sometimes affecting gene expression in a manner that is adaptively useful to the cell or organism. We also know, from transposons like Tn*3* and especially from phage λ, that the rearrangement process can be not only highly specific but also carefully regulated. Specific enzymes can be induced to catalyze rearrangement, then repressed so that the new configuration is stably inherited.

All the components needed to produce inducible adaptive heritable changes are therefore available in living organisms. As long as one focuses on base changes in coding regions as the mechanism of mutation, the distinction between acquired and innate characteristics is reinforced by the absence of known mechanisms for directed changes in base sequence. To suggest that heritable changes might be inducible then requires the gratuitous postulation of unknown mechanisms. If one includes specific rearrangements among the heritable changes to be considered, the situation is quite different. The mechanisms are there. To be a confirmed anti-Lamarckian under those circumstances requires the gratuitous postulate that known mechanisms are somehow suspended; or at least that, for unknown reasons, they never affect germinal DNA in an adaptive manner. An example of such an adaptive change that may seem less contrived than that of λ lysogenization is provided by the erythromycin resistance transposon Tn*917* (Tomich et al., 1980). Transposition of Tn*917* is specifically induced by low concentrations of the selective agent, erythromycin. This property can promote both amplification and interbacterial transfer under circumstances in which both processes may be selectively advantageous.

I think we should accept the fact that similar strategies should evolve wherever they are of use. If such mechanisms are rare and exceptional, the appropriate response for evolutionists is not to ignore them but rather to explain why they have not become more common. We may also ask what impact they are likely to have on the course

275

of evolution. I suspect that in fact they will have more consequences for our concept of heredity than they do for evolution. Both lysogenization and transposition are in a sense reversible. Elements can be added, but they also can be lost. One may regard the insertion and excision of a prophage as part of the life cycle of the phage, rather than as a step in a progression, which "evolution" implies.

The distinction is more than semantic and is one that I expect will generally hold for most or all cases of directed inheritance. The reason is simply that evolution takes place in response to selection pressure, which means that traits are selected on the basis of past experience, not in anticipation of future contingencies. The only circumstances under which a mechanism such as the erythromycin-promoted transposition of erythromycin genes is expected to evolve and to be retained is where the mechanism has been used repeatedly among the ancestors of present individuals. This can only happen with a cyclical process. To some degree, the members of a bacterial cell line undergoing one complete cycle of gain and loss might be regarded as analogous to the germ line cells of one individual higher eukaryote leading from the gametes of one generation to those of the next. The main difference is that the latter cycle is an obligatory sequence of fixed duration, whereas the former is not.

[The erythromycin transposon is not the ideal example here, because one might argue that it makes more sense to follow the transposon through a cycle that includes transfer from one host to the next, rather than following a cell line. That circumstance does not change the substance of the argument. The *Salmonella* phase variation system (Figure 4), which causes rearrangement internal to the bacterium, might be more appropriate but is not known to be regulated.]

We thus end about where we started: even if directed, adaptive germinal rearrangements sometimes occur in multicellular as well as unicellular organisms, their impact on evolution is likely to be about the same as that of rearrangements that are usually somatic but occasionally occur in the germ line. Specific rearrangements may influence the detailed course of evolution by facilitating the first steps along certain pathways of change, in which irreversible, frequent, nonspecific events follow after the reversible ones.

TRANSPOSONS IN THE EVOLUTION OF GENE REGULATION

We now leave the subject of transposons as agents for rearranging chromosomal segments with respect to one another and ask whether the influence of transposons themselves on the activity of nearby genes might have played an important part in the evolution of new patterns of gene expression.

The conspicuous regulatory effects of the maize elements led to their name "controlling elements." Insertion of an element in or near a gene may turn the gene off completely or may alter the intensity and timing of its expression during development. In mammalian cells, insertion of retroviruses in the vicinity of genes may activate their expression above normal levels. Studies of these effects may illuminate the mechanism of gene control in higher eukaryotes, which is still poorly understood.

In prokaryotes, a considerable body of data on gene regulation exists and is comprehensible within a simple conceptual framework. Transposons and insertion sequences have effects that are predictable from existing concepts. An element can either prevent gene expression (by inserting within the gene or by introducing a transcriptional stop signal upstream from it) or activate a gene (by introducing a promoter upstream from it).

Many movable elements, such as bacteriophage λ, have weak promoter activities. Some of the endogenous insertion sequences of *E. coli*, such as IS2, have strong promoter activity (Saedler et al., 1974). Deliberate selection for mutations capable of turning on an *E. coli* gene under conditions where it is poorly expressed usually nets, among other things, some IS2 insertions upstream from the selected gene (Figure 6). It is hard to doubt that similar insertions are also selected naturally, because it must sometimes happen that novel changes in the environment create a need for expression of a gene under circumstances where it was unexpressed according to the rules that previously governed it. Generally, the optimum eventuality would be replacement of the previous operator and promoter by ones following

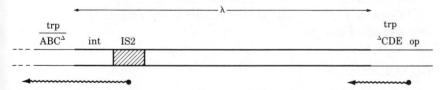

FIGURE 6. Example of laboratory selection for insertion of elements such as IS2 with strong promoter activity. The starting material was a strain in which λ prophage had inserted at a secondary attachment site within the *trpC* gene. The λ prophage blocks transcription of *trpB* from its normal promoter (*trp op*). Selection for good expression of *trpB* yields mutations in λ promoters (especially the integrase promoter) but also IS2 insertions. (Based on Zissler et al., 1977.)

different rules rather than insertion of an unregulated promoter like that of IS2. Whether IS2 insertion is a common first step in that direction, rather than an evolutionary dead end, is uncertain.

Are there any circumstances under which promoter insertion is especially likely to be the preferred route for natural changes in promoter specificity? Among the possibilities are those occasions in history where new control mechanisms developed that affected a whole battery of diverse genes, such as cyclic AMP control of certain bacterial promoters. It might have been easier to insert copies of one promoter regulated by the new effector in front of many genes rather than waiting for the appropriate binding sites to arise by random changes in the DNA of many different promoters.

If the original cyclic AMP-dependent promoters arose in that manner, long ago, most of the traces seem to have disappeared from the diverse promoter sequences found in *E. coli* today. An example of a regulated mobile promoter that may be of more recent origin is the Ty1 promoter in *Saccharomyces cerevisiae*. Insertions of the transposon Ty1 can activate expression of genes near the site of insertion. Like a number of known conjugation functions, the activity is regulated by the mating type loci; activation disappears in a/α diploids, which have no reason to conjugate (Erreda et al., 1981). It is possible that genes for normal conjugation functions are controlled by adjacent Ty1 elements (of which there are approximately 30 per haploid genome). If that proves to be correct, the implication is that either the system is of recent origin (so that the promoters have not yet diverged and lost transposability) or that transposability of these promoters serves some ongoing function, beyond its historical role of placing the promoters in the correct location. A more trivial alternative, which cannot be excluded at present, is that Ty1 has incorporated a normal cellular promoter from one of the conjugation genes.

Some models of gene control in higher eukaryotes postulate the presence of multiple control sites, which are duplicated in various combinations at other loci (Davidson and Britten, 1979). Transposons could clearly facilitate the construction of new combinations and the remodeling of old ones.

CONCLUSIONS

Movable elements are found in organisms of all types and have doubtless caused some chromosomal changes of evolutionary importance. Chromosomal rearrangements can also arise by less specific mechanisms that do not depend on specific elements. If we look for a role in evolution to which movable elements are especially suited, a likely candidate is the dissemination of specific regulatory sites around the genome.

278

Controlled gene rearrangements are observable in diverse organisms. Such rearrangements could constitute the critical mechanism whereby specialized cells of higher eukaryotes acquire their specific stable properties. Movable elements might be (1) the agents of some such rearrangements; (2) the evolutionary precursors of the cellular agents; (3) derivatives of the normal agents, which serve no useful function to the organism but are tolerated by it; or (4) all of the above.

BIBLIOGRAPHY

(Numbers in parentheses at the end of each reference indicate the chapter or chapters in which the work is cited.)

Adams, J. and E. D. Rothman (1982) Estimation of phylogenetic relationships from DNA restriction patterns and selection of endonuclease cleavage sites. Proc. Natl. Acad. Sci. 79:3560–3564. (8)

Air, G. M. (1981) Sequence relationships among the hemagglutinin genes of 12 subtypes of influenza A virus. Proc. Natl. Acad. Sci. 78:7639–7643. (9)

Air, G. M. and R. M. Hall (1981) Conservation and variation in influenza gene sequences. In: Genetic Variation among Influenza Viruses, ICN-UCLA Symposia on Molecular and Cellular Biology, D. Nayak and C. F. Fox (eds.). Academic Press, New York, Vol. 22, pp. 29–44. (9)

Anderson, S., A. T. Bankier, B. G. Barrell, M. H. L. de Bruijn, A. R. Coulson, J. Drouin, I. C. Eperon, D. P. Nierlich, B. A. Roe, F. Sanger, P. H. Schreier, A. J. H. Smith, R. Staden, and I. G. Young (1981) Sequence and organization of the human mitochondrial genome. Nature 290:457–465. (2, 4, 8, 10)

Anderson, S., M. H. L. de Bruijn, A. R. Coulson, I. C. Eperon, F. Sanger, and I. G. Young (1982) Complete sequence of bovine mitochondrial DNA: conserved features of the mammalian mitochondrial genome. J. Mol. Biol. 156:683–717. (4, 8, 10)

Anderson, S. M., M. Santos, and J. McDonald (1980) Comparative study of the thermostability of crude and purified preparations of alcohol dehydrogenase (E.C. 1.1.1.1) from *D. melanogaster*. Drosophila Information Service 55:13–14. (6)

Antonarakis, S. E., C. D. Boehm, P. J. V. Giardina, and H. H. Kazazian, Jr. (1982a) Nonrandom association of polymorphic restriction sites in the β-globin gene cluster. Proc. Natl. Acad. Sci. 79:137–141. (7)

Antonarakis, S. E., S. H. Orkin, H. H. Kazazian, Jr., S. C. Goff, C. D. Boehm, P. G. Waber, J. P. Sexton, H. Ostrer, V. F. Fairbanks, and A. Chakravarti (1982b) Evidence for multiple origins of the β^E-globin gene in South East Asia. Proc. Natl. Acad. Sci. 79:6608–6611. (7)

Aquadro, C. F. and B. D. Greenberg (1983) Human mitochondrial DNA variation and evolution: analysis of nucleotide sequences from seven individuals. Genetics 103:287–312. (4, 8)

Arber, W., M. Hümbelin, P. Caspers, H. J. Reif, S. Iida, and J. Meyer (1981) Spontaneous mutations in the *Escherichia coli* phage P1 and IS-mediated processes. Cold Spring Harbor Symp. Quant. Biol. 45:36–40 (13)

Arnheim, N. and M. Kuehn (1979) The genetic behavior of a cloned mouse ribosomal DNA segment mimics mouse ribosomal gene evolution. J. Mol. Biol. 134:743–765. (3)

Arnheim, N., M. Krystal, R. Schmickel, G. Wilson, O. Ryder, and E. Zimmer (1980) Molecular evidence for genetic exchanges among ribosomal genes on nonhomologous chromosomes in man and apes. Proc. Natl. Acad. Sci. 77:7323–7327. (3)

Arnheim, N., D. Treco, B. Taylor, and E. Eicher (1982) Distribution of ribosomal gene length variants among mouse chromosomes. Proc. Natl. Acad. Sci. 79:4677–4680. (3)

Arraj, J. A. and J. A. Campbell (1975) Isolation and characterization of the newly evolved *ebg* β-galactosidase of *Escherichia coli* K12. J. Bacteriol. 124:849–856. (12)

Attardi, G., M. Albring, F. Amalric, R. Gelfand, J. Griffith, D. Lynch, C. Merkel, W. Murphy, and D. Ojala (1976) Organization and expression of the mitochondrial genome in HeLa cells. In: Genetics and Biogenesis of Chloroplasts and Mitochondria, T. Bucher, W. Neupert, W. Sebald, and S. Werner (eds.). Elsevier, Amsterdam, pp. 573–585. (4)

Avise, J. C., R. A. Lansman, and R. O. Shade (1979a) The use of restriction endonucleases to measure mitochondrial DNA sequence relatedness in natural populations. I. Population structure and evolution in the genus *Peromyscus*. Genetics 92:279–295. (8, 9)

Avise, J. C., C. Giblin-Davidson, J. Laerm, J. C. Patton, and R. A. Lansman (1979b) Mitochondrial DNA clones and matriarchal phylogeny within and among geographic populations of the pocket gopher. *Geomys pinetis*. Proc. Natl. Acad. Sci. 76:6694–6698. (8)

Ayala, F. J. (1975) Genetic differentiation during the speciation process. Evol. Biol. 8:1–78. (5)

Ayala, F. J. (ed.) (1976) Molecular Evolution. Sinauer Associates, Sunderland, Mass. (5)

Ayala, F. J. (1982) Genetic variation in natural populations: problem of electrophoretically cryptic alleles. Proc. Natl. Acad. Sci. 79:550–554. (5)

Ayala, F. J. and Th. Dobzhansky (1974) A new subspecies of *Drosophila pseudoobscura* (Diptera: Drosophilidae). Pan-Pacific Entomologist 50:211–219. (5)

Ayala, F. J. and J. R. Powell (1972) Allozymes as diagnostic characters of sibling species of *Drosophila*. Proc. Natl. Acad. Sci. 69:1094–1096. (5)

Bach, R., I. Grummt, and B. Allet (1981a) The nucleotide sequence of the initiation region of the ribosomal transcription unit from mouse. Nucleic Acids Res. 9:1559–1569. (3)

Bach, R., B. Allet, and M. Crippa (1981b) Sequence organization of the spacer in the ribosomal genes of *Xenopus clivii* and *Xenopus borealis*. Nucleic Acids Res. 9:5311–5330. (3)

Baltimore, D. (1981) Gene conversion: some implications for immunoglobulin genes. Cell 24:592–594. (3)

Barker, W. C. and M. O. Dayhoff (1980) Evolutionary and functional relationships of homologous physiological mechanisms. BioScience 30:593–599. (2)

Barker, W. C., L. K. Ketcham, and M. O. Dayhoff (1978) Duplications in protein sequences. In: Atlas of Protein Sequence and Structure, Vol. 5, Suppl. 3, M. O. Dayhoff (ed.). Natl. Biomed. Res. Found., Washington, D.C., pp. 359–362. (2)

Barnard, E. A., M. S. Cohen, M. H. Gold, and J.-K. Kim (1972) Evolution of ribonuclease in relation to polypeptide folding mechanisms. Nature 240:395–398. (11)

Barrell, B. G., A. T. Bankier, and J. Drouin (1979) A different genetic code in human mitochondria. Nature 282:189–194. (10)

Barrell, B. G., S. Anderson, A. T. Bankier, M. H. L. de Bruijn, E. Chen, A. R. Coulson, J. Drouin, I. C. Eperon, D. P. Nierlich, B. A. Roe, F. Sanger, P. H. Schreier, A. J. H. Smith, R. Staden, and I. G. Young (1980) Different patterns of codon recognition by mammalian mitochondrial tRNAs. Proc. Natl. Acad. Sci. 77:3164–3166. (4)

Bayne, B. L., M. N. Moore, and R. K. Koehn (1981) Lysosomes and the response by *Mytilus edulis* L. to an increase in salinity. Marine Biology Letters 2:193–204. (6)

Bell, G. I., R. L. Pictet, W. J. Rutter, B. Cordell, E. Tischer, and H. M. Goodman (1980) Sequence of the human insulin gene. Nature 284:26–32. (11)

Bell, G. I., M. J. Selby, and W. J. Rutter (1982) The highly polymorphic region near the human insulin gene is composed of simple tandemly repeating sequences. Nature 295:31–35. (7, 9)

Bennetzen, J. L. and B. D. Hall (1982) Codon selection in yeast. J. Biol. Chem. 257:3026–3031. (11)

Bentley, D. L. and T. H. Rabbitts (1980) Human immunoglobulin variable region genes—DNA sequences of two V_κ genes and a pseudogene. Nature 288:730–733. (9)

Bentley, D. L. and T. H. Rabbitts (1981) Human V_κ immunoglobulin gene number: implications for the origin of antibody diversity. Cell 24:613–623. (9)

Benveniste, R. E. and G. J. Todaro (1976) Evolution of type C viral genes: evidence for an Asian origin of man. Nature 261:101–108. (4)

Benyajati, C., A. R. Place, D. A. Powers, and W. Sofer (1981) Alcohol dehydrogenase gene of *Drosophila melanogaster*: Relationship of intervening sequences to functional domains in the protein. Proc. Natl. Acad. Sci. 78:2717–2721. (9)

Benzer, S. and S. P. Champe (1961) Ambivalent *r*II mutants of phage T4. Proc. Natl. Acad. Sci. 47:1025–1038. (1)

Berg, C. M. and R. Curtiss III (1967) Transposition derivatives of an Hfr strain of *Escherichia coli* K-12. Genetics 56:503–525. (13)

Berry, R. J. and J. Peters (1977) Heterogeneous heterozygosities in *Mus musculus* populations. Proc. Roy. Soc. London (B) 197:485–503. (5)

Betz, J. L. and P. H. Clarke (1972) Selective evolution of phenylacetamide-utilizing strains of *Pseudomonas aeruginosa*. J. Gen. Microbiol. 73:161–174. (12)

Betz, J. L., P. R. Brown, M. J. Smyth, and P. H. Clarke (1974) Evolution in action. Nature 247:261–264. (12)

Bibb, M. J., R. A. Van Etten, C. T. Wright, M. W. Walberg, and D. A. Clayton (1981) Sequence and gene organization of mouse mitochondrial DNA. Cell 26:167–180. (4, 8, 10)

Bijlsma, R. (1978) Polymorphism at the glucose-6-phosphate dehydrogenase and 6-phosphogluconate dehydrogenase loci in *Drosophila melanogaster*. Part 2. Evidence for interaction in fitness. Genet. Res. 31:227–238. (6)

Birdsell, J. B., R. T. Simmons, and J. J. Graydon (1979) Microdifferentiation in blood group frequencies among twenty-eight aboriginal tribal isolates in Western Australia. Occ. Pap. Hum. Biol. 2:1–38. (5)

Birky, C. W., Jr. (1978) Transmission genetics of mitochondria and chloroplasts. Ann. Rev. Genet. 12:471–512. (8)

Birky, C. W., Jr. and R. V. Skavaril (1976) Maintenance of genetic homogeneity in systems with multiple genomes. Genet. Res. 27:249–265. (3)

Birky, C. W., Jr., A. R. Acton, R. Dietrich, and M. Carver (1982) Mitochondrial transmission genetics: replication, recombination, and segregation of mitochondrial DNA and its inheritance in crosses. In: Mitochondrial Genes, P. Slonimski, P. Borst, and G. Attardi (eds.). Cold Spring Harbor Laboratory, Cold Spring Harbor, New York, pp. 333–348. (8)

Birky, C. W., Jr., T. Maruyama, and P. A. Fuerst (1983) An approach to population and evolutionary genetic theory for genes in mitochondria and chloroplasts and some results. Genetics 103:513–527. (8)

Bishop, S. H. (1976) Nitrogen metabolism and excretion: regulation of intracellular amino acid concentrations. In: Estuarine Processes, M. Wiley (ed.). Academic Press, New York, pp. 414–429. (6)

Bishop, Y. M. M., S. E. Fienberg, and P. W. Holland (1975) Discrete Multivariate Analysis: Theory and Practice. MIT Press, Cambridge, Mass. (5)

Blake, N. M. (1979) Genetic variation of red cell enzyme systems in Australian aborigine populations. Occ. Pap. Hum. Biol. 2:39–82. (5)

Bodmer, W. F. (1970) The evolutionary significance of recombination in prokaryotes. Symp. Soc. Gen. Microbiol. 20:279–294. (5)

Bodmer, W. F. and L. L. Cavalli-Sforza (1976) Genetics, Evolution, and Man. W. H. Freeman, San Francisco. (5)

Bonen, L., T. Y. Huh, and M. W. Gray (1980) Can partial methylation explain the complex fragment patterns observed when plant mitochondrial DNA is cleaved with restriction endonucleases? FEBS Letters 111:340–346. (8)

Bonhomme, F. and R. K. Selander (1978) The extent of allelic diversity underlying electrophoretic protein variation in the house mouse. In: Origins of Inbred Mice, H. C. Morse, III (ed.). Academic Press, New York, pp. 569–589. (5)

Bonitz, S. G., G. Coruzzi, B. E. Thalenfeld, A. Tzagoloff, and G. Macino (1980) Assembly of the mitochondrial membrane system. J. Biol. Chem. 255:11927–11941. (10)

Bonner, J. J., M. Berninger, and M. L. Pardue (1977) Transcription of polytene chromosomes and of the mitochondrial genome in *Drosophila melanogaster*. Cold Spring Harbor Symp. Quant. Biol. 42:803–814. (4)

Bornstein, P. and H. Sage (1980) Structurally distinct collagen types. Ann. Rev. Biochem. 49:957–1003. (2)

Borowsky, R. (1977) Detection of the effects of selection on protein polymorphisms in natural populations by means of a distance analysis. Evolution 31:341–346. (5)

Borst, P. and R. A. Flavell (1976) Properties of mitochondrial DNAs. In: Handbook of Biochemistry and Molecular Biology, 3rd Ed., Nucleic Acids, Vol. II, G. D. Fasman (ed.). CRC Press, Ohio, pp. 363–374. (4)

Borst, P. and G. J. C. M. Ruttenberg (1969) Interaction of ribopolynucleotides with the complementary strands of chick-liver mitochondrial DNA. Biochim. Biophys. Acta 190:391–405. (4)

Botchan, M., W. Topp, and J. Sambrook (1979) Studies on SV40 excision from cellular chromosomes. Cold Spring Harbor Symp. Quant. Biol. 43:709–719. (13)

Bothwell, A. L. M., M. Paskind, M. Reth, T. Imanishi-Kari, K. Rajewsky, and D. Baltimore (1981) Heavy chain variable region contribution to the NP^b family of antibodies: somatic mutation evident in a γ2a variable region. Cell 24:625–637. (9)

Brammar, W. J., P. H. Clarke, and A. J. Skinner (1967) Biochemical and genetic studies with regulator mutants of the *Pseudomonas aeruginosa* 8602 amidase system. J. Gen. Microbiol. 47:87–102. (12)

Brasch, J. M. and D. R. Smyth (1979) Absence of silver bands in human Robertsonian translocation chromosomes. Cytogenet. Cell Genet. 24:122–125. (3)

Brenner, D. J., G. R. Fanning, F. J. Skerman, and S. Falkow (1972) Polynucleotide sequence divergence among strains of *Escherichia coli* and closely related organisms. J. Bacteriol. 109:953–965. (5)

Brennicke, A. and D. A. Clayton (1981) Nucleotide assignment of alkali-sensitive sites in mouse mitochondrial DNA. J. Biol. Chem. 256:10613–10617. (4)

Britten, R. J. and E. H. Davidson (1969) Gene regulation for higher cells: a theory. Science 165:349–357. (2)

Britten, R. J. and D. E. Kohne (1968) Repeated sequences in DNA. Science 161:529–540. (3)

Brown, B. A., R. W. Padgett, S. C. Hardies, C. A. Hutchison, III, and M. H. Edgell (1982) β-globin transcript found in induced murine erythroleukemia cells is homologous to the β*h0* and β*h1* genes. Proc. Natl. Acad. Sci. 79:2753–2757. (1)

Brown, D. D., P. C. Wensink, and E. Jordan (1972) A comparison of the ribosomal DNAs of *Xenopus laevis* and *Xenopus mulleri*: the evolution of tandem genes. J. Mol. Biol. 63:57–73. (3)

Brown, G. G. and M. V. Simpson (1981) Intra- and interspecific variation of the mitochondrial genome in *Rattus norvegicus* and *Rattus rattus*: restriction enzyme analysis of variant mitochondrial DNA molecules and their evolutionary relationships. Genetics 97:125–143. (8)

Brown, G. G. and M. V. Simpson (1982) Novel features of animal mtDNA evolution as shown by sequences of two rat cytochrome oxidase subunit II genes. Proc. Natl. Acad. Sci. 79:3246–3250. (4, 8)

Brown, J. E., P. R. Brown, and P. H. Clarke (1969) Butyramide-utilizing mutants of *Pseudomonas aeruginosa* 8602 which produce an amidase with altered substrate specificity. J. Gen. Microbiol. 57:273–298. (12)

Brown, W. M. (1976) Interspecific comparisons of animal mitochondrial DNAs. Ph.D. thesis. California Institute of Technology, Pasadena, California. (4)

Brown, W. M. (1980) Polymorphism in mitochondrial DNA of humans as revealed by restriction endonuclease analysis. Proc. Natl. Acad. Sci. 77:3605–3609. (4, 8)

Brown, W. M. (1981) Mechanisms of evolution in animal mitochondrial DNA. Ann. N.Y. Acad. Sci. 361:119–134. (4, 8)

Brown, W. M. and H. M. Goodman (1979) Quantitation of intrapopulation variation by restriction endonuclease analysis of human mitochondrial DNA. In: Extrachromosomal DNA, D. J. Cummings, P. Borst, I. B. Dawid, S. M. Weissman, and C. F. Fox (eds.). Academic Press, New York, pp. 485–499. (4, 8)

Brown, W. M. and J. Vinograd (1974) Restriction endonuclease cleavage maps of animal mitochondrial DNAs. Proc. Natl. Acad. Sci. 11:4617–4621. (4, 8)

Brown, W. M. and J. W. Wright (1979) Mitochondrial DNA analyses and the origin and relative age of parthenogenetic lizards (genus *Cnemidophorus*). Science 203:1247–1249. (4, 8)

Brown, W. M., J. Shine, and H. M. Goodman (1978) Human mitochondrial DNA: analysis of 7S DNA from the origin of replication. Proc. Natl. Acad. Sci. 75:735–739. (4)

Brown, W. M., M. George, Jr., and A. C. Wilson (1979) Rapid evolution of animal mitochondrial DNA. Proc. Natl. Acad. Sci. 76:1967–1971. (4, 8)

Brown, W. M., E. M. Prager, A. Wang, and A. C. Wilson (1982) Mitochondrial DNA sequences of primates: tempo and mode of evolution. J. Mol. Evol. 18:225–239. (4, 8)

Brutlag, D. L. (1980) Molecular arrangement and evolution of heterochromatic DNA. Ann. Rev. Genet. 14:121–144. (3)

Bryant, E. H. (1974) On the adaptive significance of enzyme polymorphisms in relation to environmental variability. Amer. Nat. 108:1–19. (5)

Bultmann, H. and C. D. Laird (1973) Mitochondrial DNA from *Drosophila melanogaster*. Biochim. Biophys. Acta 299:196–209. (4)

Burton, R. S. and M. W. Feldman (1981) Population genetics of *Tigriopus californicus*. II. Differentiation among neighboring populations. Evolution 35:1192–1205. (6)

Burton, R. S. and M. W. Feldman (1982) Changes in free amino acid concentrations during osmotic response in the intertidal copepod *Tigriopus californicus*. Comp. Biochem. Physiol. (in press) (6)

Burton, R. S. and M. W. Feldman (1983) Physiological effects of an allozyme polymorphism: glutamate-pyruvate transaminase and response to hyperosmotic stress in the copepod *Tigriopus californicus*. Biochem. Genet. (in press) (6)

Busslinger, M., S. Rusconi, and M. L. Birnstiel (1982) An unusual evolutionary behaviour of a sea urchin histone gene cluster. The EMBO Journal 1:27–33. (9)

Calder, N. (1973) The Life Game. BBC, London. (11)

Calos, M. P., D. Galas, and J. H. Miller (1978) Genetic studies of the *lac* repressor. VIII. DNA sequence charge resulting from an intragenic duplication. J. Mol. Biol. 126:865–869. (13)

Campbell, A. (1981) Evolutionary significance of accessory DNA elements in bacteria. Ann. Rev. Microbiol. 35:55–83. (13)

Campbell, A. (1983) Bacteriophage λ. In: Mobile Genetic Elements, J. A. Shapiro (ed.). Academic Press, New York (in press) (13)

Campbell, J. H., J. Lengyel, and J. Langridge (1973) Evolution of a second gene for β-galactosidase in Escherichia coli. Proc. Natl. Acad. Sci. 70:1841–1845. (12)

Cann, R. L., W. M. Brown, and A. C. Wilson (1982) Evolution of human mitochondrial DNA: molecular, genetic, and anthropological implications. In: Proc. 6th Intl. Cong. Human Genetics, Part A, B. Bonné-Tamir (ed.). Alan R. Liss, New York, pp. 157–165. (9)

Carr, D. H. and M. Gedeon (1977) Population cytogenetics of human abortuses. In: Population Cytogenetics, E. B. Hood and I. H. Porter (eds.). Academic Press, New York, pp. 1–9. (3)

Castora, F. J., N. Arnheim, and M. V. Simpson (1980) Mitochondrial DNA polymorphism: evidence that variants detected by restriction enzymes differ in nucleotide sequence rather than in methylation. Proc. Natl. Acad. Sci. 77:6415–6419. (8)

Caugant, D. A., B. R. Levin, and R. K. Selander (1981) Genetic diversity and temporal variation in the E. coli population of a human host. Genetics 98:467–490. (5)

Caugant, D., R. K. Selander, and J. S. Jones (1982) Geographic structuring of molecular and morphological polymorphisms in Pyrenean populations of the snail Cepaea nemoralis. Genetica 57:177–191. (5)

Caugant, D., B. Levin, G. Lidin Janson, T. S. Whittam, C. Svanborg Eden, and R. K. Selander (1983) Genetic diversity and relationship among strains of E. coli in the intestine and those causing urinary tract infections. Progress in Allergies 33:203–227. Karger, Basel. (5)

Cavalli-Sforza, L. L. (1966) Population structure and evolution. Proc. Roy. Soc. London (B) 164:362–379. (5)

Cavalli-Sforza, L. L. and A. W. F. Edwards (1964) Analysis of human evolution. In: Genetics Today, Proc. XI Intl. Cong. Genetics, The Hague. Pergamon Press, Oxford, pp. 923–933. (5)

Cavener, D. R. and M. T. Clegg (1981) Evidence for biochemical and physiological differences between enzyme genotypes in Drosophila melanogaster. Proc. Natl. Acad. Sci. 78:4444–4447. (6)

Chakraborty, R. (1977) Simulation results with stepwise mutation model and their interpretations. J. Mol. Evol. 9:313–322. (5)

Chakraborty, R. (1980) Gene-diversity analysis in nested subdivided populations. Genetics 96:721–723. (5)

Chakraborty, R. and M. Nei (1976) Hidden genetic variability within electromorphs in finite populations. Genetics 84:385–393. (5)

Chakraborty, R., P. A. Fuerst, and M. Nei (1978) Statistical studies on protein polymorphism in natural populations: II. Gene differentiation between populations. Genetics 88:367–390. (5, 9)

Chakraborty, R., P. A. Fuerst, and M. Nei (1980) Statistical studies on protein polymorphism in natural populations: III. Distribution of allele frequencies and the number of alleles per locus. Genetics 94:1039–1063. (5, 9)

Chang, A. C. Y., R. A. Lansman, D. A. Clayton, and S. N. Cohen (1975) Studies of mouse mitochondrial DNA in *Escherichia coli*: structure and function of eucaryotic and procaryotic chimeric plasmids. Cell 6:231–244. (8)

Chapman, R. W., J. C. Stephens, R. A. Lansman, and J. C. Avise (1982) Models of mitochondrial DNA transmission genetics and evolution in higher eucaryotes. Genet. Res. 40:41–57. (8)

Charlesworth, B., R. Lande, and M. Slatkin (1982) A neo-Darwinian commentary on macroevolution. Evolution 36:474–498. (5)

Ciarrocchi, G., J. G. Jose, and S. Linn (1979) Further characterization of a cell-free system for measuring replicative and repair synthesis with cultured human fibroblasts and evidence for the involvement of DNA polymerase in DNA repair. Nucleic Acids Res. 7:1205–1219. (4)

Clarke, B. (1975) The contribution of ecological genetics to evolutionary theory: detecting the direct effects of natural selection on particular polymorphic loci. Genetics 79s:101–113. (6)

Clarke, P. H. (1978) Experiments in microbial evolution. In: The Bacteria, L. N. Ornston and J. R. Sokatch (eds.). Academic Press, New York, pp. 137–218. (12)

Clarke, S. H., J. L. Claflin, and S. Rudikoff (1982) Polymorphisms in immunoglobulin heavy chains suggesting gene conversion. Proc. Natl. Acad. Sci. 79:3280–3284. (9)

Clayton, D. A. (1982) Replication of animal mitochondrial DNA. Cell 28:693–705. (4)

Clayton, D. A., J. N. Doda, and E. C. Friedberg (1974) The absence of a pyrimidine dimer repair mechanism in mammalian mitochondria. Proc. Natl. Acad. Sci. 71:2777–2781. (4)

Cleary, M. L., E. A. Schon, and J. B. Lingrel (1981) Two related pseudogenes are the result of a gene duplication in the goat β-globin locus. Cell 26:181–190. (2)

Cocks, G. T., J. Aguilar, and E. C. C. Lin (1974) Evolution of L-1,2-propanediol catabolism in *Escherichia coli* by recruitment of enzymes for L-fucose and L-lactate metabolism. J. Bacteriol. 118:83–88. (12)

Coen, E. S., J. M. Thoday, and G. Dover (1982) Rate of turnover of structural variants in the rDNA gene family of *Drosophila melanogaster*. Nature 295:564–568. (9)

Colantuoni, V., L. Dailey, and C. Basilico (1980) Amplification of integrated viral DNA sequences in polyoma virus-transformed cells. Proc. Natl. Acad. Sci. 77:3850–3854. (13)

Comings, D. E. (1979) Pc 1 Duarte, a common polymorphism of a human brain protein, and its relationship to depressive disease and multiple sclerosis. Nature 277:28–32. (5)

Coote, J. L., G. Szabados, and T. S. Work (1979) The heterogeneity of mitochondrial DNA in different tissues from the same animal. FEBS Letters 99:255–260. (8)

Cordell, B., G. Bell, E. Tischer, F. M. DeNoto, A. Ullrich, R. Pictet, W. J. Rutter, and H. M. Goodman (1979) Isolation and characterization of a cloned rat insulin gene. Cell 18:533–543. (11)

Coyne, J. A. (1982) Gel electrophoresis and cryptic protein variation. In: Isozymes: Current Topics in Biological and Medical Research, Vol. 5, M. C. Rattazzi, J. G. Scandalios, and G. S. Whitt (eds.). Alan R. Liss, New York, pp. 1–32. (5)

Coyne, J. A. and A. A. Felton (1977) Genic heterogeneity at two alcohol dehydrogenase loci in Drosophila pseudoobscura and Drosophila persimilis. Genetics 87:285–304. (5)

Crews, S., D. Ojala, J. Posakony, J. Nishiguchi, and G. Attardi (1979) Nucleotide sequence of a region of human mitochondrial DNA containing the precisely identified origin of replication. Nature 277:192–198. (4, 8)

Crick, F. H. C. (1966) Codon-anticodon pairing: the wobble hypothesis. J. Mol. Biol. 19:548–555. (10)

Crick, F. H. C. (1968) The origin of the genetic code. J. Mol. Biol. 38:367–379. (10)

Croce, C. M., A. Talavera, C. Basilico, and O. J. Miller (1977) Suppression of production of mouse 28S ribosomal RNA in mouse-human hybrids segregating mouse chromosomes. Proc. Natl. Acad. Sci. 74:694–697. (3)

Crow, J. F. (1968) The cost of evolution and genetic load. In: Haldane and Modern Biology, K. R. Dronamraju (ed.). Johns Hopkins University Press, Baltimore, Md., pp. 165–178. (9)

Crow, J. F. (1972) The dilemma of nearly neutral mutations: how important are they for evolution and human welfare? J. Heredity 63:306–316. (11)

Crow, J. F. (1981) The neutralist-selectionist controversy: an overview. In: Population and Biological Aspects of Human Mutation, E. B. Hook (ed.). Academic Press, New York, pp. 3–14. (11)

Crow, J. F. and M. Kimura (1970) An Introduction to Population Genetics Theory. Harper & Row, New York. (5, 11)

Curtsinger, J. W. and C. C. Laurie-Ahlberg (1981) Genetic variability of flight metabolism in Drosophila melanogaster. I. Characterization of power output during tethered flight. Genetics 98:549–564. (6)

Davidson, E. H. and R. J. Britten (1973) Organization of transcription and regulation in the animal genome. Quart. Rev. Biol. 48:565–613. (3)

Davidson, E. H. and R. J. Britten (1979) Regulation of gene expression: possible role of repetitive sequences. Science 204:1052–1059. (13)

Dawid, I. B. (1972) Evolution of mitochondrial DNA sequences in Xenopus. Devel. Biol. 29:139–151. (4)

Dawid, I. B. and A. W. Blackler (1972) Maternal and cytoplasmic inheritance of mitochondrial DNA in Xenopus. Devel. Biol. 29:152–161. (8)

Dawid, I. B., C. K. Klukas, S. Ohi, J. L. Ramirez, and W. B. Upholt (1976) Structure and evolution of animal mitochondrial DNA. In: The Genetic Function of Mitochondrial DNA, C. Saccone and A. M. Kroon (eds.). Elsevier and North Holland, Amsterdam and New York, pp. 3–13. (4)

289

Dayhoff, M. O. (ed.) (1972) Atlas of Protein Sequence and Structure, Vol. 5, Natl. Biomed. Res. Found., Silver Spring, Md. (2)

Dayhoff, M. O. and W. C. Barker (1972) Mechanisms in molecular evolution: examples. In: Atlas of Protein Sequence and Structure, Vol. 5, M. O. Dayhoff (ed.). Natl. Biomed. Res. Found., Silver Spring, Md., pp. 41–45. (2)

Dean, R. T. (1977) Lysosomes. Edward Arnold, London. (6)

De Jong, G. and W. Scharloo (1976) Environmental determination of selective significance or neutrality of amylase variants in *Drosophila melanogaster*. Genetics 84:77–94. (6)

de Vries, H. (1901) Die Mutationstheorie, Vol. I. Von Veit, Leipzig. (9)

Denaro, M., H. Blanc, M. J. Johnson, K. H. Chen, E. Wilmsen, L. L. Cavalli-Sforza, and D. C. Wallace (1981) Ethnic variation in *Hpa*I endonuclease cleavage patterns of human mitochondrial DNA. Proc. Natl. Acad. Sci. 78:5768–5772. (8)

Dev, V. G., R. Tantravahi, D. A. Miller, and O. J. Miller (1977) Nucleolus organizers in *Mus musculus* subspecies and in the RAG mouse cell line. Genetics 86:389–398. (3)

Dickinson, W. J. and D. T. Sullivan (1975) Gene-enzyme Systems in *Drosophila*. Springer, New York. (6)

Dickerson, R. E. (1971) The structure of cytochrome c and the rates of molecular evolution. J. Mol. Evol. 1:26–45. (11)

DiMichele, L. and D. A. Powers (1982a) LDH-B genotype-specific hatching times of *Fundulus heteroclitus* embryos. Nature 296:563–564. (6)

DiMichele, L. and D. A. Powers (1982b) Physiological basis for swimming endurance differences between LDH-B genotypes of *Fundulus heteroclitus*. Science 216:1014–1016. (6, 9)

Doane, W. W. (1980) Selection for amylase allozymes in *Drosophila melanogaster*. Evolution 34:868–874. (6)

Doane, W. W., L. G. Treat-Clemons, A. M. Buchberg, R. M. Gemmill, J. N. Levy, and S. A. Hawley (1983) Genetic mechanism for tissue specific control of alpha amylase expression in *Drosophila melanogaster*. In: Proc. 4th Intl. Cong. Isozymes, M. J. Siciliano (ed.). Alan R. Liss, New York (in press) (6)

Dobzhansky, Th. (1955) A review of some fundamental concepts and problems of population genetics. Cold Spring Harbor Symp. Quant. Biol. 20:1–15. (9)

Dobzhansky, Th. (1970) Genetics of the Evolutionary Process. Columbia Univ. Press, New York. (9)

Dobzhansky, Th. (1974) Genetic analysis of hybrid sterility within the species *Drosophila pseudoobscura*. Hereditas 77:81–88. (5)

Doolittle, W. F. and C. Sapienza (1980) Selfish genes, the phenotype paradigm and genome evolution. Nature 284:601–603. (9)

Dover, G. (1982a) A molecular drive through evolution. BioScience 32:526–533. (3)

Dover, G. (1982b) Molecular drive: a cohesive mode of species evolution. Nature 299:111–117. (3)

Drake, J. W. (1970) The Molecular Basis of Mutation. Holden-Day, San Francisco. (1)

290

Dreyer, W. J. and J. C. Bennett (1965) The molecular basis of antibody formation: a paradox. Proc. Natl. Acad. Sci. 54:864–869. (13)

Dugaiczyk, A., S. W. Law, and O. E. Dennison (1982) Nucleotide sequence and the encoded amino acids of human serum albumin mRNA. Proc. Natl. Acad. Sci. 79:71–75. (9)

Dykhuizen, D. and D. L. Hartl (1980) Selective neutrality of 6PGD allozymes in *E. coli* and the effects of genetic background. Genetics 96:801–817. (6, 11)

Eaton, W. A. (1980) The relationship between coding sequences and function in haemoglobin. Nature 284:183–185. (1)

Edelman, G. M. and J. A. Gally (1968) Antibody structure, diversity and specificity. Brookhaven Symp. Biol. 21:328–343. (3)

Edgell, M. H., S. Weaver, C. L. Jahn, R. W. Padgett, S. J. Phillips, C. F. Voliva, M. B. Comer, S. C. Hardies, N. L. Haigwood, C. H. Langel, R. R. Racine, and C. A. Hutchison, III (1981) The mouse beta hemoglobin locus. In: Organization and Expression of Globin Genes, G. Stamatoyannopoulos and A. Nienhuis (eds.). Alan R. Liss, New York, pp. 69–88. (1)

Edwards, Y. H. and D. A. Hopkinson (1977) Developmental changes in the electrophoretic patterns of human enzymes and other proteins. In: Isozymes: Current Topics in Biological and Medical Research, Vol. 1, M. C. Rattazzi, J. G. Scandalios, and G. S. Whitt (eds.). Alan R. Liss, New York, pp. 19–78. (2)

Efstratiadis, A., F. C. Kafatos, and T. Maniatis (1977) The primary structure of rabbit β-globin mRNA as determined from cloned DNA. Cell 10:571–585. (11)

Efstratiadis, A., J. W. Posakony, T. Maniatis, R. M. Lawn, C. O'Connell, R. A. Spritz, J. K. DeRiel, B. G. Forget, S. M. Weissman, J. L. Slightom, A. E. Blechl, O. Smithies, F. E. Baralle, C. C. Shoulders, and N. J. Proudfoot (1980) The structure and evolution of the human β-globin gene family. Cell 21:653–668. (1, 7)

Eigen, M. (1971) Self organization of matter and the evolution of biological macromolecules. Naturwissenschaften 58:465–523. (10)

Elder, J. T., J. Pan, C. H. Duncan, and S. M. Weissman (1981) Transcriptional analysis of interspersed repetitive polymerase III transcription units in human DNA. Nucleic Acids Res. 9:1171–1189. (3)

Eldredge, N. and S. J. Gould (1972) Punctuated equilibria: an alternative to phyletic gradualism. In: Models in Paleobiology, T. J. M. Schopf (ed.). Freeman, Cooper and Co., San Francisco, pp. 82–115. (9)

Elgin, S. C. R. and H. Weintraub (1975) Chromosomal proteins and chromatin structure. Ann. Rev. Biochem. 44:725–774. (2)

Eliceiri, G. L. (1972) The ribosomal RNA of hamster-mouse hybrid cells. J. Cell. Biol. 53:177–184. (3)

Eliceiri, G. L. and H. Green (1969) Ribosomal RNA synthesis in human-mouse hybrid cells. J. Mol. Biol. 41:253–260. (3)

Elsevier, S. M. and F. H. Ruddle (1975) Location of genes coding for 18S and

28S ribosomal RNA within the genome of *Mus musculus*. Chromosoma 52:219–228. (3)

Engels, W. R. (1981) Estimating genetic divergence and genetic variability with restriction endonucleases. Proc. Natl. Acad. Sci. 78:6329–6333. (8)

Erreda, B., T. S. Caidello, G. Wever, and F. Sherman (1981) Studies on transposable elements in yeast. I. ROAM mutations causing increased expression of yeast genes: their activation by signals directed toward conjugation functions and their formation by insertion of Ty1 repetitive elements. Cold Spring Harbor Symp. Quant. Biol. 45:593–602. (13)

Evans, H. J., R. A. Buckland, and M. L. Pardue (1974) Location of the genes coding for 18S and 28S rRNA in the human genome. Chromosoma 48:405–426. (3)

Ewens, W. J. (1972) The sampling theory of selectively neutral alleles. Theoret. Popul. Biol. 3:87–112. (5)

Ewens, W. F. (1979) Mathematical Population Genetics. Springer-Verlag, Berlin. (5, 9)

Ewens, W. J., R. S. Spielman, and H. Harris (1981) Estimation of genetic variation at the DNA level from restriction endonuclease data. Proc. Natl. Acad. Sci. 78:3748–3750. (7)

Falconer, D. S. (1960) Introduction to Quantitative Genetics. Ronald Press Co., New York. (9)

Farabaugh, P. J., U. Schmeissner, M. Hofer, and J. H. Miller (1978) Genetic studies of the *lac* repressor. VII. On the molecular nature of spontaneous hotspots in the *lac*I gene of *Escherichia coli*. J. Mol. Biol. 126:847–863. (1, 13)

Farin, F. and P. H. Clarke (1978) Positive regulation of amidase synthesis in *Pseudomonas aeruginosa*. J. Bacteriol. 135:379–392. (12)

Farris, J. S. (1972) Estimating phylogenetic trees from distance matrices. Amer. Nat. 106:645–668. (8)

Fauron, C. M.-R. and D. R. Wolstenholme (1976) Structural heterogeneity of mitochondrial DNA molecules within the genus *Drosophila*. Proc. Natl. Acad. Sci. 73:3623–3627. (4, 8)

Fauron, C. M.-R. and D. R. Wolstenholme (1980a) Extensive diversity among *Drosophila* species with respect to nucleotide sequences within the adenine + thymine-rich region of mitochondrial DNA molecules. Nucleic Acids Res. 8:2439–2452. (4)

Fauron, C. M.-R. and D. R. Wolstenholme (1980b) Intraspecific diversity of nucleotide sequences within the adenine + thymine-rich region of mito-chondrial DNA molecules of *Drosophila mauritiana*, *Drosophila melano-gaster*, and *Drosophila simulans*. Nucleic Acids Res. 8:5391–5410. (8)

Ferguson-Smith, M. A. (1964) The sites of nucleolus formation in human pachytene chromosomes. Cytogenetics 3:124–134. (3)

Ferguson-Smith, M. A. and S. D. Handmaker (1961) Observations on the satellited human chromosomes. Lancet *i*:638–640. (3)

Ferguson-Smith, M. A. and S. D. Handmaker (1963) The association of satel-lited chromosomes with specific chromosomal regions in cultured human somatic cells. Ann. Hum. Genet. 27:143–156. (3)

Ferris, S. D. and G. S. Whitt (1979) Evolution of the differential regulation of duplicate genes after polyploidization. J. Mol. Evol. 12:267–317. (2)

Ferris, S. D., A. C. Wilson, and W. M. Brown (1981a) Evolutionary tree for apes and humans based on cleavage maps of mitochondrial DNA. Proc. Natl. Acad. Sci. 78:2432–2436. (4)

Ferris, S. D., W. M. Brown, W. S. Davidson, and A. C. Wilson (1981b) Extensive polymorphism in the mitochondrial DNA of apes. Proc. Natl. Acad. Sci. 78:6319–6323. (4, 8, 9)

Ferris, S. D., R. D. Sage, and A. C. Wilson (1982) Evidence from mtDNA sequences that common laboratory strains of inbred mice are descended from a single female. Nature 295:163–165. (4, 8)

Ferris, S. D., R. D. Sage, and A. C. Wilson (1983) Mitochondrial DNA variation among house mice and the origins of inbred mice. Genetics (in press) (8)

Fersht, A. R. (1974) Catalysis, binding and enzyme-substrate and complementarity. Proc. Roy. Soc. London (B) 187:397–407. (6)

Fersht, A. R. (1977) Enzyme Structure and Mechanism. W. H. Freeman, San Francisco. (6)

Fisher, R. A. (1935) The sheltering of lethals. Amer. Nat. 69:446–453. (2)

Fisher, W. K., A. R. Nash, and E. O. P. Thompson (1977) Haemoglobins of the shark, *Heterodontus* portusjacksoni. III. Amino acid sequence of the β-chain. Aust. J. Biol. Sci. 30:487–506. (11)

Fitch, W. M. (1980) Estimating the total number of nucleotide substitutions since the common ancestor of a pair of homologous genes: comparison of several methods and three beta hemoglobin messenger RNA's. J. Mol. Evol. 16:153–209. (11)

Fletcher, T. S., F. J. Ayala, D. R. Thatcher, and G. K. Chambers (1978) Structural analysis of the ADH[S] electromorph of *Drosophila melanogaster*. Proc. Natl. Acad. Sci. 75:5609–5612. (6)

Fontdevila, A. and H. L. Carson (1978) Spatial distribution and dispersal in a population of *Drosophila*. Amer. Nat. 112:365–380. (9)

Ford, E. B. (1964) Ecological Genetics. Methuen, London. (9)

Fossitt, D., R. P. Mortlock, R. L. Anderson, and W. A. Wood (1964) Pathways of L-arabitol and xylitol metabolism in *Aerobacter aerogenes*. J. Biol. Chem. 239:2110–2115. (12)

Fox, T. D. and C. J. Leaver (1981) The Zea mays mitochondrial gene coding cytochrome oxidase subunit II has an intervening sequence and does not contain TGA codons. Cell 26:315–323. (10)

Francisco, J. F., G. G. Brown, and M. V. Simpson (1979) Further studies on types A and B rat mtDNA's: cleavage maps and evidence for cytoplasmic inheritance in mammals. Plasmid 2:426–436. (8)

Friedlander, M. (1980) Monospermic fertilization in *Chrysopa carnia* (Neuroptera; Chrysopidae): Behaviour of the fertilizing spermatozoa prior to syngamy. Int. J. Insect Morphol. and Embryol. 9:53–57. (8)

Fuerst, P. A., R. Chakraborty, and M. Nei (1977) Statistical studies on protein polymorphism in natural populations. I. Distribution of single locus heterozygosity. Genetics 86:455–483. (5, 9)

Fukada, K. and J. Abelson (1980) DNA sequence of a T4 transfer RNA gene cluster. J. Mol. Biol. 139:377–391. (10)

293

Gaines, M. S. and T. S. Whittam (1980) Genetic changes in fluctuating vole populations: selective *vs.* nonselective forces. Genetics 96:767–778. (5)

Gally, J. A. and G. M. Edelman (1972) The genetic control of immunoglobulin synthesis. Ann. Rev. Genet. 6:1–46. (2)

Gautier, F., H. Bunemann, and L. Grotjahn (1977) Analysis of calf-thymus satellite DNA: Evidence for specific methylation of cytosine in C-G sequences. Europ. J. Biochem. 80:175–183. (8)

Geer, B. W., C. G. Woodward, and S. D. Marshall (1978) Regulation of the oxidative NADP enzyme tissue levels in *Drosophila melanogaster.* Part 2. The biochemical basis of dietary carbohydrate and D-glycerate modulation. J. Exp. Zool. 203:391–402. (6)

Geer, B. W., D. L. Lindel, and D. M. Lindel (1979) Relationship of the oxidative pentose-shunt pathway to lipid synthesis in *Drosophila melanogaster.* Biochem. Genet. 17:881–895. (6)

Gibson, J. B. (1972) Differences in the number of molecules produced by two allelic electrophoretic enzyme variants in *Drosophila melanogaster.* Experientia 27:99–100. (6)

Gilbert, W. (1979) Introns and exons: Playgrounds of evolution. In: Eukaryotic Gene Regulation, ICN-UCLA Symposium on Molecular and Cellular Biology, R. Axel, T. Maniatis, and C. F. Fox (eds.). Academic Press, New York, pp. 1–12. (2)

Giles, R. E., H. Blanc, H. M. Cann, and D. C. Wallace (1980) Maternal inheritance of human mitochondrial DNA. Proc. Natl. Acad. Sci. 77:6715–6719. (8)

Gillespie, J. H. (1978) A general model to account for enzyme variation in natural populations. V. The SAS-CFF model. Theoret. Popul. Biol. 14:1–45. (9)

Gillham, N. W. (1978) Organelle Heredity. Raven Press, New York. (8)

Gillum, A. M. and D. A. Clayton (1978) Displacement loop replication initiation sequence in animal mitochondrial DNA exists as a family of discrete lengths. Proc. Natl. Acad. Sci. 75:677–681. (4)

Gimelli, G., E. Porro, F. Santi, S. Scuppaticci, and O. Zuffardi (1976) "Jumping" satellites in three generations: a warning for paternity tests and prenatal diagnosis. Hum. Genet. 34:315–318. (3)

Glaus, K. R., H. P. Zassenhaus, H. S. Fechheimer, and P. S. Perlman (1980) Avian mtDNA: structure, organization and evolution. In: The Organization and Expression of the Mitochondrial Genome, A. M. Kroon and C. Saccone (eds.). North Holland Publ., Amsterdam, pp. 131–135. (8)

Gō, M. (1981) Correlation of DNA exonic regions with protein structural units in haemoglobin. Nature 291:90–92. (2)

Goad, W. B. and M. I. Kanehisa (1982) Pattern recognition in nucleic acid sequences. I. A general method for finding local homologies and symmetries. Nucleic Acids Res. 10:247–263. (3)

Goddard, J. M. and D. R. Wolstenholme (1978) Origin and direction of replication in mitochondrial DNA molecules from *Drosophila melanogaster.* Proc. Natl. Acad. Sci. 75:3886–3890. (4)

Goddard, J. M. and D. R. Wolstenholme (1980) Origin and direction of replication in mitochondrial DNA molecules from the genus *Drosophila.* Nucleic Acids Res. 8:741–757. (4)

Goff, S. P., E. Gilboa, O. N. Witte, and D. Baltimore (1980) Structure of the Abelson murine leukemia virus genome and the homologous cellular gene: studies with cloned viral DNA. Cell 22:777–785. (2)

Gojobori, T., W.-H. Li, and D. Graur (1982) Patterns of nucleotide substitution in pseudogenes and functional genes. J. Mol. Evol. 18:360–369. (2)

Goodman, M. (1978) Substitutional trends and non-random changes in rates during protein evolution. In: Evolution of Protein Molecules, H. Matsubara and T. Yamanaka (eds.). Japan Scientific Societies, Tokyo, pp. 17–32. (11)

Goodman, M., G. W. Moore, J. Barnabas, and G. Matsuda (1974) The phylogeny of human globin genes investigated by the maximum parsimony method. J. Mol. Evol. 3:1–48. (11)

Goodman, M., G. W. Moore, and G. Matsuda (1975) Darwinian evolution in the genealogy of haemoglobin. Nature 253:603–608. (11)

Goodpasture, C., S. E. Bloom, T. C. Hsu, and F. E. Arrighi (1976) Human nucleolus organizers: the satellites or the stalks? Amer. J. Hum. Genet. 28:556–559. (3)

Gosden, J. R., S. S. Lawrie, and C. M. Gosden (1981) Satellite DNA sequences in the human acrocentric chromosomes: information from translocations and heteromorphisms. Amer. J. Hum. Genet. 33:243–251. (3)

Gotoh, O., J.-I. Hayashi, H. Yonekawa, and Y. Tagashira (1979) An improved method for estimating sequence divergence between related DNAs from changes in restriction endonuclease cleavage sites. J. Mol. Evol. 14:301–310. (4, 8)

Gould, S. J. (1982) Darwinism and the expansion of evolutionary theory. Science 216:380–387. (9)

Gould, S. J. and R. C. Lewontin (1979) The spandrels of San Marco and the Panglossian paradigm: a critique of the adaptationist programme. Proc. Roy. Soc. London (B) 205:581–598. (9)

Grantham, R. (1980) Workings of the genetic code. Trends in Biochemical Sciences 5:327–331. (11)

Grantham, R., C. Gautier, and M. Gouy (1980a) Codon frequencies in 119 individual genes confirm consistent choices of degenerate bases according to genome type. Nucleic Acids Res. 8:1893–1912. (11)

Grantham, R., C. Gautier, M. Gouy, R. Mercier, and A. Pavé (1980b) Codon catalog usage and the genome hypothesis. Nucleic Acids Res. 8:r49-r62. (11)

Grantham, R., C. Gautier, M. Gouy, M. Jacobzone, and R. Mercier (1981) Codon catalog usage is a genome strategy modulated for gene expressivity. Nucleic Acids Res. 9:r43–r74. (11)

Greaney, G. S. and G. N. Somero (1980) Contributions of binding and catalytic rate constants for evolutionary modifications in K_m of NADH from muscle-type (M_4) lactate dehydrogenases. J. Comp. Physiol. 137:115–121. (6)

Greenberg, B. D., J. E. Newbold, and A. Sugino (1982) Intraspecific nucleotide sequence variability surrounding the origin of replication in human mitochondrial DNA. Gene (in press) (4, 8)

295

Greenblatt, I. M. (1968) The mechanism of modulator transposition in maize. Genetics 58:585–597. (13)

Greenwalt, D. E. and S. H. Bishop (1980) The effect of aminotransferase inhibitors on the pattern of free amino acid accumulation in isolated mussel hearts subjected to hyperosmotic stress. Physiol. Zool. 53:626–629. (6)

Gresson, R. A. R. (1940) Presence of the sperm middle-piece in the fertilized egg of the mouse (*Mus musculus*). Nature 145:425. (8)

Grosschedl, R. and M. Birnstiel (1980) Identification of regulatory sequences in the prelude sequences of an HZA histone gene by the study of specific deletion mutants *in vivo*. Proc. Natl. Acad. Sci. 77:1432–1436. (3)

Grummt, I., E. Roth, and M. R. Paule (1982) Ribosomal RNA transcription *in vitro* is species-specific. Nature 296:173–174. (3)

Grunstein, M. and J. E. Grunstein (1978) The histone H4 gene of *Strongylocentrotus purpuratus*: DNA and mRNA sequences at the 5' end. Cold Spring Harbor Symp. Quant. Biol. 42:1083–1092. (11)

Grunstein, M., P. Schedl, and L. Kedes (1976) Sequence analysis and evolution of sea urchin (*Lytechinus pictus* and *Strongylocentrotus purpuratus*) histone H4 messenger RNAs. J. Mol. Biol. 104:351–369. (11)

Hacking, A. J. and E. C. C. Lin (1977) Regulatory changes in the fucose system associated with the evolution of a catabolic pathway for propanediol in *Escherichia coli*. J. Bacteriol. 130:832–838 (12)

Hacking, A. J., J. Aguilar, and E. C. C. Lin (1978) Evolution of propanediol utilization in *Escherichia coli*: Mutant with improved substrate scavenging power. J. Bacteriol. 136:522–530. (12)

Haldane, J. B. S. (1933) The part played by recurrent mutation in evolution. Amer. Nat. 67:5–19. (2)

Haldane, J. B. S. (1959) Natural selection. In: Darwin's Biological Work, P. R. Bell (ed.). Cambridge Univ. Press, Cambridge, pp. 101–149. (11)

Hall, B. G. (1976) Experimental evolution of a new enzymatic function. Kinetic analysis of the ancestral (ebg^0) and evolved (ebg^+) enzymes. J. Mol. Biol. 107:71–84. (12)

Hall, B. G. (1977) Number of mutations required to evolve a new lactase function in *Escherichia coli*. J. Bacteriol. 129:540–543. (12)

Hall, B. G. (1978a) Experimental evolution of a new enzymatic function. II. Evolution of multiple functions for EBG enzyme in *E. coli*. Genetics 89:453–465. (12)

Hall, B. G. (1978b) Regulation of newly evolved enzymes. IV. Directed evolution of the *ebg* repressor. Genetics 90:673–691. (12)

Hall, B. G. (1981) Changes in the substrate specificities of an enzyme during directed evolution of new functions. Biochemistry 20:4042–4049. (12)

Hall, B. G. (1982a) Transgalactosylation activity of the ebg β-galactosidase synthesizes allolactose from lactose. J. Bacteriol. 150:132–140. (12)

Hall, B. G. (1982b) Evolution of a regulated operon in the laboratory. Genetics 101:335–344. (12)

Hall, B. G. and N. D. Clarke (1977) Regulation of newly evolved enzymes. III. Evolution of the *ebg* repressor during selection for enhanced lactase activity. Genetics 85:193–201. (12)

Hall, B. G. and D. L. Hartl (1974) Regulation of newly evolved enzymes. I. Selection of a novel lactase regulated by lactose in *Escherichia coli*. Genetics 76:391–400. (12)

Hall, B. G. and D. L. Hartl (1975) Regulation of newly evolved enzymes. II. The *ebg* repressor. Genetics 81:427–435. (12)

Hall, B. G. and T. Zuzel (1980a) Evolution of a new enzymatic function by recombination within a gene. Proc. Natl. Acad. Sci. 77:3529–3533. (12)

Hall, B. G. and T. Zuzel (1980b) The *ebg* operon consists of at least two genes. J. Bacteriol. 144:1208–1211. (12)

Hall, J. G. and R. K. Koehn (1981) Letter to the Editor on the biochemical differences between ADH allozymes in *Drosophila*. Genetics 98:669–672. (6)

Hall, J. G. and R. K. Koehn (1983) The evolution of enzyme catalytic efficiency and adaptive inference from steady-state kinetic data. Evol. Biol. (in press) (6)

Hansen, J. N., D. A. Konkel, and P. Leder (1982) The sequence of a mouse embryonic β-globin gene. J. Biol. Chem. 257:1048–1052. (1)

Harpending, H. C. (1974) Genetic structure of small populations. Ann. Rev. Anthropol. 3:229–243. (5)

Harris, H. (1966) Enzyme polymorphisms in man. Proc. Roy. Soc. London (B) 164:298–310. (9)

Harris, H. (1971) Polymorphism and protein evolution: the neutral mutation-random drift hypothesis. J. Med. Genet. 8:444–452. (9)

Harris, H., D. A. Hopkinson, and Y. H. Edwards (1977) Polymorphism and the subunit structure of enzymes: A contribution to the neutralist-selectionist controversy. Proc. Natl. Acad. Sci. 74:698–701. (5)

Harrison, G. A. (1977) Introduction: structure and function in the biology of human populations. In: Population Structure and Human Variation, G. A. Harrison (ed.). Cambridge Univ. Press, Cambridge, pp. 1–8. (5)

Hartl, D. L. and D. E. Dykhuizen (1981) Potential for selection among nearly neutral allozymes of 6-phosphogluconate dehydrogenase in *Escherichia coli*. Proc. Natl. Acad. Sci. 78:6344–6348. (11)

Hartl, D. L. and B. G. Hall (1974) Second naturally occurring β-galactosidase in *E. coli*. Nature 248:152–153. (12)

Hartman, H. (1975) Speculation on the origin and evolution of metabolism. J. Mol. Evol. 4:359–370. (2)

Hauswirth, W. W. and P. J. Laipis (1982) Mitochondrial DNA polymorphism in a maternal lineage of Holstein cows. Proc. Natl. Acad. Sci. 79:4686–4690. (4)

Hayashi, J. I., H. Yonekawa, O. Gotoh, J. Watanabe, and Y. Tagashira (1978) Strictly maternal inheritance of rat mitochondrial DNA. Biochem. Biophys. Res. Comm. 83:1032–1038. (8)

Hazel, J. R. and C. L. Prosser (1974) Molecular mechanisms of temperature compensation poikilotherms. Physiol. Rev. 54:620–677. (6)

Heckman, J. E., J. Sarnoff, B. Alzner-DeWeerd, S. Yin, U. L. RajBhandary

(1980) Novel features in the genetic code and codon reading patterns in *Neurospora crassa* mitochondria based on sequences of six mitochondrial tRNAs. Proc. Natl. Acad. Sci. 77:3159–3163. (10)

Heffron, F., R. Kostriken, C. Morita, and R. Parker (1981) Tn*3* encodes a site-specific recombination system: identification of essential sequences, genes, and the actual site of recombination. Cold Spring Harbor Symp. Quant. Biol. 45:259–268. (13)

Heinstra, P. W. H., K. T. Eisses, W. G. E. J. Schoonen, W. Aben, A. J. de Winter, D. J. van der Horst, W. J. A. van Marrewijk, A. M. T. Beenakkers, W. Scharloo, and G. E. W. Thorig (1982) A dual function of alcohol dehydrogenase in *Drosophila*. Genetica (in press) (6)

Henderson, A. S., D. Warburton, and K. C. Atwood (1972) Location of ribosomal DNA in the human chromosome complement. Proc. Natl. Acad. Sci. 69:3394–3398. (3)

Henderson, A. S., D. Warburton, and K. C. Atwood (1973) Ribosomal DNA connectives between human acrocentric chromosomes. Nature 245:95–97. (3)

Henderson, A. S., E. M. Eicher, M. T. Yu, and K. C. Atwood (1974) The chromosomal location of rDNA in the mouse. Chromosoma 49:155–160. (3)

Henderson, A. S., K. C. Atwood, and D. Warburton (1976) Chromosomal distribution of rDNA in *Pan paniscus, Gorilla gorilla beringei* and *Symphalangus syndactylus*: comparison to related primates. Chromosoma 59:147–155. (3)

Hickey, D. A. (1977) Selection for amylase allozymes in *Drosophila melanogaster*. Evolution 31:800–804. (6)

Hickey, D. A. (1979) Selection on amylase allozymes in *Drosophila melanogaster*: selection experiments using several independently derived pairs of chromosomes. Evolution 33:1128–1137. (6)

Hieter, P. A., E. E. Max, J. G. Seidman, J. V. Maizel, Jr., and P. Leder (1980) Cloned human and mouse kappa immunoglobulin constant and J region genes conserve homology in functional segments. Cell 22:197–207. (9)

Higgs, D. R., S. E. Y. Goodbourn, J. S. Wainscoat, J. B. Clegg, and D. J. Weatherall (1981) Highly variable regions of DNA flank the human globin genes. Nucleic Acids Res. 9:4213–4224. (7)

Hilbish, T. J., L. E. Deaton, and R. K. Koehn (1982) Effect of an allozyme polymorphism on regulation of cell volume. Nature 298:688–689. (6)

Hirsch, M., A. Spradling, and S. Penman (1974) The messenger-like poly(A)-containing RNA species from the mitochondria of mammals and insects. Cell 1:31–36. (4)

Hoffman, R. J. (1981) Evolutionary genetics of *Metridium senile*. I. Kinetic differences in phosphoglucose isomerase allozymes. Biochem. Genet. 19:129–144. (6)

Holland, J., K. Spindler, F. Horodyski, E. Grabau, S. Nichol, and S. VandePol (1982) Rapid evolution of RNA genomes. Science 215:1577–1585. (9)

Hollis, G. F., P. A. Hieter, O. W. McBride, D. Swan, and P. Leder (1982) Processed genes: a dispersed human immunoglobulin gene bearing evidence of RNA-type processing. Nature 296:321–325. (2)

Holmquist, R. (1979) The method of parsimony: an experimental test and

298

theoretical analysis of the adequacy of molecular restoration studies. J. Mol. Biol. 135:939–958. (11)

Holmquist, R., T. H. Jukes, and S. Pangburn (1973) Evolution of transfer RNA. J. Mol. Biol. 78:91–116. (10)

Hood, L. and D. W. Talmage (1970) Mechanism of antibody diversity: germ line basis for variability. Science 168:325–334 (3)

Hood, L., J. H. Campbell, and S. C. R. Elgin (1975) The organization, expression and evolution of antibody genes and other multigene families. Ann. Rev. Genet. 9:305–353. (2, 3)

Hoorn, A. J. W. and W. Scharloo (1978) The functional significance of amylase polymorphism in Drosophila melanogaster. I. Properties of two amylase variants. Genetica 49:173–180. (6)

Horak, I., H. G. Coon, and I. B. Dawid (1974) Interspecific recombination of mitochondrial DNA molecules in hybrid somatic cells. Proc. Natl. Acad. Sci. 71:1828–1832. (4)

Horowitz, N. H. (1965) The evolution of biochemical syntheses—retrospect and prospect. In: Evolving Genes and Proteins, V. Bryson and H. J. Vogel (eds.). Academic Press, New York, pp. 15–23. (2)

Houck, C. M., F. P. Rinehart, and C. W. Schmid (1979) A ubiquitous family of repeated DNA sequences in the human genome. J. Mol. Biol. 132:289–306. (3)

Hsu, T. C., S. E. Spirito, and M. L. Pardue (1975) Distribution of 18S and 28S rRNA genes in mammalian genomes. Chromosoma 53:25–37. (3)

Hubby, J. L. and L. H. Throckmorton (1965) Protein differences in Drosophila. II. Comparative species genetics and evolutionary problems. Genetics 52:203–215. (5)

Hughes, M. B. and J. C. Lucchesi (1977) Genetic rescue of a lethal "null" activity allele of 6-phosphogluconate dehydrogenase in Drosophila melanogaster. Science 196:1114–1115. (6)

Hunkapiller, T., H. Huang, L. Hood, and J. H. Campbell (1982) The impact of modern genetics on evolutionary theory. In: Perspectives on Evolution, R. Milkman (ed.). Sinauer Associates, Sunderland, Mass., pp. 164–189. (2)

Hutchison, C. A., III, J. E. Newbold, S. S. Potter, and M. H. Edgell (1974) Maternal inheritance of mammalian mitochondrial DNA. Nature 251:536–538. (8)

Iino, T. (1977) Genetics of structure and function of bacterial flagella. Ann. Rev. Genet. 11:161–182. (13)

Ikemura, T. (1980) The frequency of codon usage in E. coli genes: correlation with abundance of cognate tRNA. In: Genetics and Evolution of RNA Polymerase, tRNA and Ribosomes, S. Osawa, H. Ozeki, H. Uchida, and T. Yura (eds.). Univ. of Tokyo Press, Tokyo, pp. 519–523. (11)

Ikemura, T. (1981a) Correlation between the abundance of Escherichia coli transfer RNAs and the occurrence of the respective codons in its protein genes. J. Mol. Biol. 146:1–21. (11)

299

Ikemura, T. (1981b) Correlation between the abundance of *Escherichia coli* transfer RNAs and the occurrence of the respective codons in its protein genes: a proposal for a synonymous codon choice that is optimal for the *E. coli* translational system. J. Mol. Biol. 151:389–409. (11)

Ikemura, T. (1982) Correlation between the abundance of yeast transfer RNAs and the occurrence of the respective codons in protein genes. J. Mol. Biol. 158:573–597. (11)

Jacobs, P. A., M. Mayer, and N. E. Morton (1976) Acrocentric chromosome associations in man. Amer. J. Hum. Genet. 28:567–576. (3)

Jacq, C., J. R. Miller, and G. G. Brownlee (1977) A pseudogene structure in 5S DNA of *Xenopus laevis*. Cell 12:109–120. (2)

Jacquard, A. (1974) The Genetic Structure of Populations. Springer-Verlag, New York. (5)

Jagadeeswaran, P., B. G. Forget, and S. M. Weissman (1981) Short interspersed repetitive DNA elements in eucaryotes: transposable DNA elements generated by reverse transcription of RNA pol III transcripts? Cell 26:141–142. (3)

Jahn, C. L., C. A. Hutchison, III, S. J. Phillips, S. Weaver, N. L. Haigwood, C. F. Voliva, and M. H. Edgell (1980) DNA sequence organization of the β-globin complex in the BALB/c mouse. Cell 21:159–168. (1)

Jeffreys, A. (1979) DNA sequence variants in the ^{G}γ-, ^{A}γ-, δ- and β-globin genes of man. Cell 18:1–10. (7)

Jensen, R. A. (1976) Enzyme recruitment in evolution of new function. Ann. Rev. Microbiol. 30:409–425. (2)

Jensen, R. A. and G. S. Byng (1982) The partitioning of biochemical pathways with isozymes systems. In: Isozymes: Current Topics in Biological and Medical Research, Vol. 5, M. C. Rattazi, J. G. Scandalios, and G. S. Whitt (eds.). Alan R. Liss, New York, pp. 143–175. (2)

Jobe, A. and S. Bourgeois (1972) Lac repressor-operator interactions. VI. The natural inducer of the *lac* operon. J. Mol. Biol. 69:397–408. (12)

Johnson, G. and M. W. Feldman (1973) On the hypothesis that polymorphic enzyme alleles are selectively neutral: I. The evenness of allele frequency distribution. Theoret. Popul. Biol. 4:209–221. (5)

Jones, J. S., B. H. Leith, and P. Rawlings (1977) Polymorphism in *Cepaea*: a problem with too many solutions? Ann. Rev. Ecol. Syst. 8:109–143. (5)

Jones, J. S., R. K. Selander, and G. D. Schnell (1980) Patterns of morphological and molecular polymorphism in the land snail *Cepaea nemoralis*. Biol. J. Linn. Soc. 14:359–387. (5)

Jorde, L. B. (1980) The genetic structure of subdivided human populations. A review. In: Current Developments in Anthropological Genetics. Theory and Methods, Vol. 1, J. H. Mielke and M. H. Crawford (eds.). Plenum Press, New York, pp. 135–208. (5)

Jukes, T. H. (1966) Molecules and Evolution. Columbia Univ. Press, New York, pp. 69–70. (10)

Jukes, T. H. (1973) Possibilities for the evolution of the genetic code from a preceding form. Nature 246:22–26. (10)

Jukes, T. H. (1979) Dr. Best, insulin, and molecular evolution. Canadian J. Biochem. 57:455–458. (11)

300

Jukes, T. H. (1981) Amino acid codes in mitochondria as possible clues to primitive codes. J. Mol. Evol. 18:15–16. (4, 10)

Jukes, T. H., W. R. Holmquist, and H. Moise (1975) Amino acid composition of proteins: Selection against the genetic code. Science 189:50–51. (10)

Kacser, H. and J. A. Burns (1973) The control of flux. Symp. Soc. Exp. Biol. 27:65–104. (6)

Kacser, H. and J. A. Burns (1979) Molecular democracy: who shares the controls? Biochem. Soc. Trans. 7:1149–1160. (6)

Kacser, H. and J. A. Burns (1981) The molecular basis of dominance. Genetics 97:639–666. (6)

Kammen, H. O. and S. Spengler (1980) The biosynthesis of inosinic acid in transfer RNA. Biochim. Biophys. Acta 213:352–354. (10)

Kan, Y. W. and A. M. Dozy (1978) Polymorphism of DNA sequence adjacent to the human β globin structural gene: relationship to sickle cell mutation. Proc. Natl. Acad. Sci. 75:5631–5635. (7)

Kan, Y. W., K. Y. Lee, M. Furbetta, A. Angius, and A. Cao (1980) Polymorphism of DNA sequence in the β globin gene region. New Eng. J. Med. 302:185–188. (7)

Kaplan, N. and C. H. Langley (1979) A new estimate of sequence divergence of mitochondrial DNA using restriction endonuclease mapping. J. Mol. Evol. 13:295–304. (8)

Kaplan, N. and K. Risko (1981) An improved method for estimating sequence divergence of DNA using restriction endonuclease mappings. J. Mol. Evol. 17:156–162. (8)

Karlin, S. and B. Levikson (1974) Temporal fluctuations in selection intensities: Case of small population size. Theoret. Popul. Biol. 6:383–412. (9)

Karn, M. N. and L. S. Penrose (1951) Birth weight and gestation time in relation to maternal age, parity, and infant survival. Ann. Eugenics 16:147–164. (11)

Kasamatsu, H., D. L. Robberson, and J. Vinograd (1971) A novel closed-circular mitochondrial DNA with properties of a replicating intermediate. Proc. Natl. Acad. Sci. 68:2252–2257. (4)

Kato, I., W. J. Kohr, and M. J. Laskowski (1978) Evolution of avian ovomucoids. In: Proceedings of 11th FEBS Meeting, 47, S. Magnuson, M. Ottesen, B. Taltmann, K. Danø, and H. Neurath (eds.). Pergamon Press, New York, pp. 197–206. (2)

Kedes, L. H. (1979) Histone genes and histone messengers. Ann. Rev. Biochem. 48:837–870. (11)

Kim, S., M. Davis, E. Sinn, P. Patten, and L. Hood (1981) Antibody diversity: somatic hypermutation of rearranged V_H genes. Cell 27:573–581. (9)

Kimura, M. (1964) Diffusion models in population genetics. J. Appl. Probab. 1:177–232. (11)

Kimura, M. (1968a) Evolutionary rate at the molecular level. Nature 217:624–626. (5, 9, 11)

Kimura, M. (1968b) Genetic variability maintained in a finite population due to mutational production of neutral and nearly neutral isoalleles. Genet. Res. 11:247–269. (5, 11)

Kimura, M. (1969a) The rate of molecular evolution considered from the standpoint of population genetics. Proc. Natl. Acad. Sci. 63:1181–1188. (11)

Kimura, M. (1969b) The number of heterozygous nucleotide sites maintained in a finite population due to steady flux of mutations. Genetics 61:893–903. (11)

Kimura, M. (1971) Theoretical foundation of population genetics at the molecular level. Theoret. Popul. Biol. 2:174–208. (11)

Kimura, M. (1977) Preponderance of synonymous changes as evidence for the neutral theory of molecular evolution. Nature 267:275–276. (11)

Kimura, M. (1979a) Model of effectively neutral mutations in which selective constraint is incorporated. Proc. Natl. Acad. Sci. 76:3440–3444. (5, 9)

Kimura, M. (1979b) The neutral theory of molecular evolution. Scientific American 241(5):98–126. (11)

Kimura, M. (1980) A simple method for estimating evolutionary rate of base substitutions through comparative studies of nucleotide sequences. J. Mol. Evol. 16:111–120. (2, 11)

Kimura, M. (1981a) Estimation of evolutionary distances between homologous nucleotide sequences. Proc. Natl. Acad. Sci. 78:454–458. (11)

Kimura, M. (1981b) Was globin evolution very rapid in its early stages?: a dubious case against the rate-constancy hypothesis. J. Mol. Evol. 17:110–113. (11)

Kimura, M. (1981c) Doubt about studies of globin evolution based on maximum parsimony codons and the argumentation procedure. J. Mol. Evol. 17:121–122. (11)

Kimura, M. (1981d) Possibility of extensive neutral evolution under stabilizing selection with special reference to non-random usage of synonymous codons. Proc. Natl. Acad. Sci. 78:5773–5777. (9, 11)

Kimura, M. (1982a) The neutral theory as a basis for understanding the mechanism of evolution and variation at the molecular level. In: Molecular Evolution, Protein Polymorphism and the Neutral Theory, M. Kimura (ed.). Japan Scientific Societies Press, Tokyo and Springer-Verlag, Berlin, pp. 3–56. (11)

Kimura, M. (ed.) (1982b) Molecular Evolution, Protein Polymorphism and the Neutral Theory. Japan Scientific Societies Press, Tokyo and Springer-Verlag, Berlin. (5)

Kimura, M. and J. F. Crow (1964) The number of alleles that can be maintained in a finite population. Genetics 49:725–738. (7, 9, 11)

Kimura, M. and T. Ohta (1969a) The average number of generations until fixation of a mutant gene in a finite population. Genetics 61:763–771. (11)

Kimura, M. and T. Ohta (1969b) The average number of generations until extinction of an individual mutant gene in a finite population. Genetics 63:701–709. (11)

Kimura, M. and T. Ohta (1971a) Protein polymorphism as a phase of molecular evolution. Nature 229:467–469. (11)

Kimura, M. and T. Ohta (1971b) Theoretical Aspects of Population Genetics. Princeton Univ. Press, Princeton. (5, 9, 11)

Kimura, M. and T. Ohta (1973) Mutation and evolution at the molecular level. Genetics 73(Suppl.):19–35. (11)

King, B. O., R. O. Shade, and R. A. Lansman (1981) The use of restriction endonucleases to compare mitochondrial DNA sequences in *Mus musculus*: a detailed restriction map of mitochondrial DNA from mouse L-cells. Plasmid 5:313–328. (8)

King, J. L. (1967) Continuously distributed factors affecting fitness. Genetics 55:483–492. (9)

King, J. L. and T. H. Jukes (1969) Non-Darwinian evolution: Random fixation of selectively neutral mutations. Science 164:788–798. (5, 9, 11)

Kirk, R. L. (1979) Genetic differentiation in Australia and its bearing on the origin of the first Americans. In: The First Americans: Origins, Affinities, and Adaptations, W. S. Laughlin and A. B. Harper (eds.). Gustav Fischer, New York, pp. 211–237. (5)

Kiss, A., B. Sain, and P. Venetianer (1977) The number of rRNA genes in *Escherichia coli*. FEBS Letters 79:77–79. (2)

Klaer, R., S. Kuhn, H.-J. Fritz, E. Tilman, I. Saint-Girons, P. Haberman, D. Pfeifer, and P. Starlinger (1981) Studies on transposition mechanisms and specificity of IS4. Cold Spring Harbor Symp. Quant. Biol. 45:215–224. (13)

Klar, G. T., C. B. Stalnaker, and T. M. Farley (1979) Comparative physical and physiological performance of rainbow trout, *Salmo gairdneri*, of distinct lactate dehydrogenase B² phenotypes. Comp. Biochem. Physiol. 63A:229–235. (6)

Klemenz, R. and E. P. Geiduschek (1980) The 5' terminus of the precursor ribosomal RNA of *Saccharomyces cerevisiae*. Nucleic Acids Res. 8:2679–2689. (3)

Klein, H. L. and T. Petes (1981) Intrachromosomal gene conversion in yeast: a new type of genetic exchange. Nature 289:144–148. (3)

Klukas, C. K. and I. B. Dawid (1976) Characterization and mapping of mitochondrial ribosomal RNA and mitochondrial DNA in *Drosophila melanogaster*. Cell 9:615–625. (4)

Koch, A. L. (1972) Enzyme evolution: I. The importance of untranslatable intermediates. Genetics 72:297–316. (2)

Köchel, H. G., C. M. Lazarus, N. Basak, and H. Küntzel (1981) Mitochondrial tRNA gene clusters in *Aspergillus nidulans*: Organization and nucleotide sequence. Cell 23:625–633. (10)

Koehn, R. K. (1969) Esterase heterogeneity: Dynamics of a polymorphism. Science 163:943–944. (9)

Koehn, R. K. (1978) Physiology and biochemistry of enzyme variation: the interface of ecology and population genetics. In: Ecological Genetics: the Interface, P. Brussard (ed.). Springer-Verlag, New York, pp. 51–72. (6)

Koehn, R. K. and W. F. Eanes (1978) Molecular structure and protein variation within and among populations. Evol. Biol. 11:39–100. (5)

Koehn, R. K. and F. W. Immermann (1981) Biochemical studies of aminopeptidase polymorphism in *Mytilus edulis*. I. Dependence of enzyme activity on season, tissue, and genotype. Biochem. Genet. 19:1115–1142. (6)

Koehn, R. K. and J. F. Siebenaller (1981) Biochemical studies of aminopeptidase polymorphism in *Mytilus edulis*. II. Dependence of reaction rate on physical factors and enzyme concentration. Biochem. Genet. 19:1143–1162. (6)

Koehn, R. K., R. Milkman, and J. B. Mitton (1976) Population genetics of marine pelecypods. IV. Selection, migration and genetic differentiation in the blue mussel, *Mytilus edulis*. Evolution 30:2–32. (6)

Koehn, R. K., R. I. E. Newell, and F. Immermann (1980a) Maintenance of an aminopeptidase allele frequency cline by natural selection. Proc. Natl. Acad. Sci. 77:5385–5389. (6)

Koehn, R. K., B. L. Bayne, M. N. Moore, and F. J. Siebenaller (1980b) Salinity related physiological and genetic differences between populations of *Mytilus edulis*. Biol. J. Linnean Soc. 14:319–334. (6)

Kohne, D. E., J. A. Chiscon, and B. H. Hoyer (1972) Evolution of primate DNA sequences. J. Hum. Evol. 1:627–644. (4)

Konkel, D. A., S. M. Tilghman, and P. Leder (1978) The sequence of the chromosomal mouse β-globin major gene: homologies in capping, splicing and poly(A) sites. Cell 15:1125–1132. (1)

Konkel, D. A., J. V. Maizel, Jr., and P. Leder (1979) The evolution and sequence comparison of two recently diverged mouse chromosomal β-globin genes. Cell 18:865–873. (1)

Kreitman, M. (1982) DNA sequence variation at the Adh locus in *D. melanogaster*. Presented at the Annual Meeting of the Society for the Study of Evolution, Stony Brook, New York, June, 1982. (9)

Krystal, M., P. D'Eustachio, F. H. Ruddle, and N. Arnheim (1981) Human nucleolus organizers on non-homologous chromosomes can share the same ribosomal gene variants. Proc. Natl. Acad. Sci. 78:5744–5748. (3)

Kuchino, Y., S. Watanabe, F. Harada, and S. Nishimura (1980) Primary structure of AUA-specific isoleucine transfer ribonucleic acid from *Escherichia coli*. Biochemistry 19:2085–2089. (10)

Kunkel, T. A. and L. A. Loeb (1981) Fidelity of mammalian DNA polymerases. Science 213:765–767. (4)

Küntzel, H. and H. G. Köchel (1981) Evolution of rRNA and origin of mitochondria. Nature 293:751–755. (4)

Kurtén, B. and E. Anderson (1980) Pleistocene mammals of North America. Columbia Univ. Press, New York. (9)

Kuter, K. and A. Rogers (1975) The synthesis of ribosomal protein and ribosomal RNA in a rat-mouse hybrid cell line. Exp. Cell. Res. 91:317–325. (3)

Lacy, E. and T. Maniatis (1980) The nucleotide sequence of a rabbit β-globin pseudogene. Cell 21:545–553. (2, 11)

Lagerkvist, U. (1981) Unorthodox codon reading and the evolution of the genetic code. Cell 23:305–306. (10)

Laipis, P. J. and W. W. Hauswirth (1980) Variation in bovine mitochondrial DNAs between maternally related animals. In: The Organization and Expression of the Mitochondrial Genome, A. M. Kroon and C. Saccone (eds.). North Holland Publ., Amsterdam, pp. 125–130. (8)

Landsteiner, K. (1900) Zur Kentinis der antifermentativen lytischen und agglutinierenden Wirkung des Blutserum und der Lymph. Zentr. Bakteriol. Parasitenk. 27:357–362. (5)

Langley, C. H. and W. M. Fitch (1974) An examination of the constancy of the rate of molecular evolution. J. Mol. Evol. 3:161–177. (11)

Langley, C. H., E. A. Montgomery, and W. F. Quattlebaum (1982) Restriction map variation in the Adh region of Drosophila. Proc. Natl. Acad. Sci. 79:5631–5635. (7, 9)

Lansman, R. A. and D. A. Clayton (1975) Selective nicking of mammalian mitochondrial DNA in vivo: photosensitization by incorporation of 5-bromodeoxyuridine. J. Mol. Biol. 99:761–776. (4)

Lansman, R. A., R. O. Shade, J. F. Shapira, and J. C. Avise (1981) The use of restriction endonucleases to measure mitochondrial DNA sequence relatedness in natural populations. III. Techniques and potential applications. J. Mol. Evol. 17:214–226. (8)

Lansman, R. A., J. C. Avise, C. F. Aquadro, J. F. Shapira, and S. W. Daniel (1982) Extensive genetic variation in mitochondrial DNA's among geographic populations of the deer mouse, Peromyscus maniculatus. Submitted. (8)

Lansman, R. A., J. C. Avise, and M. D. Huettel (1983) "Paternal leakage" of mitochondrial DNA—a critical experimental test. Submitted. (8)

Latter, B. D. H. (1972) Selection in finite populations with multiple alleles. III. Genetic divergence with centripetal selection and mutation. Genetics 70:475–490. (9)

Latter, B. D. H. (1981) The distribution of heterozygosity in temperate and tropical species of Drosophila. Genet. Res. 38:137–156. (5).

Lauer, J., C.-K. J. Shen, and T. Maniatis (1980) The chromosomal arrangement of human α-like globin genes: sequence homology and α-globin gene deletions. Cell 20:119–130. (3).

Laurie-Ahlberg, C. C., G. Maroni, G. C. Bewley, J. C. Lucchesi, and B. S. Weir (1980) Quantitative genetic variation of enzyme activities in natural populations of Drosophila melanogaster. Proc. Natl. Acad. Sci. 77:1073–1077. (6)

Lawn, R. M., J. Adelman, S. C. Bock, A. E. Franke, C. M. Houck, R. C. Najarian, P. H. Seeburg, and K. L. Wion (1981) The sequence of human serum albumin cDNA and its expression in E. coli. Nucleic Acids Res. 9:6103–6114. (9)

Learner, S. A., T. T. Wu, and E. C. C. Lin (1964) Evolution of a catabolic pathway in bacteria. Science 146:1313–1315. (12)

LeBlanc, D. J. and R. P. Mortlock (1971) Metabolism of D-a-rabinose. A new pathway in Escherichia coli. J. Bacteriol. 106:90–96. (12)

LeCam, L. M., J. Neyman, and E. L. Scott (eds.) (1972) Darwinian, Neo-Darwinian, and Non-Darwinian Evolution. Proc. 6th Berkeley Symp. Math. Stat. Prob., Vol. 5, Univ. of California Press, Berkeley. (5)

Leder, P. (1982) The genetics of antibody diversity. Scientific American 246(5):102–115. (2, 9)

Leder, A., D. Swan, F. Ruddle, P. D'Eustachio, and P. Leder (1981) Dispersion of α-like globin genes of the mouse to three different chromosomes. Nature 293:196–200. (2)

Lee, W.-H., M. Nunn, and P. H. Duesberg (1981) *src* genes of ten Rous Sarcoma Virus strains, including two reportedly transduced from the cell, are completely allelic; putative markers of transduction are not detected. J. Virol. 39:758–776. (13)

Leigh Brown, A. J. and D. Ish-Horowicz (1981) Evolution of the 87A and 87C heat-shock loci in *Drosophila*. Nature 290:677–682. (3, 7)

Leigh Brown, A. J. and C. H. Langley (1979) Reevaluation of genic heterozygosity in natural populations of *Drosophila melanogaster* by two-dimensional electrophoresis. Proc. Natl. Acad. Sci. 76:2381–2384. (5)

Lemischka, I. R., S. Farmer, V. R. Racaniello, and P. A. Sharp (1981) Nucleotide sequence and evolution of a mammalian α-tubulin messenger RNA. J. Mol. Biol. 151:101–120. (11)

Levin, B. R. (1981) Periodic selection, infectious gene exchange and the genetic structure of *E. coli* populations. Genetics 99:1–23. (5)

Levin, B. R., F. M. Stewart, and L. Chao (1977) Resource-limited growth, competition and predation: A model and some experimental studies with bacteria and bacteriophage. Amer. Nat. 111:3–24. (5)

Levin, B. R., F. M. Stewart, and V. A. Rice (1979) The kinetics of conjunctive plasmid transmission: Fit of simple mass action models. Plasmid 2:247–260. (5)

Levins, R. (1968) Evolution in Changing Environments. Princeton Univ. Press, Princeton, New Jersey. (9)

Lewin, B. (1980) Gene Expression, Vol. 2, Wiley, New York. (3)

Lewis, N. and J. Gibson (1978) Variation in amount of enzyme protein in natural populations. Biochem. Genet. 16:159–170. (6)

Lewontin, R. C. (1974) The Genetic Basis of Evolutionary Change. Columbia Univ. Press, New York. (5, 6, 11, 12)

Lewontin, R. C. and J. L. Hubby (1966) A molecular approach to the study of genetic heterozygosity in natural populations. II. Amount of variation and degree of heterozygosity in natural populations of *Drosophila pseudoobscura*. Genetics 54:595–609. (9)

Lewontin, R. C. and J. Krakauer (1973) Distribution of gene-frequency as a test of the theory of the selective neutralism of polymorphisms. Genetics 74:175–195. (5)

Li, W.-H. (1978) Maintenance of genetic variability under the joint effect of mutation, selection, and random genetic drift. Genetics 90:349–382. (9)

Li, W.-H. (1980) Rate of gene silencing at duplicate loci: A theoretical study and interpretation of data from tetraploid fishes. Genetics 95:237–258. (2)

Li, W.-H. (1981) A simulation study on Nei and Li's model for estimating DNA divergence from restriction enzyme maps. J. Mol. Evol. 17:251–255. (8)

Li, W.-H. (1982) Evolutionary change of duplicate genes. In: Isozymes: Current Topics in Biological and Medical Research, Vol. 6, M. C. Rattazzi, J. G. Scandalios, and G. S. Whitt (eds.). Alan R. Liss, New York, pp. 55–92. (2)

Li, W.-H., T. Gojobori, and M. Nei (1981) Pseudogenes as a paradigm of neutral evolution. Nature 292:237–239. (2, 11)

Lim, S. T., R. M. Kay, and G. S. Bailey (1975) Lactate dehydrogenase isozymes of salmonid fish. J. Biol. Chem. 250:1790–1800. (2)

Little, P. F. R. (1982) Globin pseudogenes. Cell 28:683–684. (2)

Lomedico, P., N. Rosenthal, A. Efstratiadis, W. Gilbert, R. Kolodner, and R.

Tizard (1979) The structure and evolution of the two nonallelic rat pre-proinsulin genes. Cell 18:545–558. (2, 11)

Long, E. O. and I. B. Dawid (1980) Repeated genes in eucaryotes. Ann. Rev. Biochem. 49:727–764. (3)

Long, E., M. L. Rebbert, and I. B. Dawid (1981) Nucleotide sequence of the initiation site for ribosomal RNA transcription in Drosophila melanogaster: comparison of genes with and without insertions. Proc. Natl. Acad. Sci. 78:1513–1517. (3)

Loukas, M., Y. Vergini, and C. B. Krimbas (1981) The genetics of Drosophila subobscura populations. XVII. Further genic heterogeneity within electro-morphs by urea denaturation and the effect of the increased genic variability on linkage disequilibrium studies. Genetics 97:429–441. (5)

Lucchesi, J. C., M. B. Hughes, and B. W. Geer (1979) Genetic control of pentose phosphate pathway enzymes in Drosophila. In: Current Topics in Cellular Regulation, B. L. Horecker and E. R. Stadtman (eds.). Academic Press, New York, Vol. 15, pp. 143–154. (6)

Lueders, K., A. Leder, P. Leder, and E. Kuff (1982) Association between a transposed α-globin pseudogene and retrovirus-like elements in the BALB/c mouse genome. Nature 295:426–428. (2)

Marshall, C. J., S. D. Handmaker, and M. E. Bramwell (1975) Synthesis of ribosomal RNA in synkaryons and heterokaryons formed between human and rodent cells. J. Cell. Sci. 17:307–325. (3)

Martial, J. A., R. A. Hallewell, J. D. Baxter, and H. M. Goodman (1979) Human growth hormone: complementary DNA cloning and expression in bacteria. Science 205:602–607. (11)

Maruyama, T. and M. Kimura (1980) Genetic variability and effective population size when local extinction and recolonization of subpopulations are frequent. Proc. Natl. Acad. Sci. 77:6710–6714. (5, 9)

Maruyama, T. and M. Nei (1981) Genetic variability maintained by mutation and overdominant selection in finite populations. Genetics 98:441–459. (9)

Mather, K. (1953) The genetical structure of populations. Symp. Soc. Exp. Biol. 7:66–95. (11)

Matsumoto, L., H. Kasamatsu, L. Piko, and J. Vinograd (1974) Mitochondrial DNA replication in sea urchin oocytes. J. Cell. Biol. 63:146–159. (4)

Mattei, M. G., J.-F. Mattei, S. Aymedel, and F. Giraud (1979) Dicentric Robertsonian translocation in man: 17 cases studied by R, C, and N banding. Hum. Genet. 50:33–38. (3)

May, B. (1980) The salmonid genome: Evolutionary restructuring following a tetraploid event. Ph.D. thesis. Pennsylvania State University, University Park, Pennsylvania. (2)

Mayr, E. (1963) Animal Species and Evolution. Harvard Univ. Press, Cambridge, Mass. (9)

Mayr, E. (1982) Growth of Biological Thought. Harvard Univ. Press, Cambridge, Mass. (9)

McClintock, B. (1952) Chromosome organization and gene expression. Cold Spring Harbor Symp. Quant. Biol. 16:13–47. (13)

McClintock, B. (1957) Controlling elements and the gene. Cold Spring Harbor Symp. Quant. Biol. 21:197–216. (13)

McDonald, J. F., S. M. Anderson, and M. Santos (1980) Biochemical differences between products of the *Adh* locus in *Drosophila*. Genetics 95:1013–1022. (6)

McDonald, J. F., G. K. Chambers, J. David, and F. J. Ayala (1977) Adaptive response due to changes in gene regulation: A study with *Drosophila melanogaster*. Proc. Natl. Acad. Sci. 74:4562–4566. (6)

McKechnie, S. W., M. Kohane, and S. C. Phillips (1981) A search for interacting polymorphic enzyme loci in *Drosophila melanogaster*. In: Genetic Studies of Drosophila populations, J. B. Gibson and J. G. Oakeshott (eds.). Australian National Univ. Press, Canberra, Australia, pp. 121–138. (6)

McLachlan, A. D. (1977) Growth and evolution of proteins by gene duplication. In: Molecular Evolution and Polymorphism, Proc. 2nd Taniguchi International Symposium on Biophysics, M. Kimura (ed.). National Institute of Genetics, Mishima, Japan, pp. 208–239. (2)

Menozzi, P., A. Piazza, and L. L. Cavalli-Sforza (1978) Synthetic maps of human gene frequencies in Europe. Science 201:786–792. (5)

Miesfeld, R. and N. Arnheim (1982) Identification of the *in vivo* and *in vitro* origin of transcription in human rDNA. Nucleic Acids Res. 10:3933–3949. (3)

Mikkelsen, M., A. Basli, and H. Poulsen (1980) Nucleolar organizer regions in translocations involving acrocentric chromosomes. Cytogenet. Cell. Genet. 26:14–21. (3)

Milkman, R. D. (1967) Heterosis as a major cause of heterozygosity in nature. Genetics 55:493–495. (9)

Milkman, R. D. (1973) Electrophoretic variation in *E. coli* from natural sources. Science 182:1024–1026. (5)

Milkman, R. D. (1975) Electrophoretic variation in *E. coli* of diverse natural origins. In: Isozymes, Vol. IV, C. L. Markert (ed.). Academic Press, New York, pp. 273–285. (5)

Milkman, R. D. (1982) Toward a unified selection theory. In: Perspectives on Evolution, R. D. Milkman (ed.). Sinauer Associates, Sunderland, Mass., pp. 105–118. (9, 11)

Miller, H. I., J. Abraham, M. Benedik, A. Campbell, D. Court, H. Echols, R. Fischer, J. M. Galindo, G. Guarneros, T. Hernandez, D. Mascarenhas, C. Montañez, D. Schindler, U. Schmeissner, and L. Sosa (1981) Regulation of the integration-excision reaction by bacteriophage λ. Cold Spring Harbor Symp. Quant. Biol. 45:439–445. (13)

Miller, K. G. and B. Sollner-Webb (1981) Transcription of mouse rRNA genes by RNA polymerase I: *in vitro* and *in vivo* initiation and processing sites. Cell 27:165–174. (3)

Miller, O. J., D. A. Miller, V. G. Dev, R. Tantravahi, and C. M. Croce (1976) Expression of human and suppression of mouse nucleolus organizer activity in mouse-human somatic cell hybrids. Proc. Natl. Acad. Sci. 73:4531–4535. (3)

Miller, S., R. W. Pearcy, and E. Berger (1975) Polymorphism at the α-glycer-

308

ophosphate dehydrogenase locus in *Drosophila melanogaster*. I. Properties of adult allozymes. Biochem. Genet. 13:175–188. (6)

Mirre, C., M. Hartung, and A. Stahl (1980) Association of ribosomal genes in the fibrillar center of the nucleolus: factors influencing translocation and non-disjunction in the human meiotic oocyte. Proc. Natl. Acad. Sci. 77:6017–6021. (3)

Miyata, T. and H. Hayashida (1981) Extraordinarily high evolutionary rate of pseudogenes: evidence for the presence of selective pressure against changes between synonymous codons. Proc. Natl. Acad. Sci. 78:5739–5743. (2, 11)

Miyata, T. and T. Yasunaga (1981) Rapidly evolving mouse α globin-related pseudo gene and its evolutionary history. Proc. Natl. Acad. Sci. 78:450–453. (2, 11)

Miyata, T., T. Yasunaga, and T. Nishida (1980) Nucleotide sequence divergence and functional constraint in mRNA evolution. Proc. Natl. Acad. Sci. 77:7328–7332. (11)

Modiano, G., G. Battistuzzi, and A. G. Motulsky (1981) Nonrandom patterns of codon usage and of nucleotide substitutions in human α- and β-globin genes: an evolutionary strategy reducing the rate of mutations with drastic effects? Proc. Natl. Acad. Sci. 78:1110–1114. (11)

Montoya, J., D. Ojala, and G. Attardi (1981) Distinctive features of the 5'-terminal sequences of the human mitochondrial mRNAs. Nature 290:465–470. (4)

Moore, M. N., R. K. Koehn, and B. L. Bayne (1980) Leucine aminopeptidase (aminopeptidase-I), N-acetyl-β-hexosaminidase and lysosomes in the mussel, *Mytilus edulis* L., in response to salinity changes. J. Exp. Zool. 214:239–249. (6)

Morgan, T. H. (1925) Genetics and Evolution. Princeton Univ. Press, Princeton, New Jersey. (9)

Morgan, T. H. (1932) The Scientific Basis of Evolution. W. W. Norton, New York. (9)

Mortlock, R. P. and W. A. Wood (1964a) Metabolism of pentoses and pentitols by *Aerobacter aerogenes*. I. Demonstration of pentose isomerase, pentulokinase, and pentitol dehydrogenase enzyme families. J. Bacteriol. 88:838–844. (12)

Mortlock, R. P. and W. A. Wood (1964b) Metabolism of pentoses and pentitols by *Aerobacter aerogenes*. II. Mechanism of acquisition of kinase, isomerase, and dehydrogenase activity. J. Bacteriol. 88:845–849. (12)

Mortlock, R. P., D. D. Fossitt, and W. A. Wood (1965) A basis for utilization of unnatural pentoses and pentitols by *Aerobacter aerogenes*. Proc. Natl. Acad. Sci. 54:572–579. (12)

Muller, H. J. (1929) The method of evolution. The Scientific Monthly 29:481–505. (9)

Muller, H. J. (1950) Our load of mutations. Amer. J. Hum. Genet. 2:111–176. (9)

Nagylaki, T. and T. D. Petes (1982) Intrachromosomal gene conversion and the maintenance of sequence homogeneity among repeated genes. Genetics 100:315–337. (3)

Nash, H. (1981) Integration and excision of bacteriophage λ: the mechanism of conservative site-specific recombination. Ann. Rev. Genet. 15:143–167. (13)

Nass, M. M. K. (1973) Differential methylation of mitochondrial and nuclear DNA in cultured mouse, hamster, and virus-transformed hamster cells: *in vivo* and *in vitro* methylation. J. Mol. Biol. 80:155–175. (8)

Nathans, D. and H. O. Smith (1975) Restriction endonucleases in the analysis and restructuring of DNA molecules. Ann. Rev. Biochem. 44:273–293. (8)

Neel, J. V., M. Layrisse, and F. M. Salzano (1977) Man in the tropics: the Yanomama Indians. In: Population Structure and Human Variation, G. A. Harrison (ed.). Cambridge Univ. Press, Cambridge, pp. 109–142. (5)

Nei, M. (1965) Variation and covariation of gene frequencies in subdivided populations. Evolution 19:256–258. (5)

Nei, M. (1969) Gene duplication and nucleotide substitution in evolution. Nature 221:40–41. (2)

Nei, M. (1975) Molecular Population Genetics and Evolution. North Holland, Amsterdam and New York. (5, 9, 11)

Nei, M. (1976) The cost of natural selection and the extent of enzyme polymorphism. Trends in Biochem. Sci. 1:N247–N248. (5)

Nei, M. (1977) F-statistics and analysis of gene diversity in subdivided populations. Ann. Hum. Genet. 41:225–233. (5)

Nei, M. (1980) Stochastic theory of population genetics and evolution. (Vito Volterra Symp. on Math. Models in Biology), C. Barigozzi (ed.), Springer-Verlag, Berlin, pp. 17–47. (5, 9)

Nei, M. and R. Chakraborty (1976) Electrophoretically silent alleles in a finite population. J. Mol. Evol. 8:381–385. (5)

Nei, M. and Y. Imaizumi (1966a) Genetic structure of human populations. I. Local differentiation of blood group gene frequencies in Japan. Heredity 21:9–25. (5)

Nei, M. and Y. Imaizumi (1966b) Genetic structure of human populations. II. Differentiation of blood group gene frequencies among isolated populations. Heredity 26:183–190. (5)

Nei, M. and W.-H. Li (1979) Mathematical model for studying genetic variation in terms of restriction endonucleases. Proc. Natl. Acad. Sci. 76:5269–5273. (4, 8, 9)

Nei, M. and A. K. Roychoudhury (1973) Probability of fixation of nonfunctional genes at duplicate loci. Amer. Nat. 107:362–372. (2)

Nei, M. and A. K. Roychoudhury (1982) Genetic relationship and evolution of human races. Evol. Biol. 14:1–59. (5)

Nei, M. and F. Tajima (1981) DNA polymorphism detectable by restriction endonucleases. Genetics 97:145–163. (7, 8, 9)

Nei, M. and S. Yokoyama (1976) Effects of random fluctuations of selection intensity on genetic variability in a finite population. Japan. J. Genet. 51:355–369. (9)

Nei, M., T. Maruyama, and R. Chakraborty (1975) The bottleneck effect and genetic variability in populations. Evolution 29:1–10. (5, 9)

Nei, M., P. A. Fuerst, and R. Chakraborty (1976) Testing the neutral mutation hypothesis by distribution of single locus heterozygosity. Nature 262:491–493. (9)

Nevo, E. (1978) Genetic variation in natural populations: patterns and theory. Theoret. Popul. Biol. 13:121–177. (6, 9)

Nishimura, S. (1972) Minor components in transfer RNA: Their characterization, location and function. Progress in Nucleic Acid Research and Molecular Biology 12:49–85. (10)

Nishioka, Y., A. Leder, and P. Leder (1980) Unusual α-globin-like gene that has cleanly lost both globin intervening sequences. Proc. Natl. Acad. Sci. 77:2806–2809. (2, 11)

Oakeshott, J. G., J. B. Gibson, P. R. Anderson, W. R. Knibb, D. G. Anderson, and G. K. Chambers (1982) Alcohol dehydrogenase and glycerol-3-phosphate dehydrogenase clines in *Drosophila melanogaster* on different continents. Evolution 36:86–96. (6)

O'Brien, S. J. and R. J. MacIntyre (1972a) The α-glycerophosphate cycle in *Drosophila melanogaster*. I. Biochemical and developmental aspects. Biochem. Genet. 7:141–161. (6)

O'Brien, S. J. and R. J. MacIntyre (1972b) The α-glycerophosphate cycle in *Drosophila melanogaster*. II. Genetic aspects. Biochem. Genet. 7:127–128. (6)

O'Brien, T. W., N. D. Denslow, T. O. Harville, R. A. Hessler, and D. E. Matthews (1980) Functional and structural roles of proteins in mammalian mitochondrial ribosomes. In: The Organization and Expression of the Mitochondrial Genome, A. M. Kroon and C. Sacconi (eds.). Elsevier/North Holland, New York, pp. 301–305. (4)

Ohkubo, H., E. Avvedimento, Y. Yamada, G. Vogeli, M. Sobel, G. Meilino, I. Pastan, and B. de Grombrugghe (1982) The collagen gene. In: (ICN-UCLA) Developmental Genetics (in press) (2)

Ohnishi, S., A. J. Leigh Brown, R. A. Voelker, and C. H. Langley (1982) Estimation of genetic variability in natural populations of *Drosophila simulans* by two-dimensional and starch gel electrophoresis. Genetics 100:127–136. (5)

Ohno, S. (1970) Evolution by Gene Duplication. Springer-Verlag, Berlin. (2, 9)

Ohno, S. (1972) So much "junk" DNA in our genome. In: Evolution of Genetic Systems, H. H. Smith (ed.). Brookhaven Symp. Biol. 23:366–370. (2)

Ohno, S., J. Trujillo, W. D. Kaplan, and R. Kinosita (1961) Nucleolus-organizers in the causation of chromosomal anomalies in man. The Lancet ii:123–126. (3)

Ohta, T. (1973) Slightly deleterious mutant substitutions in evolution. Nature 246:96–98. (9, 11)

Ohta, T. (1974) Mutational pressure as the main cause of molecular evolution and polymorphism. Nature 252:351–354. (9, 11)

Ohta, T. (1977a) On the gene conversion model as a mechanism for maintenance of homogeneity in systems with multiple genomes. Genet. Res. 30:89–91. (3)

Ohta, T. (1977b) Extension to the neutral mutation random drift hypothesis. In: Molecular Evolution and Polymorphism, M. Kimura (ed.), Proc. 2nd Taniguchi Intl. Symp. on Biophysics. National Institute of Genetics, Mishima, Japan, pp. 148–167. (9)

Ohta, T. (1978) Sequence variability of immunoglobulins considered from the standpoint of population genetics. Proc. Natl. Acad. Sci. 75:5108–5112. (9)

Ohta, T. (1980) Evolution and Variation of Multigene Families. Springer-Verlag, Berlin. (2, 3)

Ohta, T. (1982a) Linkage disequilibrium due to random genetic drift in finite subdivided populations. Proc. Natl. Acad. Sci. 79:1940–1944. (5)

Ohta, T. (1982b) Linkage disequilibrium with the island model. Genetics 101:139–155. (5)

Ohta, T. and M. Kimura (1971a) Functional organization of genetic material as a product of molecular evolution. Nature 233:118–119. (11)

Ohta, T. and M. Kimura (1971b) On the constancy of the evolutionary rate of cistrons. J. Mol. Evol. 1:18–25. (11)

Ohtsubo, E., R. C. Deonier, H. J. Lee, and N. Davidson (1974) Electron microscope heteroduplex studies of sequence relations among plasmids of *Escherichia coli*. IV. The F sequences in F14. J. Mol. Biol. 89:565–584. (13)

Ojala, D., C. Merkel, R. Gelfand, and G. Attardi (1980) The tRNA genes punctuate the reading of genetic information in human mitochondrial DNA. Cell 22:393–403. (4)

Ojala, D., J. Montoya, and G. Attardi (1981) tRNA punctuation model of RNA processing in human mitochondria. Nature 290:470–474. (4)

Oliver, E. J. and R. P. Mortlock (1971) Growth of *Aerobacter aerogenes* on D-arabinose: origin of the enzyme activities. J. Bacteriol. 108:287–292. (12)

Oliver, N. A. and D. C. Wallace (1982) Assignment of two mitochondrially synthesized polypeptides to human mitochondrial DNA and their use in the study of intracellular mitochondrial interaction. Mol. Cell. Biol. 2:30–41. (4)

Orgel, L. E. and F. H. C. Crick (1980) Selfish DNA: the ultimate parasite. Nature 284:604–607. (9)

Orkin, S. H., H. H. Kazazian, S. E. Antonarakis, S. C. Goff, C. D. Boehm, J. Sexton, P. Waber, and P. V. J. Giardina (1982) Linkage of β thalassemia mutations and β globin gene polymorphisms with DNA polymorphisms in the human β globin gene cluster. Nature 296:627–631. (7, 9)

Ozeki, H. (1980) The organization of transfer RNA genes in *Escherichia coli*. In: RNA-polymerase, tRNA and Ribosomes: Their Genetics and Evolution, S. Osawa, H. Ozeki, H. Uchida, and T. Yura (eds.). Tokyo University Press, Tokyo, pp. 173–183. (2)

Palmiter, R. D., R. L. Brinster, R. E. Hammer, M. E. Trumbauer, M. G. Rosenfeld, N. C. Birnberg, and R. M. Evans (1982) Dramatic growth of mice that develop from eggs microinjected with metallothionein-growth hormone fusion genes. Nature 300:611–615. (9)

Pardue, M. L. (1974) Localization of repeated DNA sequences in *Xenopus*

312

chromosomes. Cold Spring Harbor Symp. Quant. Biol. 38:475–482. (3)

Parkin, D. T. (1979) An Introduction to Evolutionary Genetics. Edward Arnold, London. (11)

Paterson, H. E. (1981) The continuing search for the unknown and unknowable: A critique of contemporary ideas on speciation. S. Afri. J. Sci. 77:113–119. (5)

Peacock, W. J., D. Brutlag, E. Goldring, R. Appels, C. W. Hinton, and D. L. Lindsley (1973) The organization of highly repeated DNA sequences in *Drosophila melanogaster* chromosomes. Cold Spring Harbor Symp. Quant. Biol. 38:405–416. (4)

Pech, M., J. Hochtl, H. Schnell, and H. G. Zachau (1981) Differences between germ-line and rearranged immunoglobulin V_κ coding sequences suggest a localized mutation mechanism. Nature 291:668–670. (9)

Perler, F., A. Efstratiadis, P. Lomedico, W. Gilbert, R. Kolodner, and J. Dodgson (1980) The evolution of genes: the chicken preproinsulin gene. Cell 20:555–566. (4)

Perutz, M. F. and H. Lehman (1968) Molecular pathology of human haemoglobin. Nature 219:902–909. (11)

Petes, T. D. (1980) Unequal meiotic recombination within tandem arrays of yeast ribosomal DNA genes. Cell 19:765–774. (3)

Piazza, A., P. Menozzi, and L. L. Cavalli-Sforza (1981a) The making and testing of geographic gene-frequency maps. Biometrics 37:635–659. (5)

Piazza, A., P. Menozzi, and L. L. Cavalli-Sforza (1981b) Synthetic gene frequency maps of man and selective effects. Proc. Natl. Acad. Sci. 78:2638–2642. (5)

Pietruszko, R. (1980) Alcohol and aldehyde dehydrogenase isozymes from mammalian liver—their structural and functional differences. In: Isozymes: Current Topics in Biological and Medical Research, Vol. 4, M. C. Rattazzi, J. C. Scandalios, and G. S. Whitt (eds.). Alan R. Liss, New York, pp. 107–130. (2)

Place, A. R. and D. A. Powers (1979) Genetic variation and relative catalytic efficiencies: Lactate dehydrogenase B allozymes of *Fundulus heteroclitus*. Proc. Natl. Acad. Sci. 76:2354–2358. (6)

Pogue-Beile, K. L., S. Dassarma, S. R. King, and S. R. Jaskunas (1980) Recombination between bacteriophage lambda and plasmid pBR322 in *Escherichia coli.* J. Bacteriol. 142:992–1003. (13)

Polan, M. L., S. Friedman, J. G. Gall, and W. Gehring (1973) Isolation and characterization of mitochondrial DNA from *Drosophila melanogaster.* J. Cell. Biol. 56:580–589. (4)

Post, L. E., G. D. Strycharz, M. Nomura, H. Lewis, and P. P. Dennis (1979) Nucleotide sequence of the ribosomal protein gene cluster adjacent to the gene for RNA polymerase subunit β in *Escherichia coli.* Proc. Natl. Acad. Sci. 76:1697–1701. (11)

Potter, S. S., J. E. Newbold, C. A. Hutchison, III and M. H. Edgell (1975) Specific cleavage analysis of mammalian mitochondrial DNA. Proc. Natl. Acad. Sci. 72:4496–4500. (8)

Powell, J. R. (1975) Protein variation in natural populations. Evol. Biol. 8:79–119. (6)

Powers, D. A. and A. R. Place (1978) Biochemical genetics of *Fundulus heteroclitus* (L.). I. Temporal and spatial variation in gene frequencies of *Ldh-B, Mdh,-A, Gpi-B,* and *Pgm-A*. Biochem. Genet. 16:593–607. (6)

Powers, D. A., G. S. Greaney, and A. R. Place (1979) Physiological correlation between lactate dehydrogenase genotype and haemoglobin function in killifish. Nature 277:240–241. (6)

Prager, E. M. and A. C. Wilson (1978) Construction of phylogenetic trees for proteins and nucleic acids: empirical evaluation of alternative matrix methods. J. Mol. Evol. 11:129–142. (8)

Prakash, S. (1972) Origin of reproductive isolation in the absence of apparent genic differentiation in a geographic isolate of *Drosophila pseudoobscura*. Genetics 72:143–155. (5)

Prakash, S. (1977) Further studies on gene polymorphism in the main body and geographically isolated populations of *Drosophila pseudoobscura*. Genetics 85:713–719. (5)

Proudfoot, N. (1980) Pseudogenes. Nature 286:840–841. (11)

Proudfoot, N. J. and T. Maniatis (1980) The structure of a human α-globin pseudogene and its relationship to α-globin gene duplication. Cell 21:537–544. (11)

Rabinowitz, M. and H. Swift (1970) Mitochondrial nucleic acids and their relation to the biogenesis of mitochondria. Physiol. Rev. 50:376–427. (4, 8)

Racine, R. R. and C. H. Langley (1980) Genetic heterozygosity in a natural population of *Mus musculus* assessed using two-dimensional electrophoresis. Nature 283:855–857. (5)

Radding, C. M. (1978) Genetic recombination: Strand transfer and mismatch repair. Ann. Rev. Biochem. 47:847–880. (3)

Ramirez, J. L. and I. B. Dawid (1978) Mapping of mitochondrial DNA in *Xenopus laevis* and *X. borealis*: the positions of ribosomal genes and D-loops. J. Mol. Biol. 119:133–146. (4)

Ramshaw, J. A. M., J. A. Coyne, and R. C. Lewontin (1979) The sensitivity of gel electrophoresis as a detector of genetic variation. Genetics 93:1019–1037. (5)

Rastl, E. and I. B. Dawid (1979a) Structure and evolution of animal mitochondrial DNA. In: Extrachromosomal DNA, D. J. Cummings, P. Borst, I. B. Dawid, S. M. Weissman, and C. F. Fox (eds.). Academic Press, New York, pp. 395–407. (4)

Rastl, E. and I. B. Dawid (1979b) Expression of the mitochondrial genome in *Xenopus laevis*: a map of transcripts. Cell 18:501–510. (4)

Richards, R. I., J. Shine, A. Ulrich, J. R. I. Wells, and H. M. Goodman (1979) Molecular cloning and sequence analysis of adult chicken β globin cDNA. Nucleic Acids Res. 7:1137–1146. (11)

Richmond, R. C. (1972) Enzyme variability in the *Drosophila willistoni* group. III. Amounts of variability in the subspecies *D. paulistorum*. Genetics 71:87–112. (5)

314

Richmond, R. C., D. G. Gilbert, K. B. Sheehan, M. H. Gromko, and F. M. Butterworth (1980) Esterase-6 and reproduction in *Drosophila melanogaster*. Science 297:1483–1485. (6)

Rieger, R., H. Nicoloff, and M. Anastassova-Kristeva (1979) "Nucleolar Dominance" in interspecific hybrids and translocation lines—a review. Biol. ZBL 98:385–398. (3)

Rigby, P. W. J., B. D. Burleigh, Jr., and B. S. Hartley (1974) Gene duplication in experimental evolution. Nature 251:200–204. (2)

Roberts, R. B. (1962a) Alternative codes and templates. Proc. Natl. Acad. Sci. 48:897–900. (10)

Roberts, R. B. (1962b) Further implications of the doublet code. Proc. Natl. Acad. Sci. 48:1245–1250. (10)

Roberts, R. J. (1982) Restriction and modification enzymes and their recognition sequences. Nucleic Acids Res. 10:117–144. (8)

Robertson, A. (1967) The nature of quantitative genetic variation. In: Heritage from Mendel, R. A. Brink (ed.). Univ. of Wisconsin Press, Madison, Wisconsin, pp. 265–280. (9)

Robertson, M. (1982) The evolutionary past of the major histocompatibility complex and the future of cellular immunology. Nature 297:629. (9)

Robins, D. M., S. Ripley, A. S. Henderson, and R. Axel (1981) Transforming DNA integrase into the host chromosome. Cell 23:29–39. (13)

Rolseth, S. J., V. A. Fried, and B. G. Hall (1980) A mutant *ebg* enzyme that converts lactose into an inducer of the *lac* operon. J. Bacteriol. 142:1036–1039. (12)

Romer, A. S. (1968) The Procession of Life. Weidenfeld and Nicolson, London. (11)

Romero-Herrera, A. E., N. Lieska, and S. Nassar (1979) Characterization of the myoglobin of the lamprey *Petromyzon marinus*. J. Mol. Evol. 14:259–266. (11)

Rosenzweig, M. L. (1978) Competitive speciation. Biol. J. Linn. Soc. 10:275–289. (5)

Rossman, M. G., A. Liljas, C.-I. Bränd'en, and L. J. Banaszak (1975) Evolutionary and structural relationships among dehydrogenases. In: The Enzymes, Vol. 11, 3rd Ed., P. D. Boyer (ed.). Academic Press, New York, pp. 61–101. (2)

Rothstein, S. and W. S. Reznikoff (1981) The functional differences in the inverted repeats of Tn*5* are caused by single basepair nonhomology. Cell 23:191–200. (13)

Rubenstein, J. L. R., D. Brutlag, and D. A. Clayton (1977) The mitochondrial DNA of *Drosophila melanogaster* exists in two distinct and stable superhelical forms. Cell 12:471–482. (4)

Ruddle, F. H. and R. P. Creagan (1975) Parasexual approaches to the genetics of man. Ann. Rev. Genet. 9:407–486. (3)

Russell, E. S. and E. C. McFarland (1974) Genetics of mouse hemoglobins. Ann. N. Y. Acad. Sci. 241:25–38. (1)

Rychkov, Y. G. and V. A. Sheremetyeva (1977) The genetic process in the system of ancient human isolates in North Asia. In: Population Structure and Human Variation, G. A. Harrison (ed.). Cambridge Univ. Press, Cambridge, pp. 47–108. (5)

Saedler, H., H. J. Reif, S. Hu, and N. Davidson (1974) IS2, a genetic element for turn-off and turn-on of gene activity in *E. coli*. Mol. Gen. Genet. 132:265–289. (13)

Sacktor, B. (1975) Biochemistry of insect flight. Part 1—Utilization of fuels by muscle. In: Insect Biochemistry and Function, D. J. Candy and B. A. Kilby (eds.). Chapman and Hall, London. pp. 1–88. (6)

Sampsell, B. and S. Sims (1982) Interaction of *Adh* genotype and heat stress on alcohol tolerance in *Drosophila melanogaster*. Nature 296:853–855. (6)

Sawada, M., S. Osawa, H. Kobayashi, H. Hori, and A. Muto (1981) The number of ribosomal RNA genes in *Mycoplasma capricolum*. Mol. Gen. Genet. 182:502–504. (2)

Scangos, G. A. and A. M. Reiner (1978) Acquisition of ability to utilize xylitol: Disadvantages of a constitutive catabolic pathway in *Escherichia coli*. J. Bacteriol. 134:501–505. (12)

Schaffner, W., G. Kunz, H. Daetwyler, J. Telford, H. O. Smith, and M. L. Birnstiel (1978) Genes and spacers of cloned sea urchin histone DNA analyzed by sequencing. Cell 14:655–671. (11)

Schneider, W. P., B. P. Nichols, and C. Yanofsky (1981) Procedure for production of hybrid genes and proteins and its use in assessing significance of amino acid differences in homologous tryptophan synthetase α polypeptides. Proc. Natl. Acad. Sci. 78:2169–2173. (9)

Schreier, P. H., A. L. M. Bothwell, B. Mueller-Hill, and D. Baltimore (1981) Multiple differences between the nucleic acid sequences of the $IgG2a^a$ and $IgG2a^b$ alleles of the mouse. Proc. Natl. Acad. Sci. 78:4495–4499. (9)

Sederoff, R. R., C. S. Levings, III, D. H. Timothy, and W. W. L. Hu (1981) Evolution of DNA sequence organization in mitochondrial genomes of *Zea*. Proc. Natl. Acad. Sci. 78:5953–5957. (8)

Seeburg, P. H., J. Shine, J. A. Martial, J. D. Baxter, and H. M. Goodman (1977) Nucleotide sequence and amplification in bacteria of structural gene for rat growth hormone. Nature 270:486–494. (11)

Segal, H. L. (1975) Lysosomes and intracellular protein turnover. In: Lysosomes in Biology and Pathology, J. T. Dingle and R. T. Dean (eds.). North Holland, Amsterdam, Vol. 4, pp. 295–302. (6)

Sekiya, J., M. Kobayashi, T. Seki, and K. Koike (1980) Nucleotide sequence of a cloned fragment of rat mitochondrial DNA containing the replication origin. Gene 11:53–62. (4)

Selander, R. K. (1975) Stochastic factors in the genetic structure of populations. In: Proceedings of the 8th Intl. Cong. Numerical Taxonomy, G. F. Estabrook (ed.). W. H. Freeman, San Francisco, pp. 284–332. (5)

Selander, R. K. and D. W. Kaufman (1975) Genetic structure of populations of the brown snail (*Helix aspersa*). I. Microgeographic variation. Evolution 29:385–401. (5)

Selander, R. K. and B. R. Levin (1980) Genetic diversity and structure in *Escherichia coli* populations. Science 210:545–547. (5, 9)

316

Shah, D. M. and C. H. Langley (1979a) Inter- and intraspecific variation in restriction maps of *Drosophila* mitochondrial DNAs. Nature 281:696–699. (5, 9)

Shah, D. M. and C. H. Langley (1979b) Electron microscope heteroduplex study of *Drosophila* mitochondrial DNA's: evolution of the A+T-rich region. Plasmid 2:69–78. (8)

Shenk, T. (1981) Transcriptional control regions: nucleotide sequence requirements for initiation by RNA polymerase II and III. In: Initiation Signals in Viral Gene Expression, A. Shatkin (ed.). Springer-Verlag, New York, pp. 25–45. (3)

Sheppard, H. W. and G. A. Gutman (1981) Allelic forms of rat κ chain genes: evidence for strong selection at the level of nucleotide sequence. Proc. Natl. Acad. Sci. 78:7064–7068. (9)

Shumaker, K. M., R. W. Allard, and A. L. Kahler (1982) Cryptic variability at enzyme loci in three plant species, *Avena barbata, Hordeum vulgaris,* and *Zea mays.* J. Heredity 73:86–90. (5)

Simons, E. L. (1972) Primate Evolution: An Introduction to Man's Place in Nature. Macmillan, New York. (4)

Simpson, G. G. (1944) Tempo and Mode in Evolution. Columbia Univ. Press, New York. (5)

Singer, J., J. Robert-Ems, and A. D. Riggs (1979) Methylation of mouse liver DNA studied by means of restriction enzymes *Msp*I and *Hpa*II. Science 203:1019–1021. (8)

Singer, M. F. (1982) SINEs and LINEs: Highly repeated short and long interspersed sequences in mammalian genomes. Cell 28:433–434. (9)

Singh, R. S. (1979) Genic heterogeneity within electrophoretic "alleles" and the pattern of variation among loci in *Drosophila pseudoobscura.* Genetics 93:997–1018. (5)

Singh, R. (1983) Genetic differentiation for allozyme and fitness characters between mainland and Bogota populations of *Drosophila pseudoobscura.* Can. J. Cytol. Genet. (in press) (5)

Singh, R. S., R. C. Lewontin, and A. A. Felton (1976) Genic heterogeneity within electrophoretic "alleles" of xanthine dehydrogenase in *Drosophila pseudoobscura.* Genetics 84:609–629. (5)

Singh, S. M. and E. Zouros (1978) Genetic variation associated with growth rate in the American oyster (*Crassostrea virginica*). Evolution 32:342–353. (9)

Skibinski, D. O. F. and R. D. Ward (1982) Correlations between heterozygosity and evolutionary rate of proteins. Nature 298:490–492. (5, 9)

Slater, J. H. and A. T. Bull (1982) Environmental microbiology: biodegradation. Philos. Trans. of the Royal Soc. B: 297:575–597. (12)

Slatkin, M. (1982) Testing neutrality in subdivided populations. Genetics 100:533–545. (5)

Slightom, J. L., A. E. Blechl, and O. Smithies (1980) Human fetal $^G\gamma$- and $^A\gamma$-globin genes: complete nucleotide sequences suggest that DNA can be exchanged between these duplicated genes. Cell 21:627–638. (1, 2, 3, 7, 9)

Smith, G. P. (1973) Unequal crossover and the evolution of multigene families. Cold Spring Harbor Symp. Quant. Biol. 38:507–513. (3)

Smith, G. P. (1976) Evolution of repeated DNA sequences by unequal crossovers. Science 191:528–534. (3)

Smith, S. C., R. R. Racine, and C. H. Langley (1981) Lack of genic variation in the abundant proteins of human kidney. Genetics 96:967–974. (5)

Smouse, P. E., V. J. Vitzthum, and J. V. Neel (1981) The impact of random and lineal fission on the genetic divergence of small human groups: a case study among the Yanomama. Genetics 98:179–197. (5)

Sneath, P. H. A. and R. R. Sokal (1973) Numerical Taxonomy. W. H. Freeman, San Francisco. (8)

Sokal, R. R. and P. Menozzi (1982) Spatial autocorrelations of HLA frequencies in Europe support demic diffusion of early farmers. Amer. Nat. 119:1–17. (5)

Sokal, R. R. and N. L. Oden (1978) Spatial autocorrelation in biology. I. Methodology. Biol. J. Linn. Soc. 10:199–228. (5)

Sollner-Webb, B. and R. H. Reeder (1979) The nucleotide sequence of the initiation and termination sites for ribosomal RNA transcription in *Xenopus laevis*. Cell 18:485–499. (3)

Soulé, M. (1976) Allozyme variation: its determinants in space and time. In: Molecular Evolution, F. J. Ayala (ed.). Sinauer Associates, Sunderland, Mass., pp. 60–77. (9)

Speyer, J. F., P. Lengyel, C. Basilio, and S. Ochoa (1962) Synthetic polynucleotides and the amino acid code. II. Proc. Natl. Acad. Sci. 48:63–68. (10)

Spritz, R. A. (1981) Duplication-deletion polymorphism 5′ to the human β globin gene. Nucleic Acids Res. 9:5037–5047. (7)

Stearns, R. E. C. (1881) On *Helix aspersa* in California and the geographic distribution of certain West American land snails, and previous errors relating thereto. Ann. N.Y. Acad. Sci. 2:129–139. (5)

Stein, J. P., J. F. Catterall, P. Kristo, A. R. Means, and B. W. O'Malley (1980) Ovomucoid intervening sequences specify functional domains and generate protein polymorphism. Cell 21:681–687. (2)

Sures, I., J. Lowry, and L. H. Kedes (1978) The DNA sequence of sea urchin (*S. purpuratus*) H2A, H2B and H3 histone coding and spacer regions. Cell 15:1033–1044. (11)

Sures, I., D. V. Goeddel, A. Gray, and A. Ullrich (1980) Nucleotide sequence of human preproinsulin complementary DNA. Science 208:57–59. (11)

Sved, J. A., T. E. Reed, and W. F. Bodmer (1967) The number of balanced polymorphisms that can be maintained in a natural population. Genetics 55:469–481. (9)

Szostak, J. W. and R. Wu (1980) Unequal crossing over in the ribosomal DNA of *Saccharomyces cerevisiae*. Nature 284:426–430. (3)

Takahata, N. (1981) Genetic variability and rate of gene substitution in a finite population under mutation and fluctuating selection. Genetics 98:427–440. (9)

Takahata, N. and M. Kimura (1981) A model of evolutionary base substitutions and its application with special reference to rapid change of pseudogenes. Genetics 98:641–657. (11)

Takahata, N. and T. Maruyama (1979) Polymorphism and loss of duplicate expression: A theoretical study with application to tetraploid fish. Proc. Natl. Acad. Sci. 76:4521–4525. (2)

Takahata, N. and T. Maruyama (1981) A mathematical model of extranuclear genes and the genetic variability maintained in a finite population. Genet. Res. 37:291–302. (8)

Tantravahi, R., D. A. Miller, V. G. Dev, and O. J. Miller (1976) Detection of nucleolus organizer regions in chromosomes of human, chimpanzee, gorilla, orangutan and gibbon. Chromosoma 56: 15–27. (3)

Tantravahi, R., D. A. Miller, G. D'Ancona, C. M. Croce, and O. J. Miller (1979) Location of rRNA genes in three inbred strains of rat and suppression of rat rRNA activity in rat-human somatic cell hybrids. Exp. Cell Res. 119:387–392. (3)

Tappel, A. L. (1969) Lysosomal enzymes and other components. In: Lysosomes in Biology and Pathology, J. T. Dingle and H. B. Fell (eds.). North Holland, Amsterdam, Vol. 2, pp. 207–244. (6)

Tartof, K. D. (1975) Redundant genes. Ann. Rev. Genet. 9:355–385. (2, 3)

Tateno, Y. and M. Nei (1978) Goodman et al.'s method for augmenting the number of nucleotide substitutions. J. Mol. Evol. 11:67–73. (11)

Templeton, A. R. (1980) The theory of speciation and the founder principle. Genetics 94:1011–1038. (5)

Templeton, A. R. (1983) Phylogenetic inference from restriction endonuclease cleavage site maps with particular reference to the evolution of man and the apes. (submitted) (8)

Therman, E. (1980) Human Chromosomes. Springer-Verlag, New York. (3)

Therman, E. and E. M. Kuhn (1976) Cytological demonstration of mitotic crossing-over in man. Cytogenet. Cell Genet. 17:254–267. (3)

Thomas, M. and R. Davis (1975) Studies on the cleavage of bacteriophage λ with EcoRI restriction endonuclease. J. Mol. Biol. 91:315–328. (8)

Throckmorton, L. H. (1977) Drosophila systematics and biochemical evolution. Ann. Rev. Ecol. Syst. 8:235–254. (5)

Tilghman, S. M., D. C. Tiemeier, F. Polsky, M. H. Edgell, J. G. Seidman, A. Leder, L. W. Enquist, B. Norman, and P. Leder (1977) Cloning specific segments of the mammalian genome: bacteriophage λ containing mouse globin and surrounding gene sequences. Proc. Natl. Acad. Sci. 74:4406–4410. (1)

Tomich, P. K., F. Y. An, and D. B. Clewell (1980) Properties of erythromycin-inducible transposon Tn917 in Streptococcus faecalis. J. Bacteriol. 141:1366–1374. (13)

Tomkins, D. J. (1981) Unstable familial translocation: A t(11;22) mat inherited as a t(11;15). Amer. J. Hum. Genet. 33:745–751. (3)

Tracey, M. L., N. F. Bellet, and C. B. Graven (1975) Excess of allozyme homozygosity and breeding population structure in the mussel Mytilus californianus. Marine Biology 32:303–311. (9)

Treat-Clemons, L. G. and W. W. Doane (1982) Purified α-amylases from Drosophila: specific activities and isoelectric focusing points. Isozyme Bull. 15:90–91. (6)

Tsujimoto, Y., S. Hirose, M. Tsuda, and Y. Suzuki (1981) Promoter sequence of fibroin gene assigned by in vitro transcription system. Proc. Natl. Acad. Sci. 78:4838–4842. (3)

Tuan, D., P. A. Biro, J. K. de Riel, H. Lazarus, and B. G. Forget (1979) Restriction endonuclease mapping of the human β globin gene loci. Nucleic Acids Res. 6:2519. (7)

Turberville, C. and P. H. Clarke (1981) A mutant of *Pseudomonas aeruginosa* PAC with an altered amidase inducible by the novel substrate. FEMS Microbiol. Letters 10:87–90. (12)

Ullrich, A., T. J. Dull, A. Gray, J. Brosius, and I. Sures (1980) Genetic variation in the human insulin gene. Science 209:612–615. (9)

Ullrich, A., T. J. Dull, A. Gray, J. A. Phillips, S. Peter (1982) Variation in the sequence and modification state of the human insulin gene flanking region. Nucleic Acids Res. 10:2225–2240. (7)

Upholt, W. B. (1977) Estimation of DNA sequence divergence from comparison of restriction endonuclease digests. Nucleic Acids Res. 4:1257–1265. (4, 8)

Upholt, W. B. and I. B. Dawid (1977) Mapping of mitochondrial DNA of individual sheep and goats: rapid evolution in the D-loop region. Cell 11:571–583. (4, 8)

Urano, Y., R. Kominami, T. Mishima, and M. Muramatsu (1980) The nucleotide sequence of the putative transcription initiation site of a cloned ribosomal RNA gene of the mouse. Nucleic Acids Res. 8:6043–6058. (3)

Ursprung, H. and E. Schabtach (1965) Fertilization in tunicates: Loss of the paternal mitochondrion prior to sperm entry. J. Exp. Zool. 159:379–384. (8)

Uyeno, T. and G. R. Smith (1972) Tetraploid origin of the karyotype of catostomid fishes. Science 175:644–646. (2)

Valenzuela, P., G. I. Bell, A. Venegas, E. T. Sewell, F. R. Masiarz, L. J. DeGennaro, F. Weinberg, and W. J. Rutter (1977) Ribosomal RNA genes of *Saccharomyces cerevisiae*. J. Biol. Chem. 252:8126–8135. (3)

Valenzuela, P., M. Quiroga, J. Zaldivar, W. J. Rutter, M. W. Kirschner, and D. W. Cleveland (1981) Nucleotide and corresponding amino acid sequences encoded by α and β tubulin mRNAs. Nature 289:650–655. (11)

van Arsdell, S. W., R. A. Denison, L.-B. Bernstein, A. Weiner, T. Manser, and R. F. Gesteland (1981) Direct repeats flank three small nuclear RNA pseudogenes in the human genome. Cell 26:11–17. (3)

van Delden, W. (1982) The alcohol dehydrogenase polymorphism in *Drosophila melanogaster*: selection at an enzyme locus. Evol. Biol. (in press) (6)

van Ooyen, A., J. van den Berg, N. Mantel, and C. Weissmann (1979) Comparison of total sequence of a cloned rabbit β-globin gene and its flanking regions with a homologous mouse sequence. Science 206:337–344. (11)

Vanin, E. F., G. I. Goldberg, P. W. Tucker, and O. Smithies (1980) A mouse α-globin-related pseudogene lacking intervening sequences. Nature 286:222–226. (11)

Vaughn, K. C., L. R. De Bonte, and K. G. Wilson (1980) Organelle alteration as a mechanism for maternal inheritance. Science 208:196–197. (8)

Walberg, M. W. and D. A. Clayton (1981) Sequence and properties of the human KB cell and mouse L cell D-loop regions of mitochondrial DNA. Nucleic Acids Res. 9:5411–5421. (4, 8)

Waley, S. G. (1969) Some aspects of the evolution of metabolic pathways. Comp. Biochem. Physiol. 30:1–11. (2)

Walsh, P. J. (1981) Purification and characterization of two allozymic forms of octopine dehydrogenase from California populations of Metridium senile. J. Comp. Physiol. 143:213–222. (6)

Warburton, D., K. C. Atwood, and A. S. Henderson (1976) Variation in the number of genes for rRNA among human acrocentric chromosomes: correlation with frequency of satellite association. Cytogenet. Cell Genet. 17:221–230. (3)

Ward, W. F. and G. E. Mortimore (1978) Compartmentation of intracellular amino acids in rat liver. Evidence for an intralysosomal pool derived from protein degradation. J. Biol. Chem. 253:3581–3587. (6)

Watt, W. B. (1977) Adaptation at specific loci. I. Natural selection on phosphoglucose isomerase of Colias butterflies: biochemical and population aspects. Genetics 87:177–194. (6)

Watterson, G. A. (1978) The homozygosity test of neutrality. Genetics 88:405–417. (5)

Watterson, G. A. (1982) On the time for gene silencing at duplicate loci. Statistics Research Report No. 12:1–24, Monash University, Australia. (2)

Weatherall, D. J. and J. B. Clegg (1979) Recent developments in the molecular genetics of human hemoglobin. Cell 16:467–479. (2)

Weaver, S., M. B. Comer, C. L. Jahn, C. A. Hutchison, III, and M. H. Edgell (1981) The adult β-globin genes of the "single" type mouse C57BL. Cell 24:403–411. (1)

Wickens, M. P., S. Woo, B. W. O'Malley, and J. B. Gurdon (1980) Expression of a chicken chromosomal ovalbumin gene injected into frog oocyte nuclei. Nature 285:628–634. (3)

Wilde, C. D., C. E. Crowther, and N. J. Cowan (1982) Diverse mechanisms in the generation of human β-tubulin pseudogenes. Science 217:549–552. (2)

Wiley, D. C., I. A. Wilson, and J. J. Skehel (1981) Structural identification of the antibody-binding sites of Hong Kong influenza haemagglutinin and their involvement in antigenic variation. Nature 289:373–378. (9)

Wilson, A. C., S. S. Carlson, and T. J. White (1977) Biochemical evolution. Ann. Rev. Biochem. 46:573–639. (4, 5, 11)

Woese, C. R. (1981) Archaebacteria. Scientific American 244(6):98–122. (10)

Wolstenholme, D. R. (1973) Replicating DNA molecules from eggs of Drosophila melanogaster. Chromosoma 43:1–18. (4)

Wolstenholme, D. R., J. M. Goddard, and C. M.-R. Fauron (1979) Structure and replication of mitochondrial DNA from the genus Drosophila. In: Extrachromosomal DNA, D. J. Cummings, P. Borst, I. B. Dawid, S. M. Weissman, and C. F. Fox (eds.). Academic Press, New York, pp. 409–425. (4)

321

Wood, W. G., J. B. Clegg, and D. J. Weatherall (1977) Developmental biology of human hemoglobins. In: Progress in Hematology, X. E. B. Brown (ed.). Grune and Stratton, New York, pp. 43–90. (2)

Workman, P. L. and J. D. Niswander (1970) Population studies on southwestern Indian tribes. II. Local genetic differentiation in the Papago. Amer. J. Hum. Genet. 22:24–49 (5)

Wozney, J., D. Hanahan, V. Tate, H. Boedtker, and P. Doty (1981) Structure of the pro α2(I) collagen gene. Nature 294:129–135. (2)

Wright, J. E., Jr., B. May, M. Stoneking, and G. M. Lee (1980) Pseudolinkage of the duplicate loci for supernatant aspartate aminotransferase in brook trout, Salvelinus fontinalis. J. Heredity 71:223–228. (2)

Wright, S. (1931) Evolution in Mendelian populations. Genetics 16:97–159. (5, 9, 11)

Wright, S. (1940) Breeding structure of populations in relation to speciation. Amer. Nat. 74:232–248. (5)

Wright, S. (1943) Isolation by distance. Genetics 28:114–138. (5)

Wright, S. (1965) The interpretation of population structure by F-statistics with special regard to systems of mating. Evolution 19:395–420. (5)

Wright, S. (1970) Random drift and the shifting balance theory of evolution. In: Mathematical Topics in Population Genetics, K. Kojima (ed.). Springer-Verlag, Berlin, pp. 1–31. (9)

Wright, S. (1977) Evolution and the Genetics of Populations, Vol. III, Experimental Results and Evolutionary Deductions. University of Chicago Press, Chicago. (5)

Wright, S. (1978) Evolution and the Genetics of Populations, Vol. IV, Variability Within and Among Natural Populations. University of Chicago Press, Chicago. (5)

Wright, S. (1982) Character change, speciation, and the higher taxa. Evolution 36:427–443. (5)

Wu, T. T., E. C. C. Lin, and S. Tanaka (1968) Mutants of Aerobacter aerogenes capable of utilizing xylitol as a novel carbon source. J. Bacteriol. 96:447–456. (12)

Yamada, Y., V. E. Avvedimento, M. Mudryj, H. Ohkubo, G. Vogeli, M. Irani, I. Pastan, and B. de Crombrugghe (1980) The collagen gene: evidence for its evolutionary assembly by amplication of a DNA segment containing an exon of 54 bp. Cell 22:887–892. (2)

Yamazaki, T. and T. Maruyama (1972) Evidence for the neutral hypothesis of protein polymorphism. Science 178:56–58. (5)

Yanofsky, C., T. Platt, I. P. Crawford, B. P. Nichols, G. E. Christie, H. Horowitz, M. van Cleemput, and A. M. Wu (1981) The complete nucleotide sequence of the tryptophan operon of Escherichia coli. Nucleic Acids Res. 9:6647–6668. (2)

Yardley, D. G., W. W. Anderson, and H. E. Schaffer (1977) Gene frequency changes at the α-amylase locus in experimental populations of Drosophila pseudoobscura. Genetics 87:357–369. (5)

Yonekawa, H., K. Moriwaki, O. Gotoh, J. Watanabe, J.-I. Hayashi, N. Miyashita, M. L. Petras, and Y. Tagashira (1980) Relationship between laboratory mice and the subspecies Mus musculus domesticus based on restriction endonuclease cleavage patterns of mitochondrial DNA. Japan. J. Genetics 55:289–296. (8)

Yonekawa, H., K. Moriwaki, O. Gotoh, J.-I. Hayashi, J. Watanabe, N. Miyashita, M. L. Petras, and Y. Tagashira (1981) Evolutionary relationships among five subspecies of *Mus musculus* based on restriction enzyme cleavage patterns of mitochondrial DNA. Genetics 98:801–816. (8)

Young, J. P. W., R. K. Koehn, and N. Arnheim (1979) Biochemical characterization of 'LAP', a polymorphic aminopeptidase from the blue mussel, *Mytilus edulis*. Biochem. Genet. 17:305–323. (6)

Zakian, V. A. (1976) Electron microscopic analysis of DNA replication in main band and satellite DNA's of *Drosophila virilis*. J. Mol. Biol. 108:305–331. (4)

Zera, A. J., R. K. Koehn, and J. G. Hall (1983) Allozymes and biochemical adaptation. In: Comprehensive Insect Physiology, Biochemistry and Pharmacology, G. A. Kerkut and L. I. Gilbert (eds.). Pergamon Press, New York (in press). (6)

Zieg, J. and M. Simon (1980) Analysis of the nucleotide sequence of an invertible controlling element. Proc. Natl. Acad. Sci. 77:4196–4200. (13)

Zimmer, E. A., S. L. Martin, S. M. Beverley, Y. W. Kan, and A. C. Wilson (1980) Rapid duplication and loss of genes coding for the α chains of hemoglobin. Proc. Natl. Acad. Sci. 77:2158–2162. (3)

Zissler, J., E. Mosharuafa, W. Pilacinski, M. Fiandt, and W. Szybalski (1977) Position effects of insertion sequences IS2 near the genes for prophage λ insertion and excision. In: DNA Insertion Elements, Episomes, and Plasmids, A. I. Bukhari, J. A. Shapiro, and S. L. Adhya (eds.) Cold Spring Harbor Laboratory, Cold Spring Harbor, New York. (13)

Zuckerkandl, E. (1975) The appearance of new structure and functions in proteins during evolution. J. Mol. Evol. 7:1–57. (2)

Zuckerkandl, E. and L. Pauling (1965) Evolutionary divergence and convergence in proteins. In: Evolving Genes and Proteins, V. Bryson and H. J. Vogel (eds.). Academic Press, New York, pp. 97–166. (9, 11)

INDEX

acid phosphatase, 104
adaptive and nonadaptive evolution;
  see also neutral theory, 56, 116–
  118, 184–187, 255–257
adipose gene, 123–124
advantageous mutations, 60, 255
Aerobacter aerogenes, 237
alanine synthesis, 131
albumin, 18, 20, 214
albumin gene, 177
alcohol dehydrogenase, 121, 177
  activity, 122
  Adh[6] in *Drosophila,* 108
  E and S chain divergence, 27
  geographic variation, 122
aldolase, 240
allolactose synthesis, 254
α globin, 22
α2 type I collagen gene, 16, 18
α-amylase in *D. melanogaster,* 123
α-glycerophosphate dehydrogenase,
  124–125
Alu I family, 38, 46
amino acid code; *see also* codon, 192
  archetypal, 191–195
  mitochondrial, 196
  origin, 203–204
  universal, 197–202
amino acid conservation, 81
aminoacylation, 196–198
aminopeptidase-I, 131–134
anthropological genetics, 102
anticodon, 193
  archetypal, 194
  in tRNA, 206
  in universal code, 197
anticodon deaminase, 201
antigenic shift, 180
ape, 50
archetypal code, 191–195
artificial selection, 187
Ascaris lumbricoides, 72
Asians, 142
ATP, 129
autotetraploidization, 27
average heterozygosity, 210

bacteriophage λ, 267
β-globins, 1–13, 137, 177, 214, 219
  β[thal] alleles, 142
  β thalassemia mutations, 143
  chicken, 221
  haplotypes, 141
  phylogeny, 10, 11
β-galactosidase evolution, 242–245
Blacks, 142
blood type, 104
Bohr effect, 25
Bombyx mori, 58
bottleneck effect, 169
Bush baby (*Galago*), 67
butyramide, 241

calcium-dependent regulator protein,
  20
Capra hircus, 153
carbonic anhydrases, 27
catalytic efficiency, 120, 128, 137
catostomid fishes, 28, 185
Catostomus clarkii, 185
cell volume regulation, 130–135
centripetal selection, 186
Cepaea, 95–102
  population structure, 95–98
  shell characteristics, 100
Cercopithecus aethiops, 74
charge-change substitutions, 91
chicken α tubulin, 221
chicken β globin, 221
Chinese hamster, 55
chromosomal rearrangements, 274
  and transposons, 270
chymotrypsin, 26, 27
cis-vaccenyl acetate, 124
clines, 104, 128, 133
Cnemidophorus sexlineatus mtDNA,
  66, 158
code, 191–195, 227–228
  amino acid, 192
  archetypal, 191–195
  pretermination, 228
  two letter, 203

usage, 227
universal, 193
coding strategy, 226
coenzyme binding domain, 20
coincidental evolution; *see also* concerted evolution, 43
cointegrate, 262
collagen gene, 22
concerted evolution, 22, 38, 43, 45
  model of, 45
conjugative-plasmid transfer of genes, 106
conservative site-specific recombination, 260
copepods, 131
cow, 67, 69, 77
  mtDNA, 77
cryptic variation, 90–92
cultural diffusion, 104
cytochrome *c*, 214

D-arabitol metabolism, 236
D-arabitol permease, 237
deleterious mutations, 15
deletion, 6, 7, 138, 183, 271
demic spread, 104
depletion of energy reserves, 134
2,3-diphosphoglycerate, 22
distribution of single-locus heterozygosity (*h*), 170
D-loop, 64, 68, 147–148
DNA, 14, 39, 74, 173, 183
  content, 14
  heteroduplexes, 74
  hybridization, 39
  length polymorphism, 183
  mechanisms of polymorphism, 183
  polymorphism, 137, 173
Down's syndrome, 55–56
Drosophila engyochracea, 168, 171
D. mauritiana, 66
D. melanogaster, 21, 66–67, 71–72, 92, 120–128, 183
  alcohol dehydrogenase, 121, 177
  α-amylase, 123
  enzymes, 121
  esterase-6, 124
  mtDNA, 66, 69
D. neohydei, 67
D. pseudoobscura, 92
D. simulans, 66, 92
D. subobscura, 92
duplicative transposition, 262

EBG operon, 243, 249
EBG enzyme, 246, 250
EBG repressor, 252
EBG system, 255
effective number of alleles, 91
element expression, 266
embryonic hemoglobin, 22
emergence of new genes, 14–37
enzyme polymorphism, 90–94, 115–136, 166–173
episomes, 258–259
ε globin, 22, 139
Escherichia coli, 14, 106, 136, 227
  electrophoretic types, 108–109
  strain C, 238
  strain K12, 242
esterase-6, 124
estimation of matriarchal phylogeny, 160
eukaryotic gene structure, 15
evolution
  aliphatic amidase, 240
  amino acid code, 191
  antibody gene, 23, 24
  β-galactosidase, 242–245
  chromosomal rearrangement, 270–274
  D-arabinose utilization, 238
  EBG enzyme, 246, 250
  EBG repressor, 252
  EBG system, 255
  lactose utilization, 246, 256
  mitochondria, 195
  new genes, 14–37
  new metabolic functions, 234–257
  phenylalanine codons, 199
  propanediol utilization, 239
  pseudogenes, 30, 223
  regulatory functions, 251
  transcription, 60
  transposons, 264
  universal code, 196
  xylitol utilization, 236
evolutionary convergence, 163
evolutionary rates, 85, 217–218
  α and β hemoglobin, 218
  insulin, 217
evolutionary tree
  hominoid primates, 78
  β-globins, 8–11
  immunoglobin trees, 24
excisionase, 266
exons, 15–16

326

immunoglobulin, 18, 20, 22, 27,
  177–178
 evolution, 24
 gene, 22, 177–178
 IgG molecule, 23
 pseudogenes, 30–32
independent evolution model, 40
influenza A viruses, 180
inheritance of mtDNA, 159
insertion/deletion, 138, 183
insertion sequences; see also transposon, 260
insulin, 18, 27, 138, 221
insulin gene, 177, 184
integrase, 266
interchromosomal rDNA interactions, 51
interdemic selection, 111
intergroup selection, 113
intervening sequence (IVS); see also
  intron, 4, 9, 15, 30, 145
intragenic recombination, 250
intron, 15, 30

junk DNA, 37

k_{cat}, 120, 128, 132
kanamycin resistance, 266
Klebsiella aerogenes, 236
K. pneumoniae, 237–238
K_M, 120, 122, 125, 128, 136

lac operon, 6, 253
lactalbumin-lysozyme divergence, 27
lactate dehydrogenase, 26–28, 128,
  185
lactobionate, 248, 249
lactose utilization, 242
lacZ gene, 242
L-fucose pathway, 238
L-fucose permease, 240
linkage disequilibrium, 107
lipid synthesis, 127
lysosomes, 130
Lytechinus pictus, 67, 218

mammalian hemoglobin; see also
  globin, 25
mammalian mitochondrial code, 195
matriarchal phylogeny, 160

mechanism of adaptation, 116–119,
  135–137
mechanisms of concerted evolution,
  43
mechanisms of DNA polymorphism,
  138
mechanism of transposition, 260
Mediterraneans, 142
meiotic nondisjunction, 55
methylation of mtDNA, 158
mitochondrial genome; see also
  mtDNA, 21, 62, 64–65, 147, 151,
  154, 196
 codon, 64, 196
 DNA, 62, 66, 147, 151, 154
 DNA homogeneity, 151
 DNA polymorphism, 154
 genome size, 66
 protein genes, 82
 rRNA, 84
 tRNA, 83
models of unequal crossing-over, 46
moderately repetitive genes, 22
molecular drive, 47
molecular evolutionary clock, 214
molecular mechanisms of adaptation, 116–119
morphological evolution, 185
mouse; see also *Mus,* 3, 8–9, 30, 32,
  58, 67, 69, 72, 77, 92, 147, 153,
  223
 pseudogenes, 30–32, 223
 mtDNA, 77
movable elements, 258–279
 classification of, 259
mRNA, 21, 64
mtDNA; see also mitochondrial genome, 62–88, 147–164
 evolution, 62–88
 mutation, 86
 nucleomorphs, 161
 nucleotide sequence evolution, 74
 nucleotide substitutions, 77
 polymorphism, 147–164
 size, 62
 turnover time, 86
multigene families, 22, 38
multimeric proteins, 24
multiple β gene origins, 144
Mus caroli; see also mouse, 9
M. castaneous, 9
M. cervicolor, 9
M. musculus, 3, 8–9, 92, 147, 153

328

M. phari, 9
mutation, 30, 59, 85, 137, 180, 236
  faulty replication, 85
  frame shift, 30, 32
  neutral, 59, 137
  nonsense, 30
  regulatory genes, 236
  RNA viruses, 180
  somatic, 179
mutationism, 189
Mycoplasma capricolum, 21
myeloma, 179
myoglobin, 25–26
Mytilus californianus, 72
Mytilus edulis, 130

NADPH, 127
negative selection, 208
neoclassical theory, 189
neolithic spread of farming, 103
neuraminidase, 180
Neurospora crassa, 201
neutral mutations; *see also* neutral
    theory, 59, 137
neutral theory, 59, 89, 170, 208–233
  molecular evolution, 208
  tests, 170
new genes, 14–37
non-Darwinian evolution, 209
nonsense mutations, 30
nontranscribed spacer; *see also* in-
    tervening sequence, intron, 41–
    42
nucleic acid synthesis, 127
nucleolar association, 55
nucleolar dominance, 59
nucleolus organizer, 47, 50, 52
nucleomorph, 154
nucleotides, 9, 16, 71, 81, 145, 174
  base composition of mtDNA, 71
  changes, 9, 16
  conservation, 81
  diversity (π), 145, 174
  diversity (π), estimates of, 175
  substitution, 32, 34, 79, 81
  substitutions in introns, 222
  variability in β globin gene clus-
    ter, 144
null mutation, 35

octopine dehydrogenase, 130
organellar redundancy, 86
origin of the amino acid code, 203
origin of β gene frameworks, 145
orthologous sequences, 2
Ovis aries, 153
ovomucoid gene, 16–17
oxidoreductase, 240

Pan paniscus, 47, 153
Pan troglodytes, 47, 153
pancreatic ribonucleases, 217
pandemic (antigenic shift), 180
parallel evolution, 43–44
parvalbumin, 20
pentose shunt, 125–128
periodic selection, 106
Peromyscus bairdii, 9
P. maniculatus, 8, 153, 161–163
  mtDNA, 159
P. miliaris, 222
P. polionotus, 153
phage φX174, 14
phenotypic evolution, 185, 230
phenylacetamide, 241
phosphoglucomutase, 104
6-phosphogluconate dehydrogenase,
    125, 127, 136
phosphoglucose isomerase, 108, 130
3-phosphoglycerate kinase, 28
plant mtDNA polymorphism, 164
plasminogen, 20
poly(A), 9
poly(G) sequence, 6
poly(T) sequence, 4
polymorphism, 87, 113, 115–136,
    139, 153, 184–190, 208–233
  β globin gene, 139
  enzyme, 115–136
  mtDNA, 87
  neutral, 113, 184–190, 208–233
Pongo pygmaeus, 153
population structure, 89–114
  Cepaea, 96–102
  E. coli, 106
  Helix, 96–102
  Mytilus, 133
pregrowth hormones, 219–221
preproinsulins, 219–220

presomatotropin, 221
pretermination codons, 228
primordial gene, 16
production of NADPH, 127
proinsulin, 216
promoter insertion, 278
protease inhibitor, 20
protein polymorphism, 115–136
protein y, 2
prototype sequence, 9
pseudogenes, 8, 9, 15, 29–37, 223
  defects, 31
Pseudomonas aeruginosa, 240
purifying selection, 39, 40
pyruvate kinase, 28

rabbit, 30, 67
  mtDNA, 67
  pseudogene ψβ2, 12, 30
Rana pipiens, 72
random differentiation of subpopulations, 35, 112
random drift, 35
random fluctuation of selection intensity, 169
rat; *see also Rattus,* 221
  α tubulin, 221
  preproinsulin, 221
rate of adaptive evolution, 112
rate of evolution, 211
rate of fixation of a pseudogene, 35
rate of molecular evolution, 212
rate of nucleotide substitution, 32
Rattus norvegicus, 153, 155
R. rattus, 153, 155, 159, 221
rDNA, 47–49, 51–56
  chromosomal distribution, 49
  chromosomal exchange, 53
  concerted evolution, 56
  multigene family, 47
  primates, 48
  restriction map, 49
  role in nucleolar organizer, 51–55
reactivation, 29
recombination, 87, 106
  E. coli, 106
  mtDNA, 87
regulation, 26, 253
  divergence of, 26
  lactose permease, 253
repeated DNA, 22, 37, 39
repetitive sequence, 7, 22, 38

replication of mammalian mtDNA,
  64, 148
replicon, 262
replicon fusion, 271
restriction maps, 48, 149
  mtDNA, 149–150
  ribosomal gene, 48–49
retrograde evolution, 29
retrovirus, 30, 267
ribitol dehydrogenase, 236
ribosomes, 41
  repeating gene units, 41
  RNA genes, 21, 22
  RNAs, 21, 30, 38–39, 63
RNA, 41, 48, 57–58
  gene sequences, 48
  polymerase, 57–58
  precursor, 41
  5S rRNA, 30
Robertsonian translocation, 54, 56
role of mutation in evolution, 187

Saccharomyces, 21, 226
Salmo gairdneri, 72
salmonid fishes, 28, 72
sea urchin; *see Strongylocentrotus*
selection, 4, 115–136
  enzyme polymorphism, 115–136
selective constraint, 216–218
sequence divergence of mtDNA, 76
sequence rearrangements, 80
sequence variation in β globin gene
  cluster, 140
serine protease, 16
serum albumin, 18, 20
shell characters, 111
shifting balance theory, 113, 188
Shigella, 107
silent polymorphism, 81, 90, 176
silent/replacement substitution ratio, 83
silent substitutions, 81
silk fibroin, 58
small multigene families, 42
somatic mutation, 179
spacer sequences; *see also* intervening sequence, 63
spatial autocorrelation, 105
spread of agriculture in Europe, 105
stabilizing selection, 231
stereochemical theory, 204
strand bias, 73

This book was set in Linotron 202 Century Schoolbook at DEKR Corporation. The production team consisted of Jodi Simpson, copy editor, Joseph J. Vesely, designer and production coordinator, and Fredrick J. Schoenborn, illustrator. The book was manufactured by R. R. Donnelley & Sons.